Computed Radiation Imaging

Computed Radiation Imaging

Physics and Mathematics of Forward and Inverse Problems

Esam M.A. Hussein

Department of Mechanical Engineering
University of New Brunswick
Fredericton, N.B. Canada

ELSEVIER

AMSTERDAM • BOSTON • HEIDELBERG • LONDON • NEW YORK • OXFORD
PARIS • SAN DIEGO • SAN FRANCISCO • SINGAPORE • SYDNEY • TOKYO

Elsevier
32 Jamestown Road London NW1 7BY
225 Wyman Street, Waltham, MA 02451, USA

First edition 2011

British Library Cataloguing in Publication Data
A catalogue record for this book is available from the British Library

Library of Congress Cataloging-in-Publication Data
A catalog record for this book is available from the Library of Congress

ISBN 978-0-323-16529-7

For information on all Elsevier publications
visit our website at *www.elsevierdirect.com*

This book has been manufactured using Print On Demand technology. Each copy is produced to order and is limited to black ink. The online version of this book will show color figures where appropriate.

Transferred to Digital Printing in 2016

Contents

Preface

Imaging with atomic/nuclear radiation takes two forms: (1) a direct imprint as in radiography, or (2) a computed image as in computed tomography (CT) scans. Both forms of imaging are an integral part of modern medical-diagnostic practice, and are utilized in research and industrial imaging. Photons, in the form of x-rays or gamma-rays, are typically employed in such imaging systems, but neutrons are also utilized when photons cannot provide the needed indications. Radiography gives a direct imprint of the radiation that succeeds in penetrating the interrogated object. Therefore, radiographic imaging does not require much post-exposure numerical processing to extract the image features. In the case of computed radiography (CR),[1] some computations are done to enhance the image quality and enable image manipulation, but such computations are not essential for producing the raw image. In digital radiography (DR),[2] x-rays are captured directly, providing a digital image (without intermediate storage) that can also be numerically manipulated for optimum viewing.

Any numerical handling of a readily available image is an image enhancement and/or manipulation process. Computed imaging refers here strictly to images produced only after numerical processing of measurements, because the detected radiation, even when digitized, cannot give the required image features. For instance, a tomograph[3] can only be obtained by numerical reconstruction using many radiation projections of radiation-transmitted through an object, while a radiograph gives a single projection without numerical reconstruction. Therefore, one speaks of computed tomography (CT) or computer-assisted tomography (CAT) scans, which implies that the image involves both computations and the acquisition of many projections through a scan. Computed imaging is also involved in positron emission tomography (PET), single photon emission tomography (SPECT), as well as scattering-based imaging methods.

This book provides a framework for the computed imaging problem by presenting it as a "forward" (direct) problem, and as an "inverse" problem. The forward problem tackles the physical aspects of imaging via mathematical models that relate the to-be-imaged physical parameters to the collected measurements. The inverse problem

[1] Computed radiography (CR) employs a phosphorus plate that captures x-ray photons in gaps and stores their energy in its crystal structure. The stored energy is later extracted, in the form of light, by a low-energy laser, which is collected and converted into electrical signals, stored and processed for viewing.

[2] Digital radiography (DR) involves the use of CCD (charged-coupled device) cameras or photodiodes to capture light from a scintillation screen when exposed to x-rays, or a direct photoconductor material that produces electric signals.

[3] The word "tomo" is Greek for a slice or a section.

reconstructs numerically the imaged parameters from the measurements. This categorization is not new and is used in the solution of the inverse problem of image reconstruction. However, in addressing the inverse problem, the forward problem is often taken as a *fait accompli*. This is in spite of the fact that the forward problem inherently hides many of the physical simplifications of imaging methods, and contributes to some of the artifacts in the reconstructed images that are not attributable to measurement uncertainties. A good investigation of the forward problem of a particular imaging technique is necessary, not only to provide a foundation for understanding the physics of the technique, but also to help in elucidating the often overlooked limitations of imaging methods. Part I of this book covers the formulation of the forward problem for various radiation imaging modalities. Part II presents a general formulation of the inverse problem.

This book is directed toward graduate students and researchers. Therefore, it takes a critical view of all material presented, so that readers become fully aware of the inherent physical and numerical capabilities and limitations of the methods presented for both the forward and inverse problems. This necessitates an in-depth discussion of the founding physical and mathematical principles upon which the examined methods are based. Parts of this book can be adopted in senior undergraduate courses, by selecting a few chapters of each part of the book, so that the essential fundamentals are covered.

The book is written in a manner that makes it understandable by readers with a background in physics, mathematics, computer science, or engineering. This is done by avoiding, as much as possible, the use of specialized terminology without clear introduction and definition. Mathematicians interested in the numerical aspects of the inverse problem will be exposed to the physical foundations of the forward problem. On the other hand, physicists and engineers should be able to follow the mathematical logic of the presented methods without being bewildered by the jargon of abstract mathematics. By making the material readable by a wide variety of scientists, it is hoped to encourage those outside the imaging research community to involve themselves in medical or industrial imaging, while providing those within the imaging field with a thorough look at the physical and mathematical challenges of the problem.

The author expresses his gratitude to John Bowles of Inversa Systems Ltd. and Esmaeil Enjilela of the University of New Brunswick for reading draft chapters of this book, and providing valuable comments.

Esam M.A. Hussein
Fredericton, October 2010

1 Radiation Imaging

1.1 Why Radiation?

The word "radiation" is utilized to cover electromagnetic (EM) waves, as well as nuclear particulate radiation (alpha, beta, positrons, and neutrons). The most rudimentary form of imaging is that provided by natural vision, which relies on the deflection of visible light to the eye. Thermal imaging is a form of visualization based on the detection of infrared radiation that emanates from the surface of a hot body. Radar (radio detection and ranging) can also be viewed as a crude imaging process of "seeing" far-away objects. Radar utilizes EM waves in the microwave (above 1 GHz) range and observes their reflection off surfaces. At the extreme end of the radiofrequency band (a few 100 GHz), where the wavelength is in the order of millimeters, electromagnetic radiation can penetrate some organic materials, such as clothing. These so-called millimeter waves are reflected, upon encountering material interfaces, due to the change in the dielectric constant. The intensity of the reflected waves can be used in surface imaging, as done in some body screening security systems.

Imaging with the above methods requires the occurrence of a discontinuity in specific "bulk" physical properties: light's index of refraction for optical viewing, thermal conductivity and emissivity for thermal imaging, and dielectric constant or electric conductivity in the case of radar. This limits their visualization capabilities to certain applications.

One imaging method that relies on microscopic (rather than bulk) properties is magnetic resonance imaging (MRI). In this method, a powerful magnetic field is applied to align nuclei with non-zero spin[1] (such as ^{1}H, ^{3}H, ^{13}C, ^{19}F, ^{23}Na, ^{31}P, and ^{39}K) in a particular direction. Some of these nuclei are then temporarily disoriented by radiofrequency (RF) pulses. As the disturbed nuclei realign, they emit RF signals that are used to determine their location and characterize the surrounding medium. Obviously, if non-zero spin nuclei are present, no imaging information is obtainable. The presence of a magnetic material also interferes with the operation of MRI. As such, MRI is limited in use to certain materials. This restricts the visualization capability of MRI to certain applications. Therefore, MRI can also be viewed as an EM imaging method that relies on microscopic (rather than bulk) properties.

[1] A nucleus that has an odd number of nucleons has an intrinsic magnetic moment and angular momentum, i.e. a spin > 0.

Computed Radiation Imaging. DOI: 10.1016/B978-0-12-387777-2.00001-X

At the high frequency end of the EM spectrum, the wavelength of radiation becomes so low that it ceases to be sensitive to bulk properties and begins to be affected by individual atoms. For soft (low energy) x-rays, the wavelength range (in the nanometer range) becomes comparable to the spacing between atoms in a crystallized structure, and the phenomenon of x-ray diffraction begins to occur, providing a useful tool for examining crystal formations. As the frequency increases farther (into the hard x-ray and gamma-ray range), the wavelength becomes so small that the EM radiation resembles a "quantum" of radiation and possess corpuscular capabilities. Each photon quantum becomes like a "bullet" that moves in the vast vacuum space between atoms, until it encounters an atom with which it can interact. This corpuscular nature of x- and gamma-rays gives them their penetrating properties, and makes them attractive probes for "seeing" deep within matter. Since this high-frequency radiation interacts with individual atoms, it is not affected by bulk optical, electrical, or thermal properties and can be used for imaging all types of materials.

Acoustic waves can also penetrate deep inside matter and can be used for imaging purposes. Short-wavelength sound waves are, however, needed to capture minute details within matter. Therefore, ultrasonic waves (>1 MHz) are employed for this purpose. Sonic waves are reflected when they encounter an interface between two different materials, due to the change in acoustic impedance (which depends on the speed of sound in the material and the material's density). By sending pulses of ultrasonic waves via a small probe and recording the time of travel and intensity of reflected signals, the position of the reflected surface and the change in acoustic impedance at the surface can be determined. This information is used to construct an image of interfaces as a probe scans an object. Ultrasonic imaging is, therefore, a method for visualizing interfaces, but, when too many interfaces are present, its indications become quite difficult to decipher into an image. It also requires contact between the probe and the surface of the inspected object to provide the acoustic coupling needed for efficient transfer of the ultrasonic waves into and out of the medium. Imaging ocean floors and submerged objects with sonic waves (known as sonar imaging) functions also by monitoring the reflection of sound waves at interfaces.

Nuclear-particle radiation is capable of penetrating matter regardless of its bulk properties. However, charged particles, such as alpha and beta particles, are limited in their penetrability, though highly accelerated charged particles can reach some considerable depth in certain materials. On the other extreme, there are the neutrinos and cosmic rays that are so penetrating that they are not affected much by matter and are difficult to detect. Neutrons, being neutral particles with some mass, are good candidates for imaging, since they are affected by matter and are readily detectable. Therefore, neutron imaging is used in some applications.

1.2 Imaging Modalities

The modality used in imaging with non-ionizing radiation (light and microwaves) and sound (ultrasonic and sonar) is mainly that of *scattering* (reflection or refraction) of waves bouncing off surfaces and interfaces. Magnetic resonance imaging relies on the *emission* modality of imaging, as it records emitted secondary RF waves. However, the most common modality of imaging with x-rays, gamma-rays, and neutrons, relies on the *transmission* modality, by measuring the radiation that succeeds in penetrating the imaged object, passing through it, without interacting with the intervening medium. The dominance of the transmission modality in imaging with penetrating radiation is due to the fact that this radiation interacts with matter at the microscopic (the atomic/nucleus) level. Since most of the matter is made of empty space, with the nuclei and atomic electrons occupying a very small portion of the overall space, most of the radiation passes through matter without interaction. Therefore, the strength of the non-interacting transmitted radiation is quite high, while the interacting radiation that does not succeed in passing through carries information on the nature of the interrogated object. As such, transmitted radiation provides a "negative" impression of the material it passes through; the denser the material, the weaker the transmitted signal and vice versa. This is the basis of radiography, schematically shown in Figure 1.1(a),

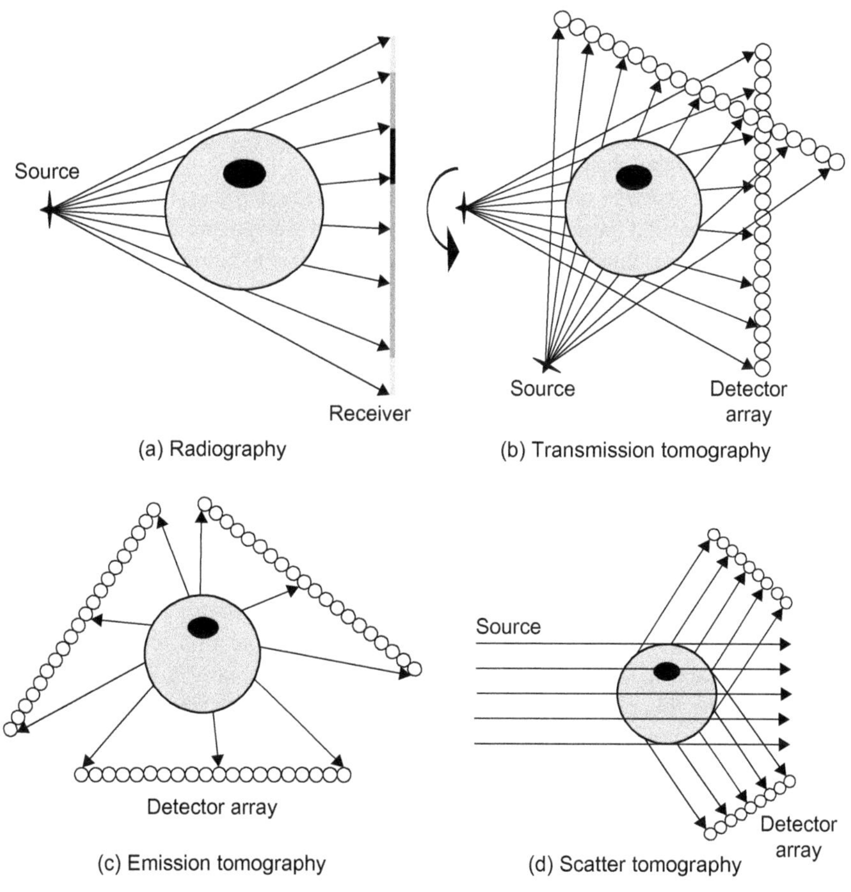

Figure 1.1 Various radiation imaging modalities.

where the internal details of an object are projected by the penetrating radiation on a receiving plate (a film or a scintillating screen). The projected images can be directly interpreted by the viewer in terms of relative material density (dark/light) and their geometric mapping (projection) in the density distribution.

Radiography enables the projection of the internal details of an object on a screen. However, like any other type of projection, it compresses the internal details in the direction of projection, i.e. all material traversed by an incident radiation ray is projected as a single point on the receiving plate. In order to decipher the material information along each ray, many intersecting projections must be taken. Then, with the aid of numerical manipulations, the material attributes at each point (or more precisely, each elementary area) can be determined. This is the essence of transmission tomography, schematically depicted in Figure 1.1(b); more widely know as computed tomography (CT) or computer-assisted tomography (CAT). The word "tomos" is Greek for a section or a slice. Therefore, tomographic imaging is usually performed slice-by-slice. The word "computed" in CT signifies the fact that an image is obtained after performing certain computations. This also makes it necessary to acquire measurements in a "digital" form, unlike in radiography where "analog" (e.g. film) recording of the measurements is sufficient. We also speak of CT scans, since the source and the detector assembly has to scan the object at various directions to provide the necessary multiple and overlapping projections; although in some cases, the source and detectors are kept fixed, and the object itself is rotated.

When distinct features are expected in the interrogated object, the effect of intervening material around these features can be reduced by acquiring digital radiography at different source orientations while fixing the receiver's location. A series of image slices, parallel to the imaging plane, can subsequently be numerically reconstructed from the acquired radiographs to depict the distribution of these features at different planes. This is due to the fact that features at different depths in the reflected object are projected differently at various angles of source orientation. This imaging technique is known as digital tomosynthesis (Grant, 1972), or simply tomosynthesis. It is quite attractive for use in mammography, where tumors may be masked by dense breast tissue. The technique is also useful in angiography, chest diagnosis, dental imaging, and orthopaedic imaging of joints (Dobbins III and Godfrey, 2003, Haaker et al., 1985). The concept of tomosynthesis is a numerical form of earlier analog or geometric tomography (also called laminography) systems, which involved a linear and opposing motion of the an x-ray source and a radiographic screen around a pivot to generate an image of the fulcrum plane parallel to the film. A point in the object on the fulcrum plane appears on the same location of the film as the source-film assembly moves, while points above and below this plane appear at different locations and as such their imprint on the film is blurred. In tomosynthesis, this is accomplished numerically by performing a shift-and-add operation of the logarithm of measurements (to linearize the response of transmitted radiation). The shifting process aligns the imprints of distant features on each tomographic plane. Features in the focal (fulcrum) plane then line up exactly on top of each other, so that the addition of the projections enhances their presence. On the other hand, features above or below the fulcrum plane are distributed in the added image to different locations, and their imprints are blurred.

Transmission imaging methods, as Figure 1.1(a) and (b) show, are global in nature; i.e. they image everything in between the source and receiver with the same resolution. Sometimes, however, the resulting image does not portray well particular regions in the object, due to their low density or similarity to the surroundings. If such feeble regions are accessible, a contrast material with higher absorption ability can be introduced into such regions to enhance their presence in the image; as done in the x-ray imaging of the gastrointestinal tract by the introduction of barium liquid. In other cases, particularly for examining physiological functions, a radiation source is introduced as a pharmaceutical, and the emitted radiation is monitored. This modality of emission imaging, schematically shown in Figure 1.1(c), can provide a radiogram by direct viewing of the recorded emission at a particular direction. Alternatively, emissions recorded in a digital form at different directions can be numerically manipulated to provide a tomograph. Single photon emission tomography (SPECT) and double photon emission tomography, better known as positron emission tomography (PET), are the medical forms of this emission imaging modality. While photon (gamma-ray) emitters, such as ^{99m}T, ^{123}I, and ^{133}Xe, are used in SPECT; positron emitters, such as, ^{15}O, ^{13}N, ^{11}C and ^{18}O, are utilized in PET. Each emitted positron is combined with an electron in the medium, with the two annihilating each other to produce two photons that emanate in opposite directions; hence the "double" photon emission terminology. The radioisotopes used in emission imaging have short half-lives to enable high decay rate. In this imaging modality, the strength and spatial distribution of the embedded source is deciphered. As radiation from the source travels toward the detector, it is attenuated by the intervening material, which reduces the intensity of the recorded radiation and, in turn, shadows the image. Such attenuation should be accounted for using *a priori* anatomical information, obtained perhaps from a CT image. While emission tomography is compatible with medical imaging, it is not always easy to embed a radioisotope within an industrial medium. However, if the medium is readily radioactive, as in spent nuclear fuel, emission imaging can be utilized.

The scatter imaging modality is a good candidate for industrial imaging, particularly when the inspected object is too thick or too dense to allow transmission imaging and when emission imaging is not possible. Moreover, scatter imaging can also be utilized when access to opposite sides of the object is not available. Though scatter imaging is the modality of imaging surfaces and interfaces with light, microwaves and ultrasonic waves, scatter imaging with penetrating radiation is much more complex, because radiation experiences no abrupt change as it crosses from one medium to another. The radiation scattering signal can also be quite convoluted, since source radiation can scatter many times before leaving the object and reaching a detector. Nevertheless, imaging based on once-scattered radiation (i.e. single scattering) have emerged. Figure 1.1(d) shows a schematic of a scatter imaging modality. Notice in this arrangement that radiation scattered to a particular direction from a unidirectional source is shown to be detected. This can be done with the aid of a collimated detector and is necessary in order to know from where the detected radiation emerges. Many directions of scattering can be monitored, and various arrangements can be adopted. Scatter imaging can be viewed as a form of emission imaging, with a virtual source at the scattering point. Therefore, scatter imaging produces images of the intensity of scattering, hence material density. This is unlike transmission imaging, which provides an indirect measure of density, since it detects radiation attenuation. The attenuation coefficient is a function of the material density, but can also be affected by material composition, while the scattering coefficient can be made, by proper choice of radiation type and energy, to be a function of only material density. Therefore, a combination of scatter and transmission imaging can provide both density and composition maps. In fact, the motivation for developing scatter imaging systems arose in medical imaging to obtain composition information, which is important for understanding tumor histology, and density maps are needed for planning for radiotherapy, where radiation deposition is strongly density dependent.

1.3 Direct and Reconstructed Imaging

Two types of images can be extracted from any of the imaging modalities shown in Figure 1.1: direct and computed. Direct images are those depicted by the measurements, as in the case of transmission radiography or the radiograms obtained from emission imaging. Direct images obtained from the scattering modality are usually quite convoluted, since in essence the obtained image is modulated by transmission, from the source to the scattering point, and by emission, from the scattering point to the detector. Direct images, when acquired in a digital form, can be supplemented by numerical processing to enhance the image quality, manipulate the image display, store an image in a digital form or render an image remotely accessible. For example, in computed radiography (CR), the image is acquired in a digital form by extracting light energy stored in a phosphorus plate as result of exposure to x-rays, which is then converted to electrical signals for storage and viewing. On the other hand, in digital radiography (DR), digital images are directly obtained by capturing light from a scintillation screen with the aid of a CCD (charged-coupled device) camera, photodiodes or a plate made of photoconductor material. In both CR and DR, the image attributes are available in a numerical form, but they are representing "measurements" of radiation intensity, rather than any physical property of the material in the imaged object. They are still, therefore, direct imprints of the measurements, since any subsequent numerical manipulation does not extract any parameters related to physical properties.

In transmission computed tomography (CT), measurements are gathered in a digital form then processed to *reconstruct* the radiation attenuation attributes within each pixel of the section of the image. Then numerical manipulation is more than just simple manipulation of the measurements. Image reconstruction takes a set of measurements and from them produces a set of physical parameters (attenuation coefficients). A similar process occurs in computed emission tomography (SPECT and PET), where radiation measurements are used to reconstruct images of the intensity (activity) of the introduced radiation source. Though then, the source's attributes are not direct physical parameters; a source's distribution in a medium depends on the physiological function in the case of living beings, the permeability of soil and rocks, and the fluid flow rate in gases and liquids. In scatter tomography, measurements are used to reconstruct images of the density of the medium (or its scattering coefficient), the attenuation coefficient of the incident radiation, the attenuation coefficient of the scattered radiation or all three physical attributes. Therefore, in all forms of *reconstructed* imaging, measurements are converted into physical attributes. This requires the availability of a measurement model that mathematically describes how a measurement value is related to the sought physical attributes. Such a measurement model enables the calculation of modeled "measurements" (synthetic data) corresponding to the actual measurements, for a known set of physical attributes, via the so-called forward mapping. Determining the physical attributes then involves the inversion of the forward model, i.e. inverse mapping of actual measurements, so that modeled measurements match, as much as possible, the actual measurements. These forward/inverse mapping processes are the workhorses of image reconstruction and are examined in further detail in Sections 1.4 and 1.5.

1.4 The Forward and Inverse Problems

In order to decipher a *parameter* depicting the physical attributes of a medium from a *measurement*, one must first establish the relationship between the two. For example, in a mercury thermometer, the physical attribute is the temperature, T, of the medium in which the thermometer is immersed, but what is measured is the height, h, of the mercury in its capillary tube; see Figure 1.2. One must then establish a relationship between T and h, in order to determine the value of T by measuring h. This can be provided by developing a *mathematical model* that relates h to T based on the *physical* principles that govern the expansion of a liquid like mercury. Upon further examination, one would realize that what measured is Δh (the change in h), which is related to the change in the volume of the mercury, and to ΔT (the change in T) by the volume expansion coefficient, β, for mercury:

$$\beta V \Delta V = \frac{1}{V}\frac{dV}{dT} \tag{1.1}$$

One would then realize that the determined physical attribute is the change in temperature, ΔT, not the temperature, T, as originally thought. Then one can relate Δh to

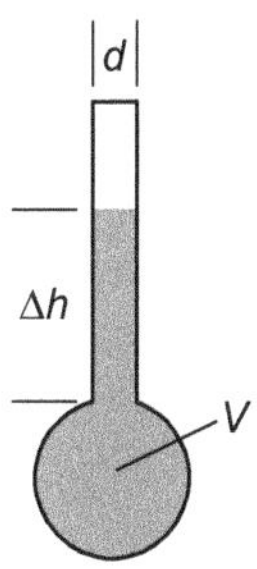

Figure 1.2 A schematic of a mercury thermometer.

ΔT by:

$$\Delta h = \frac{4V_0\beta}{\pi d^2}\Delta T \tag{1.2}$$

where use was made of $\Delta V = \pi \frac{d^2}{4}\Delta h$, with V_0 being the volume of the mercury in the spherical reservoir (bulb) at the one end, and d is the diameter of the capillary tube. The physical attribute is determined by *inverting* Eq. (1.2) to obtain:

$$\Delta T = \left(\frac{4V_0\beta}{\pi d^2}\right)^{-1} \quad \Delta h = \left(\frac{\pi d^2}{4V_0\beta}\right)\Delta h \tag{1.3}$$

The inverse problem of Eq. (1.3) is said to be *well-posed*, in accordance to Jacques Hadamard (1923), because it has a solution that is (i) unique and (ii) continuously depends on measurement; there is always a temperature change that uniquely depends on the change in h and ΔT changes continuously with Δh.

In the above simple example of the mercury thermometer, we have formulated the *forward problem* by Eq. (1.2), which relates the observed, or measured, quantity, Δh, to the physical attribute, ΔT. To arrive at the forward model of Eq. (1.2), we have relied on the physical theory that defines the volume expansion coefficient, Eq. (1.1), of liquids (mercury in this case) and the geometric relationship between a tube's volume and its height. The *inverse problem* was formulated by Eq. (1.3), which in this case is a simple reciprocal of a fraction. We had in this process already made some explicit and implicit assumptions that have implications on solving the inverse problem of determining temperature using the expansion of mercury. We realized that the intended parameter, T, was not determined, rather its change, ΔT, was found. In order to relate T to ΔT, some *calibration* is needed, to know the value of T at $\Delta h = 0$. The range of measurable temperatures is determined by finding the value of T at the maximum value of h, at the tip of the mercury tube. Then we know the space (domain) of T, or ΔT. The space of Δh is known from the geometry of the thermometer. In developing the forward model of Eq. (1.2) from the physical statement of Eq. (1.1), we have also made the assumption that $\frac{dV}{dT} \approx \frac{\Delta V}{\Delta T}$, which is a valid approximation when the change in V with T is linear. This is valid over a certain range of temperatures, before mercury changes phase from liquid to gas or to solid. This also limits the space of ΔT, as well as the applicability of the forward model of Eq. (1.2). In addition, instead of using the total volume of the mercury, V, we used the volume in the reservoir, V_0, to arrive at Eq. (1.2), which also necessitates that the volume of the capillary column be much smaller than that of the reservoir for the model to be valid. A physicist is also likely to question the formulation of the forward model of Eq. (1.2) with the view that temperature change also affects the volume of the glass body of the thermometer, which has its own expansion coefficient, β_g. Accordingly, a more accurate forward model that accounts for the glass expansion takes the form:

$$\Delta h = \frac{4V_0(\beta - \beta_g)}{\pi d^2}\Delta T \tag{1.4}$$

This model also illustrates that we will have a *singular* inverse problem, if $\beta = \beta_g$, since Δh will always be equal to zero, no matter what ΔT is. When β is very close in value to β_g, the inverse problem becomes *ill-posed*, because then a small error in observing Δh will lead to a larger error in the value of ΔT. Fortunately, most liquids have larger volume expansion coefficients than solids, by two orders of magnitude or so. Therefore, this simple inverse problem is *well conditioned.* More sophisticated physical models for the forward problem could have been derived, for instance by considering the actual volume V, instead of V_0, or by writing a two-dimensional model that takes into account the lateral expansion in the (radial) direction of the tube. In this thermometer problem, the observed parameter, Δh, has a range that extends from a minimum of zero to a maximum equal to the height of the capillary tube. Accordingly, ΔT and, consequently, T, have minimum and maximum values corresponding to those of h. Therefore, we know the domain of the parameter ΔT, or T and that of the measurements Δh. Therefore, both the forward *mapping* of Δh to ΔT and the inverse mapping of ΔT to Δh should produce values within their respective domains.

The forward and inverse problems of the mercury thermometer, though used to introduce useful concepts, are quite simple and are not solved in practice. The observer does not see a value for Δh or ΔT; a value of T is directly given. This is done via a *calibration* process that establishes the forward model without explicitly expressing it mathematically and solves the inverse problem without performing any computations. In imaging problems, the situation is not that simple. The problem is multidimensional, since an image consists of many pixels and the physical attribute has to be obtained for each pixel/voxel. Moreover, many measurements need to be acquired, and each measurement in almost all cases will depend on the physical attributes of the material content of each element corresponding to a pixel/voxel in the image. Sometimes one can relate a set of physical parameters to associated measurements to produce a *response matrix*. Such a matrix can be applied to other sets of parameters in the same imaging system, if the response is linear; i.e. the material in one image element does not affect the measurements corresponding to any other element in the image. Even then, such a measured response matrix will be limited to a particular configuration, and its generation for a large-size image may prove to be tedious and impractical. Then one has to rely on a mathematically formulated forward mapping that models the physics of the problem. As the simple example of the thermometer indicated above, the forward model is quite vital in correctly predicting the dependence of observed measurements on physical parameters. Unfortunately, more attention is usually paid to solving the inverse problem than to the formulation of the forward problem. In many cases, the solution of the inverse problem demands simplification of the forward problem to enable efficient solution of the problem. Adding details of secondary effects to the forward model, unless they have a significant impact on the observed (measured) parameter, (Δh in the thermometer's example), become an unnecessary complication. Nevertheless, one should be aware of the effect of simplification of the forward model and should attempt as much as possible to have an accurate, but amenable to inversion, modeling of the physical process. These are the reasons behind devoting Part I to the formulation of the forward problem.

1.5 Forward and Inverse Mapping

An image consists of a set of pixels (in two dimensions), or more generally a set of voxels (in three dimensions). Associated with each voxel is an attribute or a parameter reflecting a physical property. The magnitude of the parameter can be associated with a gray or color level for visualization purposes. The latter association is a marginal process and is not elaborated upon in this book. However, the association between each parameter in the set of image attributes and the spatial position and size of its pixel/voxel is implicitly assumed to be fixed in advance.

The forward problem is formulated by mapping a set of parameters, $\boldsymbol{c}$, in the space of physical parameters, $\boldsymbol{C}$, into a set of measurements, $\boldsymbol{e}$, in the space of measurements $\boldsymbol{E}$. This is schematically shown in Figure 1.3, which also depicts the inverse problem as the inverse mapping of measurements in $\boldsymbol{E}$ to parameters in $\boldsymbol{C}$. A set $\boldsymbol{c}$ in $\boldsymbol{C}$ ($\boldsymbol{c} \in \boldsymbol{C}$) is shown in Figure 1.3 to correspond to a single set $\boldsymbol{e}$ in $\boldsymbol{E}$ ($\boldsymbol{e} \in \boldsymbol{E}$). Figure 1.3 shows two forward mappings, $\boldsymbol{m}_0$ and $\boldsymbol{m}$, to indicate that physical modeling can assume many levels of sophistication. If the mapping $\boldsymbol{m}_0$ is exact, it produces an exact match with a measurement $\boldsymbol{e}_0$. But when obtaining measurements, one would acquire a measurement, $\hat{\boldsymbol{e}}$, within some uncertainty, $\pm\delta_{\boldsymbol{e}}$, due to statistical fluctuations and instrument errors. If the measuring device is functioning adequately, one should have: $\hat{\boldsymbol{e}} - \delta_{\boldsymbol{e}} \leq \boldsymbol{e}_0 \leq \hat{\boldsymbol{e}} + \delta_{\boldsymbol{e}}$. In almost all cases, an approximate mapping, $\boldsymbol{m}$, is utilized. For a parameter, $\boldsymbol{c}$, one would *calculate* a measurement $\boldsymbol{e}$; we will call this the measurement estimate, or the *modeled measurement*. If the mapping is sufficiently representative of the physical

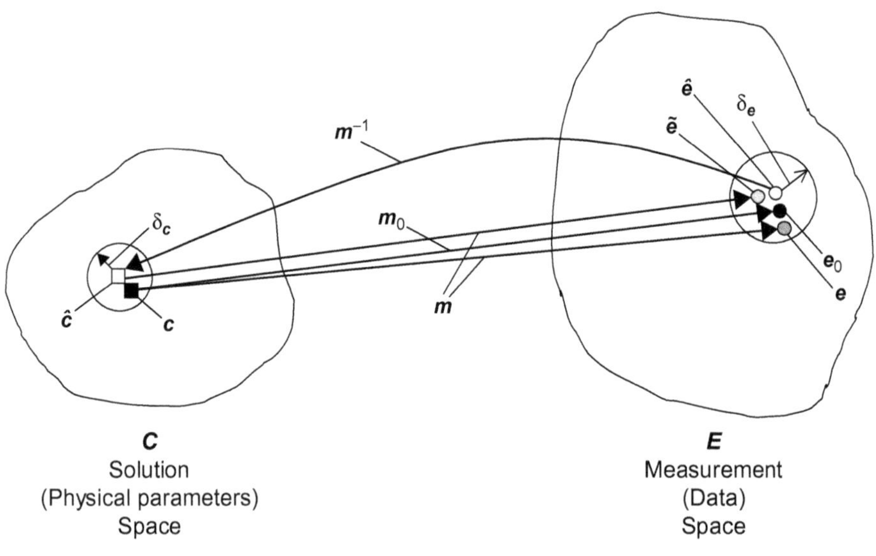

Figure 1.3 A schematic showing the mapping with $\boldsymbol{m}$ of a physical parameter, $\boldsymbol{c}$, to a measurement estimate $\boldsymbol{e}$; with $\hat{\boldsymbol{e}}$ being an actual measurement with an uncertainty ($\delta_{\boldsymbol{e}}$), $\boldsymbol{e}_0$ is the measurement determined by the exact mapping, $\boldsymbol{m}_0$ (if available), and $\hat{\boldsymbol{c}}$ is the result of inverse mapping the measurement $\hat{\boldsymbol{e}}$ with $\boldsymbol{m}^{-1}$ (the inverse of $\boldsymbol{m}$); δ_c is the uncertainty margin in $\hat{\boldsymbol{c}}$, and $\tilde{\boldsymbol{e}}$ is the measurement estimate corresponding to $\hat{\boldsymbol{c}}$.

process, one should have $\boldsymbol{e} \approx \boldsymbol{e}_0$, or at least $\boldsymbol{e} - \Delta_m \leq \boldsymbol{e}_0 \leq \boldsymbol{e} + \Delta_m$, where Δ_m is the *modeling error (inexactness)*, which is usually unknown, but can be approximated as: $\Delta_m = \boldsymbol{m}\boldsymbol{c} - \hat{\boldsymbol{e}}$.

Inverse mapping is applied to the available measurements to obtain: $\hat{\boldsymbol{c}} = \boldsymbol{m}^{-1}\hat{\boldsymbol{e}}$. Since there is an uncertainty, δ_e, associated with $\hat{\boldsymbol{e}}$ and modeling error, Δ_m, associated with $\boldsymbol{m}$, one would anticipate an uncertainty, δ_c, in the reconstructed value, $\boldsymbol{c}$. One would hope that $\delta_c < \boldsymbol{c}$, in order for $\boldsymbol{c}$ to give a realistic representation of $\boldsymbol{c}$. The value of δ_c can be controlled by bounding the values of $\boldsymbol{e} \in \boldsymbol{E}$ and $\boldsymbol{c} \in \boldsymbol{C}$. For example, one has: $\boldsymbol{e} \geq \boldsymbol{0}$ and $\boldsymbol{c} \geq \boldsymbol{0}$, since there are no negative absolute measurements, and in radiation imaging $\boldsymbol{c}$ is almost always a non-negative physical attribute. The solution space, $\boldsymbol{C}$, can also be assigned an upper-bound from *a priori* knowledge of the nature of the imaged object. One may also designate upper-bounds on the measurement space, $\boldsymbol{E}$. However, some inverse problems are ill-posed. An ill-posed problem does not meet one of the conditions Hadamard assigned to a well-posed problem, see Section 1.4. That is, an ill-posed problem may not have a unique solution, and the solution may not continuously depend on the measurements. Then a small change in $\hat{\boldsymbol{e}}$, such as that caused by statistical fluctuations, can lead to large changes in $\boldsymbol{c}$. One must then resort to solution bounding and constraining measures, regularization and smoothing and/or sharpening procedures to ensure that the reconstructed $\boldsymbol{c}$ values are physically and numerically acceptable. These procedures are discussed in Part II of this book. Note that, since in practice the inverse problem is discrete, in the sense that it relates a finite number of measurements to a finite number of image pixels/voxels, it can be formulated in a matrix form. An ill-posed problem produces an ill-conditioned matrix. Then the matrix corresponding to the forward mapping will have a large condition number; i.e. the ratio between its largest and smallest eigenvalues is high.

In summary, one can write the following mapping relationships:

$$\boldsymbol{c} \xrightarrow{\boldsymbol{m}_0} \boldsymbol{e}_0; \quad \boldsymbol{c} \in \boldsymbol{C}; \quad \boldsymbol{e}_0 \in \boldsymbol{E} \tag{1.5}$$

$$\boldsymbol{c} \xrightarrow{\boldsymbol{m}} \boldsymbol{e}; \quad \boldsymbol{c} \in \boldsymbol{C}; \quad \boldsymbol{e} \in \boldsymbol{E} \tag{1.6}$$

$$\hat{\boldsymbol{e}} \xrightarrow{\boldsymbol{m}^{-1}} \hat{\boldsymbol{c}}; \quad \hat{\boldsymbol{e}} \in \boldsymbol{E}; \quad \hat{\boldsymbol{c}} \in \boldsymbol{C} \tag{1.7}$$

$$\hat{\boldsymbol{c}} \xrightarrow{\boldsymbol{m}} \tilde{\boldsymbol{e}}; \quad \tilde{\boldsymbol{c}} \in \boldsymbol{C}; \quad \tilde{\boldsymbol{e}} \in \boldsymbol{E} \tag{1.8}$$

In addition, one should have:

$$\boldsymbol{e}_0 \in [\hat{\boldsymbol{e}} \pm \delta_e]; \quad \boldsymbol{e} \in [\hat{\boldsymbol{e}} \pm \delta_e]; \quad \boldsymbol{c} \in [\hat{\boldsymbol{c}} \pm \delta_c]; \quad \tilde{\boldsymbol{e}} \in [\hat{\boldsymbol{e}} \pm \Delta_m] \tag{1.9}$$

That is, within the interval $[\hat{\boldsymbol{e}} - \delta_e, \hat{\boldsymbol{e}} + \delta_e]$, one expects to find the *true* value, $\boldsymbol{e}_0$, of the measurement, and hopes that the forward model produces a measurement estimate, $\boldsymbol{e}$, (for a given $\boldsymbol{c}$) also within the same interval. In addition, one hopes that the true solution, $\boldsymbol{c}$, is within the interval $[\hat{\boldsymbol{c}} - \delta_c, \hat{\boldsymbol{c}} + \delta_c]$, and that the measurement estimate, $\tilde{\boldsymbol{e}}$, corresponding to the obtained solution, $\hat{\boldsymbol{c}}$, is also within $[\hat{\boldsymbol{e}} - \delta_e, \hat{\boldsymbol{e}} + \delta_e]$. The measurement, $\hat{\boldsymbol{e}}$, is sometimes referred to as the set of *observable* parameters, while the results of the inversion process, $\hat{\boldsymbol{c}}$, its estimated error, δ_c, and the corresponding modeled measurement, $\tilde{e}$, are called *a posteriori* information. Information that is known but is independent of the measurement is called *a priori information*. Though the true value of a measurement, $\boldsymbol{e}_0$, is never known, a good estimate can be obtained by accurate measurements, with a small δ_e value.

It is appropriate at this stage to distinguish between *inexactness*, *accuracy*, *precision*, and *uncertainty*. We use inexactness to refer to the difference between a computed modeled estimate of a measured value and its exact value, $\boldsymbol{e}-\boldsymbol{e}_0$; caused by neglecting some secondary aspects in the modeling process, to arrive at a numerically manageable forward mapping. Accuracy is the difference between an actual measurement and its true value, $\hat{\boldsymbol{e}}-\boldsymbol{e}_0$, caused by systematic measurement errors that are, at least in principle, correctable. Uncertainty refers to effects that are unintentional, typically statistical, that lead to variability in the value of a measurement not related to the physical or laboratory setup. Due to uncertainty, one has: $\hat{\boldsymbol{e}} \in [\hat{\bar{\boldsymbol{e}}} \pm \delta_{\bar{e}}]$, where $\delta_{\bar{e}}$ is the measure of variability of $\hat{\boldsymbol{e}}$ around its mean value, $\hat{\bar{\boldsymbol{e}}}$. The precession (reproducibility) of a measurement can be estimated by the ${\delta_{\bar{e}}}^{-1}$. The less the fluctuations in a measurement, the more precise it is. In other words, inexactness reflects systematic modeling approximations, while accuracy, uncertainty, and precession are associated with the measurement process. Of course, the exact value is never known; the "truth" is always elusive. However, one can get $\boldsymbol{e}$ to approach $\boldsymbol{e}_0$ by more accurate and detailed modeling and can have a value of $\hat{\boldsymbol{e}}$ that is closer to $\boldsymbol{e}_0$ by acquiring more accurate and precise measurements.

References

J. T. Dobbins III and D. J. Godfrey, "Digital x-ray tomosynthesis: Current state of the art and clinical potential," *Physics in Medicine and Biology*, vol. 48, pp. R65–R106, 2003.

D. G. Grant, "Tomosynthesis: A three-dimensional radiographic imaging technique," *IEEE Transactions on Biomedical Engineering*, vol. 19, pp. 20–28, 1972.

P. Haaker, E. Klotz, R. Koppe, R. Linde, and H. Moller, "A new digital tomosynthesis method with less artifacts for angiography," *Medical Physics*, vol. 12, pp. 431–436, 1985.

Part I

The Forward Problem

As far as the laws of mathematics refer to reality, they are not certain; and as far as they are certain, they do not refer to reality.

Albert Einstein*

In order to be able to correctly formulate a forward mapping, one must first be familiar with the underlying physical processes that relate the image parameters of interest to radiation measurements. For an image parameter to reveal its presence to a measurement, it must be induced to do so by a source of radiation. Such a source is either an external radiation source that bombards the surface of an object or an internal source introduced into the object. In either case, the source radiation is *transported* through the object, and some of that radiation must be able to escape from the object and reach detectors placed outside the object. These detectors provide, in turn, the measurement signal, after some electronic processing, to convert an analog signal to a digital signal suitable for numerical processing. In order to simulate these physical processes and formulate them into a forward model, one first needs to be able to simulate the transport of radiation through the object and its surroundings. Radiation transport theory is governed by the Boltzmann particle-transport equation, which is presented in Chapter 2. The issues related to the formulation of a generalized forward model in radiation imaging are introduced in Chapter 3. The forward models for each of the three radiation modalities of transmission, emission, and scattering are subsequently discussed in Chapters 4, 5, and 6, respectively.

* In: Calaprice, A., Ed. (2005). *The new quotable Einstein*. Princeton University Press, p. 228.

2 Radiation Transport

2.1 Introduction

Imaging with penetrating radiation involves mainly the use of x-ray and gamma-ray photons, and to a lesser extent neutrons. Neutrons are particles, and as such their transport is governed by the particle transport equation, which is discussed in this chapter. However, low-energy neutrons exhibit wave-like behavior, enabling their use to examine lattice structures, in a manner similar to that of soft (low energy) x-ray diffraction. Photons, on the other hand, are packets (or quanta) of electromagnetic energy, but also possess both wave behavior (at low frequency, hence low energy) and corpuscular attributes at high energy. The latter behavior is what enables photons to penetrate deep into matter and facilitates their use in imaging deep within objects. Therefore, the word "particle," is used here to refer to high-energy photons, as well as to neutrons; keeping in mind, though, that its use for photons is a metaphor. Therefore, the particle transport equation, presented below, is equally applicable to photons and neutrons. It should be noted here that both photons and neutrons have no electric charge, and as such are treated as neutral particles not affected by external electromagnetic fields. The discussion below does not initially address the transport of charged particles, which are dealt with at the end of the chapter, in Section 2.8. In essence, we have already made an implicit assumption in the radiation transport equation, presented in Section 2.4: (I) only neutral particles are considered. These assumptions are marked throughout this chapter, with uppercase Roman numerals, so that the reader can keep track of them.

2.2 Variables

In order to simulate the radiation transport process, one must first introduce a quantity that defines the intensity of the transported radiation. The most basic quantity in this regard is the particle density, i.e. the number of particles, n, per unit volume. Since radiation transport involves the "flow" of particles, a more commonly used quantity is the particle flux, ϕ, which is the number of particles per unit area per unit time. The latter definition of flux poses a problem though, since radiation intensity can vary with the orientation of radiation with respect to the area on which it impinges. Therefore, a more rigorous definition of flux is the number of radiation track-lengths per unit volume per unit time, i.e., the track-length density (Hussein, 2007). This definition

Computed Radiation Imaging. DOI: 10.1016/B978-0-12-387777-2.00002-1

avoids relating the neutron intensity to a particular area in the space and does not require the definition of a certain shape of a volume. The flux and particle density are related by:

$$\phi = nv \tag{2.1}$$

where v is the particle velocity.

In addition to defining the dependent parameters, n or ϕ, which quantify the amount of radiation transported, one must also define the independent variables of the transport process, i.e., the parameters n and ϕ depend upon. Naturally, the position in space, $\vec{r}$, and the time, t, are four basic parameters; keeping in mind that the vector $\vec{r}$ is measured from the point of origin and requires three parameters for its definition: one for its length and the other two for its direction. These four independent variables are sufficient to describe classical fluid flow, where the particles (molecules) of the flow move collectively, not independent from each other, because the molecules are bound to each other. Then, the flow velocity of fluids and the related quantities of momentum and energy are also dependent variables. In the particle transport of radiation, the particle density, n, (II) is assumed to be sufficiently large that its value is not affected by statistical fluctuations, but is (III) small enough that particles do not interact with each other. As such, a $n = 10^{15}$ particles per m^3 is not large given the fact that a mole of material contains 6.022×10^{23} atoms or molecules (Avogadro's number). The fact that radiation particles do not affect each other and do not move as a collective dictates that the velocity, $\vec{v}$, be designated as an additional independent variable. A vector, $\vec{v}$, is composed of three variables, one for magnitude and two for direction. Since photons always move at the speed of light, the magnitude of the velocity, $|v|$, is replaced by the particle energy, E, and the direction of motion is designated by the vector, $\vec{\Omega}$. Note that:

$$E = h\nu; \quad \text{for a photon} \tag{2.2}$$

where h is Planck constant (6.626×10^{-34} J s) and ν is the radiation's frequency.

It is important to note here that the energy, E, refers to the energy of one radiation particle, unlike fluid flow where energy refers to the total energy of the collective. Similarly, in particle transport, the velocity, $\vec{v}$, momentum, $\vec{p}$, and direction, $\vec{\Omega}$, are all particle-related quantities.

There are other important differences between fluid flow and radiation transport: fluid flow is confined within solid boundaries, while radiation penetrates boundaries. The movement of fluids is, therefore, limited to within directions defined by solid boundaries. Fluids do form interfaces between each other, producing clouds, droplets, and clusters, which is not the case with radiation. Therefore, it is said that radiation does *not* respect the *impenetrability* principle. The implication is that radiation can move in any direction, and as such its flux has a distribution in direction, i.e. ϕ is a function of $\vec{\Omega}$. Radiation can also have a different energy at each direction, i.e. ϕ is a function of E and $\vec{\Omega}$. Therefore, the flux, ϕ, in general is expressed as $\phi(\vec{r}, E, \vec{\Omega}, t)$, with $\phi(\vec{r}, E, \vec{\Omega}.t)dEd\vec{\Omega}$ being the flux of particles at position $\vec{r}$, with energy within E and $E + dE$ in a direction within $\vec{\Omega}$ and $\vec{\Omega} + d\Omega$. Integrating over $d\vec{\Omega}$ gives $\phi(\vec{r}, E, t)$; the flux at $\vec{r}$ and t with energy in E and $E + dE$ in all directions. Accordingly, $\phi(\vec{r}, t)$ is the flux at $\vec{r}$ at time t, incorporating all particle energies and directions. The quantity $\int \phi(\vec{r},t)dt$ is known as the particle fluence.

Neutral particles are (IV) not affected by external electromagnetic fields. Having a very small mass or no mass at all in the case of photons, radiation particles are also (V) not affected by gravity. Since the movement of radiation particles is neither confined by boundaries nor by interactions with other radiation particles, and is (VI) not affected by external fields, they (VII) move in straight lines in the vast vacuum between target atoms/molecules. This straight-line movement continues until a radiation particle is so close to the field of an atom or the short-range field of a nucleus that its path is altered. As particles approach an atom/nucleus target, quantum mechanics or electrodynamics, depending on the type of particle and the nature of the target, govern the interaction process (Hussein, 2007), in terms of determining the nature of the interaction and its probability of occurrence. The latter is quantified, in terms of interaction cross sections, as explained in Section 2.3. A particle emerging from an interaction will continue to travel in another straight line, until another interaction takes place. An observer in the laboratory will only record straight-line movement of radiation; though the change, from one straight line to another during the interaction process, may not be an abrupt change. Therefore, (VIII) quantum mechanics and electrodynamics are not considered directly in studying the transport process, but their effect is incorporated into cross sections of the medium in which particles travel.

2.3 Cross Sections

Consider a radiation beam of I_0 particles at distance $x = 0$ moving in the $\vec{x}$ direction. Let Σ be the probability per unit distance that these radiation particles will interact, one way or another, with one of the targets in the medium within which radiation travels. The probability, q, that the radiation particles will not interact within a small distance interval Δx is then $q = 1 - \Sigma\Delta x$. That is, qI_0 particles will remain in the beam, still moving in the direction $\vec{x}$, while the rest, $(1-q)I_0$, would have been removed from the beam by some interaction (either absorbed by the target atom/nucleus or changed direction by scattering). After traveling another distance, Δx, only $q^2 I_0$ particles will remain in the beam. As such, within a distance $x = n\Delta x$, the number of radiation particles remaining in the beam is $I(x) = q^n I_0$. If Δx was infinitesimally small, i.e. $\Delta x \to 0$, then $n \to \infty$, and:

$$\frac{I(x)}{I_0} = \lim_{n\to\infty} q^n = \lim_{n\to\infty}(1 - \Sigma\Delta x)^n = \lim_{n\to\infty}\left(1 - \Sigma\frac{x}{n}\right)^n = \mathrm{e}^{-\Sigma x} \tag{2.3}$$

When $I_0 = 1$, Σ becomes the probability of interaction, per unit length, of one radiation particle with the target atoms/nuclei of the medium. If there are N atoms per unit volume (atomic density), then:

$$\sigma = \frac{\Sigma}{N} \tag{2.4}$$

is the probability of interaction per unit distance per target atomic density. Note that:

$$N = \frac{\rho}{A\mathrm{u}} \tag{2.5}$$

where ρ is the material density, A is its mass number, or atomic weight, and u is the atomic mass unit ($\frac{1}{12} \times$ the mass of a ^{12}C atom), $\mathrm{u} = 1.660565 \times 10^{-27}$ kg. Since σ has dimensions of area and is for interactions at the atomic/nucleus level, it is known as the *microscopic cross section*. Accordingly, because Σ involves all target atoms/nuclei in a medium, it is called the *macroscopic cross section*. The microscopic cross section, σ, can be seen as the area of the field projected by an atom to incoming radiation, though it is not necessarily equal to the geometric area; in fact, it can be much larger. Then, Σ is the total area projected by atoms in a unit volume. Therefore, the value of σ depends on the nature of the atom and the nature of the incident radiation and its energy, while Σ depends on all the latter, in addition to the atomic density of the material.

Equation (2.3) expresses the degree of attenuation of a beam of radiation as it travels a distance x within a material of a macroscopic cross section, Σ. Accordingly, Σ is also known as the attenuation coefficient. Then, one can say that a change in beam intensity, dI, within a distance, dx, is proportional to Idx, with Σ being the proportionality constant. That is,

$$dI = -\Sigma I\, dx \tag{2.6}$$

with the negative sign indicating that dI decreases with x. Integrating Eq. (2.6) from $x = 0$ to I leads to:

$$I = I_0 \exp(-\Sigma x) \tag{2.7}$$

with $I = I_0$ at $x = 0$. Equation (2.7), derived also in Eq. (2.3), is known as the attenuation law; equivalent to Lambert's third law of optics. The total cross section is also a *relaxation distance*, since it is the distance that results in a reduction in the initial flux by $\frac{1}{\mathrm{e}} = 0.368$.

An interesting implication of Eq. (2.7) is that it can be used to determine the average distance of travel between interactions, the so-called mean-free-path, λ as:

$$\lambda = \frac{\int_0^\infty x \exp(-\Sigma x)\, dx}{\int_0^\infty \exp(-\Sigma x)\, dx} = \frac{1}{\Sigma} \tag{2.8}$$

That is, Σ is the reciprocal of the mean-free-path.

Neither Eqs. (2.3) and (2.6) nor Eq. (2.7) identify the nature of radiation interaction. They simply stipulate that radiation is removed from a radiation beam, or more precisely a narrow radiation beam. Radiation can be removed from such a beam by absorption or scattering off the beam's direction. Therefore, Σ is given the subscript "t," for total, to indicate that it incorporates both absorption and scattering.

Since the occurrence of one interaction is independent from the other, Σ_t and accordingly σ_t, are further subdivided into cross sections for specific interactions. For example, $\Sigma_t = \Sigma_a + \Sigma_s$, where Σ_a incorporates all interactions in which radiation is fully absorbed and Σ_s encompass all forms of scattering (elastic, inelastic, etc.). As such σ_a and σ_s can also be separated into corresponding interaction types. To avoid deviating from the main theme of the discussion in this chapter, interaction types are discussed in Chapters 4 to 6, when formulating the forward mapping for each imaging modality. Hussein (2007) provides a comprehensive compilation and classification of all radiation interactions.

For the scattering cross section, a differential cross section is defined to determine the probability of scattering to a specific energy and direction, so that:

$$\Sigma(E \to E'; \vec{\Omega} \to \vec{\Omega}') = \frac{d\Sigma(E, \vec{\Omega})}{dEd\Omega} \tag{2.9}$$

expresses the cross section for scattering from energy E to energy E', which is associated with a change in direction from $\vec{\Omega}$ to $\vec{\Omega}'$. Integrating Eq. (2.9) with solid angle leads to $\Sigma(E \to E')$; the energy differential cross section. A similar expression can be devised for an interaction that can generate new particles.

The attenuation coefficients for photons are usually designated by the Greek symbol μ and tabulated in terms of the mass attenuation coefficient, $\mu_m = \frac{\mu}{\rho}$, where ρ is the material density. The mass attenuation coefficients for elements, mixtures and compounds at various photon energies are available on-line via XCOM (Berger et al., 1999; `http://www.physics.nist.gov/PhysRefData/Xcom/Text/XCOM.html`). The cross sections for neutrons, on the other hand, are reported as microscopic cross sections, expressed in a unit called barn (b) with $1\,\text{b} = 10^{-28}\,\text{m}^2$. ENDFPLOT (Nuclear Data Evaluation Lab., 2000) tabulates the neutron cross sections for various elements as a function of neutron energy (`http://atom.kaeri.re.kr/endfplot.shtml`).

The definitions of the macroscopic cross section, Σ, and the flux, ϕ, enable the determination of the interaction rate per unit volume as:

$$\text{Interaction Rate} = \Sigma\phi \tag{2.10}$$

Recalling that Σ is the interaction probability per unit length, and ϕ is the track-length density, then $\Sigma\Phi$ is the interaction density. Note that with the appropriate subscript for Σ in Eq. (2.10), the interaction rate for a certain interaction can be obtained.

In discussing the interaction probability above, we implicitly assumed that (IX) the nature of the target atoms/nuclei did not change from one interaction to the next, that is, Σ remains constant as the radiation beam travels a distance x. This is a reasonable assumption, since the number of radiation particles is much smaller than the number of target atoms, as indicated above. Then, even if all radiation particles interact, their overall effect on the macroscopic physical properties of the material they interact with will be negligible.

2.4 Boltzmann Transport Equation

The assumptions, marked in the above sections by uppercase Roman numerals, point to a linear transport process of neutral particles that travel in straight lines until they encounter a target atom/nucleus to interact with. The linearity of the transport process is justified by the assumptions that radiation particles do not interact with each other and that the medium with which radiation interacts does not change during the transport process. The latter two assumptions enable the determination of the macroscopic cross sections separately and independently of the transport process. Then, the interaction rate becomes only dependent on the particle flux, ϕ, and increases linearly with ϕ, for a given interaction or a set of interactions defined by Σ.

Let us consider a finite phase volume, $\Delta\vec{r}\ \Delta E\ \Delta\Omega\ \Delta t$, in the seven-dimensional space of radiation transport. Simple conservation principles dictate that the rate of change in the number of particles per unit phase volume be equal to the amount of generation and removal of radiation within the volume. A systematic cataloging of these processes gives (Hussein, 2007):

$-\Sigma_t(\vec{r}, E, \vec{\Omega}, t)\, \phi(\vec{r}, E, \vec{\Omega}, t) =$ the *removal* rate of radiation from the element, by absorption or scattering, with Σ_t being the macroscopic total cross section of the material of the element.
$\int (\Sigma_s(\vec{r}, E', \vec{\Omega}' \rightarrow E, \vec{\Omega}, t)\, \phi(\vec{r}, E', \vec{\Omega}', t)\, dE' d\vec{\Omega}' =$ the rate of *scattering into* the element, of radiation with energy E in direction $\vec{\Omega}$, from all other energies, E' and directions $\vec{\Omega}'$, with $\Sigma_s(\vec{r}, E', \vec{\Omega}' \rightarrow E, \vec{\Omega}, t)$ being the differential scattering cross section at $\vec{r}$ from the state $(E', \vec{\Omega}')$ to the state $(E, \vec{\Omega})$.
$\int \nu \Sigma_g(\vec{r}, E', \vec{\Omega}' \rightarrow E, \vec{\Omega}, t)\, \phi(\vec{r}, E', \vec{\Omega}', t)\, dE' d\vec{\Omega}' =$ the *generation* rate of new radiation particles by some interaction producing particles with energy E and direction $\vec{\Omega}$ within the element, with $\Sigma_g(\vec{r}, E', \vec{\Omega}' \rightarrow E, \vec{\Omega}, t)$ being the differential generation cross section, and ν is the number of radiation particles produced per generation interaction.
$Q(\vec{r}, E, \vec{\Omega}, t) =$ the rate of introducing particles within the element by an external source.

The balance equation within the finite phase volume then becomes:

$$\begin{aligned}\frac{1}{v}\frac{d}{dt}\phi(\vec{r}, E, \vec{\Omega}, t) &= -\Sigma_t(\vec{r}, E, t)\phi(\vec{r}, E, \vec{\Omega}, t) \\ &\quad + \int \left(\Sigma_s(\vec{r}, E' \rightarrow E; \vec{\Omega}' \rightarrow \vec{\Omega}, t) + \nu\Sigma_g(\vec{r}, E' \rightarrow E; \vec{\Omega}' \rightarrow \vec{\Omega}, t)\right) \\ &\quad \times \phi(\vec{r}, E', \vec{\Omega}', t)\, dE'\, d\vec{\Omega}' + Q(\vec{r}, E, \vec{\Omega}, t) \end{aligned} \tag{2.11}$$

The left-hand-side of Eq. (2.11) is divided by the magnitude of the velocity, v, corresponding to E, to produce the net rate of change of the number of particles per unit volume. In addition, the left-hand-side of Eq. (2.11) is written as a *total derivative*, also called the material derivative, because the balance portrayed by the equation is what would be recorded by an observer moving with the radiation particles. An observer in the laboratory would see a *streaming* of radiation through the volume element; i.e. a spread of radiation in space, leading to:

$$\frac{1}{v}\frac{d}{dt}\phi(\vec{r}, E, \vec{\Omega}, t) = \frac{1}{v}\frac{\partial}{\partial t}\phi(\vec{r}, E, \vec{\Omega}, t) + \vec{\Omega} \cdot \nabla\phi(\vec{r}, E, \vec{\Omega}, t) \tag{2.12}$$

The *streaming* term, $\vec{v} \cdot \nabla n(\vec{r}, \vec{v}, t)$, accounts for the rate of particles entering the volume element minus those leaving it. That is, even in a vacant (material free) volume element, the number of particles leaving the volume would be less than that entering it, due to divergence (spread) of radiation particles. Therefore, the transport equation, in the laboratory frame of reference, takes the form:

$$\begin{aligned}\frac{1}{v}\frac{\partial}{\partial t}\phi(\vec{r}, E, \vec{\Omega}, t) &= -\vec{\Omega} \cdot \nabla\phi(\vec{r}, E, \vec{\Omega}, t) - \Sigma_t(\vec{r}, E, t)\phi(\vec{r}, E, \vec{\Omega}, t) \\ &+ \int \left(\Sigma_s(\vec{r}, E' \to E; \vec{\Omega}' \to \vec{\Omega}, t) + \nu\Sigma_g(\vec{r}, E' \to E; \vec{\Omega}' \to \vec{\Omega}, t)\right) \\ &\times \phi(\vec{r}, E', \vec{\Omega}', t)\, dE'\, d\vec{\Omega}' + Q(\vec{r}, E, \vec{\Omega}, t) \end{aligned} \tag{2.13}$$

This is the Boltzmann particle transport equation, which was originally derived in 1872 for the kinetic theory of gases. It is the most comprehensive equation for describing radiation transport. The transport equation is, in many ways, quite complicated, not only because of its many terms, but also because it has seven independent variables. Even at a steady state, with only six independent variables, the two variables for direction, $\vec{\Omega}$, are quite difficult to incorporate into computations. This is because, for all other variables, one can establish points of reference from which the variable is measured (point of origin in space, some zero reference time and a zero energy). A particle's direction, on the other hand, is measured with respect to the position of the particle in space, which changes as the particle is transported. This makes it quite difficult to determine a particle's direction with respect to an absolute reference. Although the spherical harmonics and discrete ordinates methods, see for example Hussein (2007), are devised to overcome this problem, they are too complicated to be implemented in formulating measurement models for the forward problem. Forward mapping relies, therefore, on simplified forms of the transport equation. But it should be kept in mind that we are simplifying an equation that itself is based on assumptions, indicated in this chapter by the uppercase Roman numerals.

2.5 Source-Free Steady-State Problem

Let us solve the transport equation at (i) steady-state within a domain (ii) away from any external sources of radiation. These conditions are easily satisfied in most imaging problems, performed under steady-state conditions, with the radiation source outside the object. The domain of the problem solution is that of the imaged object. The effect of the external source can then be defined by a boundary condition that specifies the flux at the surface of the object on which a source impinges. Even when dealing with emission imaging, where sources are embedded within the object, one can define a domain that excludes the regions containing radiation sources and assign appropriate internal boundary conditions at the surface of the source-containing regions.

Let us further (iii) narrow the domain of radiation transport for imaging to the domain seen by a detector. Since detectors used in imaging are usually collimated to monitor a narrow field-of-view, further confinement of the transport domain is not unreasonable; yet it is a simplification. With a narrow domain, one can assume that (iv) any radiation scattered within the domain leaves the beam and does not contribute to the radiation flux within the domain. Let us assume that (v) radiation scattering into the domain of interest from the outside has no significant effect on the radiation flux within the domain. Moreover, let us assume that (vi) the absorbed radiation does not result in the generation of more radiation that contributes to the flux within the studied domain. With assumptions (i) to (vi), the last three terms (for scattering, generation, and external source) in Eq. (2.13) can be dropped. In a steady-state problem, Eq. (2.13) is reduced to:

$$-\vec{\Omega}\cdot\nabla\phi(\vec{r},E,\vec{\Omega})-\Sigma_t(\vec{r},E)\phi(\vec{r},E,\vec{\Omega})=0 \tag{2.14}$$

where the time, t, was dropped, since the problem considered is at a steady state. Now Σ_t signifies complete removal of radiation from the domain, by absorption and/or scattering to the outside of the domain. Then there is (vii) no reason for the source radiation to change energy or direction, since the streaming term does not affect radiation energy or direction, while absorption gets rid of radiation altogether and scattering moves out of the domain of interest. This condition can be considered as an assumption that removes the scattering term in Eq. (2.13), since radiation changes its energy and direction only if there is scattering. With no change in direction, a radiation particle moving in a particular direction will keep moving in that direction (in a straight line) until it is fully absorbed. It is then possible to consider one direction at a time and solve Eq. (2.14) in that direction. This reduces Eq. (2.14) in effect to a one-dimensional problem in the direction of $\vec{\Omega}$. Let us designate a distance along $\vec{\Omega}$ by R, then Eq. (2.14) can be written as:

$$\begin{aligned}-\frac{d\phi(R,E,\vec{\Omega})}{dR}&=\Sigma_t(R)\phi(R)\\ \frac{d\phi(R,E,\vec{\Omega})}{\phi(R,E,\vec{\Omega})}&=-\Sigma_t(R)dR\end{aligned} \tag{2.15}$$

Here the position vector, $\vec{r}$, is removed, since it is defined by R, assuming that $R=0$ is at the surface of a detector within the domain of the radiation beam downstream of the direction, $\vec{\Omega}$. Integrating from $R=0$ to some value R along $\vec{\Omega}$, one obtains:

$$-\int\frac{d\phi(R,E,\vec{\Omega})}{\phi(R,E,\vec{\Omega})}=-\int_0^R\Sigma_t(R)dR$$

$$-\ln\frac{\phi(R,E,\vec{\Omega})}{\phi(0)}=-\int_0^R\Sigma_t(R)dR$$

$$\phi(R,E,\vec{\Omega})=\phi(0,E,\vec{\Omega})\exp\left[-\int_0^R\Sigma_t(R)dR\right] \tag{2.16}$$

$$\phi(R,E,\vec{\Omega})=\phi(0,E,\vec{\Omega})\exp[-\Sigma_t R] \tag{2.17}$$

The latter integration step is possible when (viii) the cross section, Σ_t, remains constant within the domain, i.e. when the domain contains a single homogeneous material.

Equations (2.16) and (2.17) are two different forms of the attenuation law of radiation, previously derived from probabilistic principles in Eq. (2.7) and using a basic attenuation principle in Eq. (2.3). However, the above thorough derivation from the Boltzmann transport equation elucidates the many assumptions associated with its derivation, indicated by the lowercase Roman numerals. This simple, and widely used, radiation attenuation relationship should be applied with great caution. The flux in Eqs. (2.16) and (2.17) is intentionally designated with E and $\vec{\Omega}$ to remind ourselves that this attenuation relationship is applicable only to radiation beams in a particular direction and when the radiation energy does not change. It is best suited, therefore, for simulating the transmission of (ix) narrow radiation beams. The consideration of a single direction in arriving at this law of attenuation (x) does not allow for the accounting of radiation spread with distance.

Since radiation particles cannot be confined to a narrow beam for a long distance, this attenuation law is (xi) not valid over a long distance of radiation travel. However, normalizing a measured transmitted flux to that recorded in the absence of the actuating material (i.e. in void or in air, where Σ_t can be considered to be equal to zero) eliminates the divergence effect. The reasons behind the presence of radiation divergence becomes clear when we apply the transport equation in a voided domain, as shown below.

2.6 Steady-State Problem in Void

In Section 2.5, external sources were considered to be outside the domain of the problem. When this is actually the case, the source would be located in air some distance from the object. In many situations, the attenuation of radiation in air is so small that it can be neglected, i.e. Eq. (2.17) will lead to no change in flux. Obviously, this is not realistic, since we know that the intensity of a radiation source decreases with distance, even in void. To quantify this change in flux with distance, let us consider the transport equation in void at steady state, in the presence of an external source. Then, all terms in Eq. (2.13) disappear, since there are no interactions or transient changes, except for the streaming and source terms, and one has:

$$\vec{\Omega}\cdot\nabla\phi(\vec{r},E,\vec{\Omega}) = Q(\vec{r},E,\vec{\Omega}) \tag{2.18}$$

Let us consider, as done in Section 2.5, a distance R along $\vec{\Omega}$. Then, one can express $\vec{\Omega}\cdot\nabla\phi(\vec{r},E,\vec{\Omega})$ as a partial derivative with respect to R, which reduces Eq. (2.18) to:

$$\frac{\partial}{\partial R}\phi(\vec{r},\vec{\Omega}) = Q(\vec{r},\vec{\Omega}) \tag{2.19}$$

Consider a *point isotropic source* of intensity Q_0 positioned at $\vec{r}_0$: $Q(\vec{r},E,\vec{\Omega}) = \frac{Q_0(E)}{4\pi}\delta(\vec{r}_0)$, with the delta function indicating that the source exists only at one spatial point in space. With this point source, one can integrate both sides of Eq. (2.19) along an entire infinite void space:

$$\int_{-\infty}^{\infty} \frac{\partial}{\partial R}\phi(\vec{r}, E, \vec{\Omega})\, dR = \phi(\vec{r}, E, \vec{\Omega}) = \frac{Q_0(E)}{4\pi} \int_{-\infty}^{\infty} \delta(\vec{r} - R_0\vec{\Omega})\, dR \tag{2.20}$$

The argument of the delta function in the above formulation indicates that the integral is from the source location along $\vec{\Omega}$ to the point $\vec{r}$ at which the flux is determined. Since the source is assumed to be isotropic, the flux should be the same in all directions, and there is no need to express it explicitly in terms of $\vec{\Omega}$. Integrating Eq. (2.20) over Ω gives:

$$\begin{aligned}\phi(\vec{r}, E) &= \frac{Q_0(E)}{4\pi} \int_{-\infty}^{\infty} \int_{4\pi} \delta(\vec{r} - R_0\vec{\Omega})\, dR\, d\Omega \\ &= \frac{Q_0(E)}{4\pi} \int \frac{\delta(\vec{r} - R_0\vec{\Omega})}{R^2}\, dV \\ &= \frac{Q_0(E)V}{4\pi R_0^2}\end{aligned} \tag{2.21}$$

since $d\Omega = \frac{dA}{R^2}$ and $dV = dRdA$, where dA and dV are infinitesimal intervals of area and volume, respectively.

Equation (2.21) is identical to Lambert's first law of optics, widely known as the inverse-square law of radiation divergence. The derivation from the transport equation indicates that this law is valid only for *a point isotropic source in void*. This law affirms the basic fact that the total intensity of a radiation source, $Q_0(E)V$, is constant, and as such the flux, being intensity per unit area, decreases with the increase of its projected area, consequently with R_0^2, the distance from a point on the source to the location where the flux is determined. Since the source energy does not change in void, the flux will have the same energy as that of the source, as Eq. (2.21) shows. For a volume source, one can integrate the flux evaluated for many non-overlapping points in the source volume to obtain the total flux.

2.7 Point-Kernel Method

In a non-void medium, one can superimpose the attenuation law of Eqs. (2.16) or (2.17) with the divergence law of Eq. (2.21), by accounting for the attenuation of one direction at a time. Then one obtains the so-called *uncollided* flux, since the attenuation law accounts only for particles that succeeded in surviving any interaction (all other particles are considered to be outside the domain in which the flux is evaluated, as discussed in Section 2.5). The combination of the two operations of attenuation and divergence gives the so-called flux *point kernel*, expressed as:

$$K(\vec{r};\vec{r}_0,E) = \frac{\exp[-\Sigma_t(E)|\vec{r}-\vec{r}_0|]}{4\pi|\vec{r}-\vec{r}_0|^2} \tag{2.22}$$

for the contribution of a point isotropic source $\vec{r}_0$ for the flux at $\vec{r}$. Then, the uncollided flux for a point isotropic source of strength, Q_0, is given by:

$$\phi_{uncollided}(\vec{r},E) = Q_0(\vec{r}_0,E)K(\vec{r};\vec{r}_0,E) = \frac{Q_0\exp[-\Sigma_t(E)|\vec{r}-\vec{r}_0|]}{4\pi|\vec{r}-\vec{r}_0|^2} \tag{2.23}$$

To deal with the limitation of the attenuation law, discussed in Section 2.5, a *buildup* factor (greater than one) can be introduced. This factor accounts for radiation scattering into the domain of radiation transport, which was ignored in the attenuation law. Then the flux in Eq. (2.23) becomes:

$$\phi(\vec{r},E) = B(|\vec{r}-\vec{r}_0|,E)\frac{Q_0\exp\Big[-\Sigma_t(E)|\vec{r}-\vec{r}_0|\Big]}{4\pi|\vec{r}-\vec{r}_0|^2} \tag{2.24}$$

To facilitate calculation of the buildup factor, it is often expressed as a series expansion in the form:

$$B(R,E) = 1 + a_1(E)(\Sigma_t R) + a_2(E)(\Sigma_t R)^2 + \cdots \tag{2.25}$$

with a's being coefficients evaluated experimentally or obtained from detailed calculations using the entire transport equation for a given problem.

2.8 Charged Particles

The Boltzmann transport equation, Eq. (2.13), is not applicable to charged particles, because of some of the assumptions upon which the equation is based. First, a charged particle travels in a curved path, since it is continuously affected by the Coulomb field of the atom and its nucleus. Therefore, charged particles violate the straight-line assumption of the transport equation. The continuous influence of the Coulomb field on charged particles results in a very large interaction cross section and no mean-free-path. Instead, a *range*, R, is defined, in terms of the energy-loss per unit distance, the so-called *stopping power*, $S = \frac{dE}{dr}$:

$$R(E_0) = \int_0^{R(E)} ds = \int_E^0 \frac{ds}{dE}dE = \int_0^E \frac{1}{\frac{-dE}{ds}}dE = \int_0^E \frac{dE}{S} \tag{2.26}$$

where E is the initial energy of the particle. The SRIM computer code (Ziegler, 1999) gives values for the stopping power and range of positively charged particles (ions) in solids, liquids, and gases. The range is a straight-line (crow-flight) equivalent distance, and as such can be used as an effective transport path from the start (source) point to the end (full absorption) point of a charged particle. This straight-line approximation is justified by the fact that charged-particle interactions tend to result in low momentum transfer, and as such the change in direction is not drastic. One must keep in mind that the range is only an estimate of the mean distance of travel, in the same direction as the source radiation, and as such the end point is not a definite point. A charged particle can stop either slightly before or after the distance dictated by the range value, due to the straggling effect, i.e. the statistical fluctuations of the value of the actual travel distance.

Another complicating effect of charged particles is that, as they move through the atomic field, they transfer energy to atomic electrons. The received energy can be sufficient to release atomic electrons from the shells they are bound to. The released electrons, called "delta" rays, also travel within matter depositing energy. In addition, the original charged particles, and the liberated secondary electrons, produce photons as they travel through the atomic electron field; these photons are known as the bremsstrahlung (radiation produced by braking). The emitted photons, in turn, liberate atomic electrons as they travel through matter. The newly liberated electrons release photons by the bremsstrahlung process, and so on. Therefore, a single charged particle produces in its wake a cascade of electrons, or a *shower* of electrons. This cascade of electrons eventually collapses as the released electrons become so low in energy that their associated bremsstrahlung photons cannot liberate more atomic electrons, and energy loss becomes dominated by collisions between electrons. Therefore, detailed analysis of the transport process can become so complex that it needs detailed Monte Carlo calculations. In many situations, though, it is sufficient to employ the range as a means for determining how far a charged particle would directly move.

References

M. J. Berger, J. H. Hubbell, S. M. Seltzer, J. S. Coursey, and D. Zucker, *XCOM: Photon Cross Section Database (version 1. 2), [Online]*, National Institute of Standards and Technology, NBSIR 87-3597, Gaithersburg, 1999, (http://physics.nist.gov/xcom, accessed, May 2010).

E. M. A. Hussein, *Radiation Mechanics: Principles and Practice.* Amsterdam: Elsevier, 2007.

Nuclear Data Evaluation Lab., "ENDFPLOT-2.0 online plotter for MCNP and ENDF cross section data," Korea Atomic Energy Research Institute (http://atom.kaeri.re.kr/endfplot.shtml), 2000.

F. J. Ziegler, "The stopping and range of ions in matter (SRIM-2000)," IBM-Research, Yorktown, NY, October 1999, (http://www.srim.org/, accessed May 2010).

3 Measurement Models

3.1 Formulation

The forward problem provides a mathematical formulation (a forward mapping) that produces an estimate of the value of a measurement, $e \in E$, given a set of physical parameters, $\boldsymbol{c} \in C$, as explained in Section 1.5. This mathematical formulation is sometimes referred to as a "physical model," but this terminology is avoided here, because it is also used to describe geometric or topological features. Instead, the term "measurement model" is used in this book to refer to the mathematical formulation of the forward mapping. In order to distinguish between a measurement obtained in the laboratory and the corresponding value estimated by a measurement model, we will refer to the latter as the *modeled* measurement.

3.2 Scaling

In formulating a measurement model one would aim at providing a best match with measurements, keeping in mind that measurement uncertainties are always associated with actual measurements. One must also keep in mind that the measurement model is to be used in the inverse problem, to generate the inverse mapping. If the measurement model is too complicated, so will the inversion process be. This will not only increase computational demands and slow-down the solution of the inverse problem, but it may make the generation of the inverse mapping quite difficult. A more complex modeling may also involve second order parameters that may not have much impact on the generated values. Very often it is not possible to numerically simulate all aspects of a physical system, even when powerful simulation tools are available. For instance, a radiation detector has a physical size and a response that depends on the energy of the incident radiation, and even on the direction of incidence of radiation on the detector. Moreover, when radiation deposits energy into a detector, it produces charged particles or light signals that diffuse through the detection medium in a manner that depends on the voltage applied on the detector, as well as on the detector's type and geometry. In addition, the source strength and size also affect the measured values. Such detector and source characterization details are often accounted for via calibration in the laboratory. In other words, to arrive at a measurement estimate that is comparable to an actual measurement, one must introduce a scaling constant, k, via calibration. Therefore, the forward mapping process of Figure 1.3 consists, in effect,

Computed Radiation Imaging. DOI: 10.1016/B978-0-12-387777-2.00003-3

of two sequential mappings, so that the mapping of Eq. (1.6) becomes:

$$c \xrightarrow{m_0} e_0; \quad c \in C; \quad e_0 \in E_0 \tag{3.1}$$

$$e_0 \xrightarrow{k} e; \quad e_0 \in E_0; \quad e \in E \tag{3.2}$$

where C, E_0 and E are, respectively, the spaces of physical parameters, modeled measurements and actual measurements. We used the zero subscript to indicate that the measurement model produces a numeric estimate of the measurement, via the m_0 mapping, with the calibration process introducing another mapping, k. In essence, we created an intermediate space, E_0, for the numerical estimate, e_0.

In order to facilitate numerical implementation of the forward model and its subsequent inversion, a simplified model, via the m_0 forward mapping, is usually sought. This results in, as shown in subsequent chapters for various imaging modules, idealized measurement models that ignore some of the secondary physical effects. The scaling mapping, k, helps somewhat in retrieving the effect of those secondary processes, by comparing the modeled measurement estimate, e_0, to an actual measurement, $\hat{e}$, obtained at high accuracy. However, calibration can only be done for certain reference objects, with known size and material content. Imaging is done, however, for an object that inevitably differs from calibration objects, and the scaling mapping, k, will not exactly account for the secondary physical effects incorporated in idealized measurement models; though k would generally account well for source and detector characteristics. In addition, no matter how accurate the calibration measurements are, they are never error-free. Therefore, at the end a modeled measurement, $e = ke_0$ will carry within it some modeling inexactness; see Section 1.5 for definition of inexactness.

The scaling parameter, k, is, in effect, a mapping of the incident radiation flux into a measurable electronic signal. This mapping is a function of the energy of the radiation incident on the detector and can also be affected by the direction and location of incidence as it determines the radiation's path length within the detector's sensitive volume. When the detected radiation is decomposed into components in energy and the decomposed components are utilized as individual c values in the imaging problem, the detector's measured pulse-height distribution must be "unfolded" with the detector's response function. This forms another inverse sub-problem of radiation detection that must be resolved before solving the inverse problem of image reconstruction. This book does not directly address this unfolding problem, but all the methods described in Part II for solving a linear inverse problem can be used for the detector unfolding problem, provided that the detector's response matrix is available from prior measurements or calculations.

3.3 Measured Response

Measurement modeling to obtain the mapping, m_0, can be avoided altogether by performing measurements to determine the system response, i.e. to find $m = m_0k$ directly. However, this is only doable in linear problems, where a change in a measurement, e_i, due to a change in a parameter, c_j, does not depend on changes in any other parameter, $c_{j'}$. If an imaging problem is linear or is transformed to a linear problem, one can then construct a response matrix, $\boldsymbol{A}$, that relates a vector of measurement estimates, $\boldsymbol{e}$, to a vector $\boldsymbol{c}$ by:

$$\boldsymbol{e} = \boldsymbol{A}\boldsymbol{c} \tag{3.3}$$

One can obtain a value for each element, A_{ij}, by fixing all values of $\boldsymbol{c}$ and recording a measurement, e_i, then increasing the value of c_i to $c_i + \Delta c_j$ and getting a measurement, $e_i + \Delta e_{ij}$. The two measurements are then related by:

$$\begin{aligned} e_i &= \sum_n A_{in} c_n \\ e_i + \Delta e_{ij} &= \sum_n A_{in} c_j + A_{ii} \Delta c_j \\ A_{ij} &= \frac{e_i + \Delta e_{ij} - e_i}{\Delta c_j} \end{aligned} \tag{3.4}$$

One can then construct the entire response matrix, $\boldsymbol{A}$, by varying the physical parameters one at a time, for each measurement. Obviously, this is a tedious process applicable only to linear problems, but it provides a measured response matrix and bypasses the mathematical modeling process. Detailed simulations, such as those involving Monte Carlo simulations, can be used instead of measurements, to generate the response matrix, $\boldsymbol{A}$. For linear problems, the response matrix, and consequently the forward mapping, do not change from one object to another and are not affected by changes in the material content.

3.4 Sensitivity

A simple measurement model enables one to determine the sensitivity of a measurement to changes in the physical parameters. For a measurement $e \in E$, the sensitivity to a measurement $c \in C$ defines sensitivity, S, as:

$$S = \frac{\partial e}{\partial c} \tag{3.5}$$

For a linear system, as Eq. (3.4) indicates, an element A_{ij} of the response matrix, $\boldsymbol{A}$, defines the sensitivity of a measurement, e_i, to a system parameter, c_j. The minimum value of S gives the imaging system's lowest sensitivity, and the maximum value is the best sensitivity. An imaging system that is too sensitive to changes is quite susceptible to measurement fluctuations resulting from random variations or spurious changes. On the other hand, an imaging system whose measurements are not quite sensitive to changes in physical parameters will not produce good quality images. Using a simple measurement model, one can develop an analytical formulation for the sensitivity of an imaging system that can provide guidance on how sensitivity can be improved, e.g. by changing the source's intensity, energy or even type. To arrive at such a simplified measurement model, some assumptions must be made. These assumptions may impose some restrictions on the configuration of the imaging system and the method of measurements. By understanding the measurement model's inherent assumptions, one can identify the optimal conditions under which the measurements are valid, and strive to meet them. This can also help in diagnosing the reasons for obtaining poor images.

3.5 Variability

A measurement model enables the assessment of the susceptibility of an imaging system to fluctuations in measurements that are not caused by changes in physical parameters. Such changes could be due to random statistical variability, system instabilities, motion of imaged object, or even systematic errors caused by inaccurate assignment of model parameters. The variance, σ_e^2, in a measurement, $e \in E$, affects the variance, σ_c^2, in a parameter, $c \in C$, through the relationship:

$$\sigma_c^2 = \left(\frac{\partial c}{\partial e}\right)^2 \sigma_e^2 = \frac{1}{S^2}\sigma_e^2 \tag{3.6}$$

where S, the sensitivity defined by Eq. (3.5), is assumed to be finite in value. Equation (3.6) shows that a system with poor sensitivity will also propagate measurement "error" into the reconstructed value of a physical parameter. The word "error" is used here with caution, since it does not indicate erroneous measurements. For instance, statistical fluctuations, which are inherent in any radiation measurement, are a genuine physical variability that are not caused by flaws in the measurement method. Therefore, measurement noise often refers to random fluctuations.

3.6 Components

Imaging with radiation involves three basic components:

Source: that sends radiation into an object in the case of transmission, induced emission and scatter imaging or is emanated from within the object in case of embedded-emission imaging.

Transport: of radiation from the source through the object. During this transport process, the radiation must interact, in order for it to be able to convey information about the object after it leaves it.

Detector: measures the intensity of radiation that succeeds in leaving the object and reaches pre-designated locations.

Therefore, in order to construct a measurement model, one must be able to characterize the radiation source(s), model the transport process and simulate the response of the detector(s) employed in the imaging system.

3.6.1 Source

In order for a measurement to exist, a source of radiation must be present. This is usually an active source, i.e. a source introduced for the purpose of inducing a measurement; though measurements from passive (natural or artificial radioactive materials) sources are also possible. The source is external to the object in the case of transmission, induced emission and scatter imaging, but is internal (within the object) in embedded-emission imaging. The discussion in Chapter 2 indicates that, for a source, one needs not only to know its intensity, Q, (particles emitted per unit time), but also its position in space, $\vec{r}$, its energy, E, and direction, $\vec{\Omega}$. Each of these parameters can vary: Q with time; $\vec{r}$ according to the source size and shape, E as determined by the intrinsic energy spectrum of the radiation source; and $\vec{\Omega}$ by the angular distribution of the emitted radiation.

For an external source, volume and shape are dictated by the type of source used. The intensity of photons from an x-ray source and neutrons from a neutron generator (both produce radiation by accelerating charge particles that bombard a fixed target) can be controlled by the magnitude of the current (number of particles per second) of charged particles. A pulsed source of radiation can be obtained from an accelerator-based generator by pulsing the accelerated particles at a certain rate. The intensity of a pulsed source depends on the strength of each pulse and on the pulsing rate. The intensity of a radioisotopic source is a function of its activity in becquerels (Bq); with 1 Bq being one disintegration per second, but can decrease with time as the source decays, at a rate determined by its half-life. It should be kept in mind that the intensity of a source, whether accelerator-based or isotopic, is statistically variant, due to the random nature of nuclear interactions and the decay process; see Section 15.4.3.

In the case of an x-ray machine, the applied voltage determines the maximum energy of the emitted photons, while in a neutron generator, the neutron energy is governed by the nuclear reaction that takes place when the accelerated charged particles bombard the target. Radiation (photons or neutrons) emanating from a radioisotope, on the other hand, have an energy determined by the nature of the isotope and cannot be altered.

The angular distribution from a radioisotope is isotropic, since there is no reason for radioactive decay to prefer one direction over another. The distribution of neutrons from a neutron generator tend to be, however, biased in the forward direction, i.e. the direction of charged particle that bombard the target. The angular distribution from photons produced from an x-ray machine also has a favored direction that depends on the thickness, shape and material of the target and on the energy and angle of incidence of the electrons bombarding the target. Laboratory measurements can be performed to determine the distribution of an accelerator-based source. The field-of-view of a source can, however, be confined with the aid of a collimator. Since penetrating radiation, as indicated in Chapter 2, possesses corpuscular properties, it cannot be collimated with the aid of a lens, as is the case with radiation that has wave properties. Therefore, collimation of particle radiation is accomplished by elimination, that is, by a material that absorbs radiation and removes it from all directions except the desired direction. Simple ray-tracing from the source location through the hollow part of the collimator body can then be employed to determine the source's field-of-view. However, the collimator walls also scatter radiation, and some of that radiation may contaminate the collimated (unabsorbed) radiation.

The above discussion indicates that even the simulation of a radiation source, and associated collimation, is not as straightforward as it may appear at first glance. In simplifying the modeling of the radiation source in an imaging problem, one must be aware of the above discussed factors and their influence on the measurement model. Details on the nature of various radiation sources can be found in Hussein (2003a), and the design and characterization of collimators is given in Hussein (2003b).

Based on the above discussion, one can conclude that modeling a source may require discretization or integration of source parameters in time, space, energy and angle. A discretization process involves the *lumping* of a continuous parameter into definite intervals and considering one value within each interval (typically the mean value) as representative of the interval. For example, a volume source can be represented by a number of contiguous small elementary volumes, with each volume element simulated as a point source located at the center of the element. Similarly, the energy spectrum of a source (if not monoenergetic) can be represented by a number of energy groups spanning the entire spectrum. A fan/cone-beam can also be simulated as a set of rays, each at a distinct direction. The intensity of a source, with time-varying intensity, can be taken as the time averaged intensity within the measurement time. While these discretization processes enable more inclusive representation of source parameters, care must be taken in selecting the proper intervals, so that source parameters are reasonably represented. Taking a small number of wide intervals simplifies the formulation of a measurement model, but an averaging process within a wide interval smoothes out abrupt changes. On the other hand, too many fine intervals can be computationally demanding, not only for the forward mapping, but also for the subsequent inverse mapping.

3.6.2 Detector

The transport process is quite complicated, as discussed in Chapter 2. However, this process can be simplified using the point-kernel method of Section 2.7; keeping in mind the many idealizations involved in arriving at such models. This method is applied to the different models of imaging, as discussed in the ensuing chapters. The simulation of the transport process, whether simple or sophisticated, determines the flux at a certain location, particle energy and direction at a given time. The value of the flux at the position of a detector is used to indicate the intensity of radiation. However, a detector has its own intrinsic response function that determines the amplitude of its response to an incident radiation of a certain energy. The direction with which radiation impinges on a detector is not usually a major factor, since a radiation particle "forgets" its initial direction after encountering a few collisions, within a detector, before being fully absorbed. The response function of a detector depends on its nature, i.e. the physical process it relies on to convert an absorbed neutral radiation particle into a measurable electronic signal, as well as on the shape and size of the detector.

In addition, not every radiation particle incident on a detector will be absorbed, and, hence, a detector also has an efficiency. A collimator is usually employed to "focus" the field-of-view of a detector. This field-of-view tends, however, to widen with distance due to the nature of collimation-by-elimination, which can affect the amount of radiation reaching a detector. Moreover, the internal walls of a collimator can redirect radiation, by scattering, toward the detector. Detectors may also produce spurious signals due to random exposure to cosmic radiation, thermal effects or microphonic effects. The electronic components used to amplify, shape and acquire a detector's signal can also introduce noise that can affect the detector's response. In addition, background radiation can contribute to a detector's response. Background radiation can result from natural radioactivity, scattering of source radiation on surrounding walls and floors and adjacent detectors (in the case of detector arrays), and even from direct leakage from the source to the detector. It is obvious from the above discussion that using the flux, produced even from a detailed modeling of radiation transport, may not be a direct measure of a detector's response. Therefore, one should be aware of the role a detector plays in arriving at a measurement model.

Similar to source parameters, detector parameters can also be discretized. The volume of a detector can be represented by a set of point detectors, and a detector response can be calculated within a number of energy groups and within angular intervals. Moreover, in problems where time variation is monitored, the detector response can be recorded within discrete time periods. As in source discretization, the number of selected intervals should be large enough to give proper representation of the detector response, but not too large to be computationally cumbersome.

3.7 Image Grid

A digital image consists of a number of small contiguous elements, with each element having a numerical value directly representing a physical parameter, $c \in C$. These elements usually form a cuboid grid and cover the entire imaged object; elements outside the object are considered to be empty (air filled). An image element is known as a *pixel* (picture element) in a two-dimensional (sectional or a slice) image and is called a *voxel* in a three-dimensional (volume) image. A sectional image is known as a *tomograph*. Since penetrating radiation is not affected by surfaces, it requires a volume element; as shown in the formulation of the radiation transport equation in Chapter 2. Therefore, a pixel in an image must have a thickness that defines an associated volume. This thickness is that of the section for which a tomograph is reconstructed. Square pixels are typically employed in imaging, but a corresponding voxel is not necessarily a cube, since a slice's depth may not be equal to the width of a pixel. A set of contiguous tomographs form a volume image. The size of a voxel (or pixel) defines the *spatial resolution* of an image.

The discretization of a volume into small elements for the purpose of generating an image necessitates the application of the forward model over discrete values: $c_j \in C$. Each value c_j corresponds to a voxel j in the image. This implies that each voxel, j, is considered to contain a single material with a physical property defined by c_j such that:

$$c_j = \frac{\int_0^{V_j} c \, dV}{V_j} \tag{3.7}$$

where V_j is the volume of voxel j. Obviously, if voxel j contains a homogeneous material, then c_j is an accurate indicator of the physical property of the material within the voxel. On the other hand, if a voxel happens to contain more than one material, then c_j will represent a *volume-averaged* value of the physical property. For instance, if a voxel exists at the interface between two materials, partially covering part of each, the averaging (or homogenization) produces a c_j value weighted by the volume of each material in the voxel. The result is a loss of image *sharpness*, with the interface between the two materials appearing blurred (less distinct). An extreme case occurs at the external surface of an object, where a voxel may be partially filled with the object material with the rest being air. Since air is quite low in density, its effect on a material property induced by penetrating radiation is practically nil. Then a voxel, partially covered with air, will have a c_j value less than that of the material present within the rest of the voxel.

Naturally, it is desirable to obtain a sharp image with good spatial resolution. Therefore, the voxel size should be as small as practically possible. The smaller the voxel's volume, the more voxels one would have in the reconstructed image, the more representative the c value of the voxel's contents is, and the better the image quality is. There is, however, a limit on how small the voxel volume can be. A larger number of voxels in an image requires a correspondingly large number of measurements. A very small voxel also has a very feeble effect on measurements, which in turn requires the acquisition of more accurate measurements, by using a more intense source and/or a long counting period. The forward mapping of c to e via a measurement model also becomes susceptible to the accumulation of numerical (truncation and round-off) errors, as the voxel size decreases. Such errors can then *diffuse* to the reconstructed images, reducing its quality. There is no need in practice to provide an image with a resolution better than the size of the smallest details one is interested in.

3.8 Idealization

It is obvious from the above discussion that developing a measurement model that simulates physical reality is an arduous task. The best simulation models have their own physical limitations, and, even when used, incorporating accurately all physical details can be cumbersome and often not possible. It is also desirable to have a simple measurement model for the forward problem, since it has to be inverted to reconstruct an image. Therefore, measurement models are usually idealized for simple conditions that reflect the parameter of primary interest (the image attribute), as shown in Chapters 4 to 6. The simplicity of a measurement model can be compensated for by a calibration process, as indicated in Section 3.2. Since imaging involves objects of different attributes, the calibration process should be performed using a reference image that is not very different from the imaged object.

3.9 Computer Coding

For use in image reconstruction, an analytical measurement model has to be formulated as a numerical algorithm, and subsequently coded into a computer program. This associated numerical implementation involves discretization or lumping of continuous variables. For example, a source with a small volume may be lumped into a single point at its centroid; source and/or detector continuous radiation energies may be represented as a set of discrete energy groups; an entire source/detector energy spectrum may be represented by an equivalent or an average energy; a narrow cone-beam of radiation may be treated as a pencil (parallel-ray) beam, etc. Such averaging and lumping numerical processes introduce further approximations that can affect the value of modeled measurements. Also, misunderstanding or misinterpretation of analytical formulations may lead to errors in their numerical implementation. The coding of the computer program itself is subject to human error. Therefore, one must ensure that a computer code for a measurement model is programmed as analytically and numerical intended. This process is known in computer simulation as code *verification*, as opposed to code *validation*, which is the process of demonstrating that the measurement model and its computer code reflect physical reality.[1] Roache (1998) provides detailed analysis for code validation and verification.

3.9.1 Verification

The purpose of code verification is to ensure that the computer code is not only free of *model-implementation errors* and *logic errors* in the model's specification and implementation, but also that the code is free of *data errors*; i.e. it uses the proper source and detector parameters and the correct geometry and material properties. The verification process does not end with the completion of computer coding, as one must ensure every time the code is used that there are no *input errors*; i.e. the input data is as intended. Comparing the simulation code results against hand (or spreadsheet) calculations for simple problems and testing it under bounding (extreme) conditions, as those of maximum or minimum response, are among the methods used in code verification.

3.9.2 Validation

The validation process examines how close the modeled measurements are to actual corresponding measurements. After introducing the proper calibration parameters that adjust the model's predictions to match those obtained experimentally for a reference setup, one must test the model's results against measurements obtained for a number of configurations different from those used in the reference setup. Validation is a continual process, as one must always ensure that the results of the model are reasonable and are within the expected space; i.e. $e \in E$.

[1] These definitions of verification and validation are along the same lines as those of the American Institute of Aeronautics and Astronautics: Guide for the Verification and Validation of Computational Fluid Dynamics Simulations, AIAA G-077-1998, 1998.

References

E. M. A. Hussein, *Handbook on Radiation Probing, Gauging, Imaging and Analysis: Basics and Techniques*. Dordrecht: Kluwer Academic Pub., 2003a, vol. I.

——, *Handbook on Radiation Probing, Gauging, Imaging and Analysis: Applications and Design*. Dordrecht: Kluwer Academic Pub., 2003b, vol. II.

P. J. Roache, *Verification and Validation in Computational Science and Engineering*. Socorro: Hermosa Publishers, 1998.

4 Transmission

4.1 Basic Model

The forward mapping of a radiation imaging technique that relies on transmission employs a measurement model based on the exponential attenuation law, Eq. (2.17). For a pencil-beam source, the intensity (particles per unit time), $I(x)$, of radiation that succeeds in penetrating a material of thickness x along the beam's direction, see Figure 4.1, is given by:

$$I(x) = I_0 \exp\left[-\int_0^x \Sigma(s)\, ds\right] \tag{4.1}$$

$$p(x) = -\ln\frac{I(x)}{I_0} = \int_0^x \Sigma(s)\, ds \tag{4.2}$$

where I_0 is the intensity of the incident beam, $\Sigma(s)$ is the total cross section (attenuation coefficient) of the material at position s along the beam; $0 \leq s \leq x$.

The ratio $\frac{I(x)}{I_0}$ is sometimes called the *transmittance*, and the negative of its natural logarithm, p, is known as the *ray-sum*, or the *projection*. It signifies the integration of the physical parameter of interest here, Σ, with distance along the line of integration. Mathematically, the integral in the right-hand-side of Eq. (4.2) is the Radon transform of the projection of $\Sigma(s)$. Since the measured intensity, $I(x)$, determines the value of p, with I_0 known, then p can be considered to be directly the measurement, e, in the forward mapping: $c \to e$ with $c \equiv \Sigma$ and $e \equiv p$. One advantage of the projection formulation of Eq. (4.2) is that the normalization by I_0 eliminates detector effects (size and geometry), when the same detector is used to measure both I and I_0. This is because both I and I_0 have the same energy, hence same detection efficiency, and are in the same direction (as defined by the source and detector collimation), and as such have the same source and detector fields-of-view.

Recall from Chapter 2 that the attenuation law, as presented by Eq. (4.1), is burdened with many assumptions, the most important of which is that it is only applicable to a narrow parallel (non-diverging) beam of radiation; hence the use of the word "pencil" beam, so that any radiation that scatters off the beam does not contribute to the value of $I(x)$, hence to p. One may claim that this narrow beam assumption can

Computed Radiation Imaging. DOI: 10.1016/B978-0-12-387777-2.00004-5

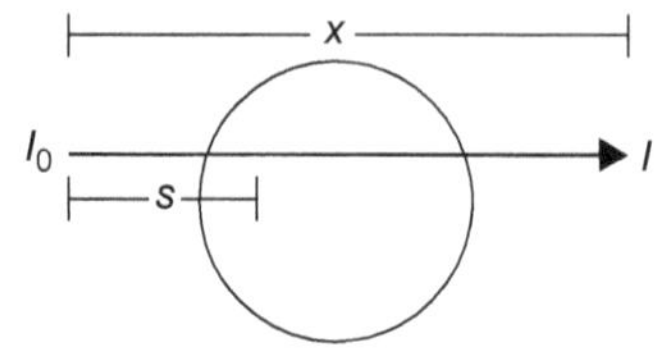

Figure 4.1 A schematic for the transmission of a pencil beam of radiation.

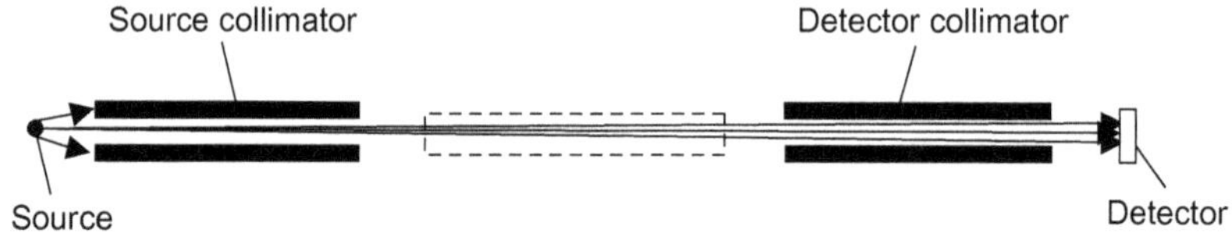

Figure 4.2 A schematic of a collimated source seen by a collimated detector.

be satisfied by collimating the source radiation into a pencil beam and confining the detector's field-of-view into a pencil-shaped configuration that matches that of the source. However, this is not practically possible, since radiation is not naturally emitted as a pencil beam and penetrating radiation collimation is accomplished by elimination, i.e. absorption of radiation. Therefore, as schematically shown in Figure 4.2, even with a narrow parallel collimator, the detected radiation will not appear to the detector as a parallel beam. Deviation from parallel collimation is more pronounced at shorter length of collimation, smaller source-to-detector distance and larger source size. On the other hand, a long collimator and a larger source-to-detector distance reduce the intensity of the detected radiation, due to the reduction in intensity by divergence, Eq. (2.21). The result is that the distance x in Eqs. (4.1) and (4.2) is underestimated if the pencil-beam distance is used. Therefore, it is desirable to perform the integration in these two equations over a set of pencil beams covering the field-of-view of the detector. This complicates, however, the solution of the inverse problem of solving for Σ, by making the measurement model of Eq. (4.2), and, in turn, its inversion more convoluted.

When the object exposed to radiation is viewed as consisting of small voxels, source divergence exposes more voxels to radiation as the beam travels through the object, due to the expansion of the source radiation field-of-view, as Figure 4.2 shows. Therefore, voxels downstream of the source beam appear to contribute more to the projection, p in Eq. (4.2). However, this apparent increase in contribution is dampened by the decrease in source intensity per unit area (flux) with divergence, in accordance to Eq. (2.21) for a point source. Moreover, by exposing the object to radiation from two opposite directions, i.e. by exchanging the source and detector positions, the effect of divergence on image voxels reverses its trend, providing a form of averaging that dampens the effect of divergence. Therefore, though opposite exposures to radiation may appear to be redundant, they in effect improve the quality of measurements by smoothing out the effect of divergence.

Before leaving the discussion on radiation divergence, one should consider incorporating it directly into Eq. (4.1). The latter equation was deduced from Eq. (2.17), which is in terms of flux, ϕ, rather than intensity, I, as in Eq. (4.1). Let us consider the flux and the point-kernel model of Eq. (2.23) and replace Q_0 in the latter equation with I_0 to be consistent with Eq. (4.1). Then the flux at some distance x from a source of intensity I_0 can be expressed as:

$$\phi(x) = \frac{I_0}{4\pi x^2} \exp\left[-\int_0^x \Sigma(s)\, ds\right] \tag{4.3}$$

$$= \phi(x)|_{\Sigma=0} \exp\left[-\int_0^x \Sigma(s)\, ds\right] \tag{4.4}$$

where $\phi|_{\Sigma=0}$ is the flux at distance x, if the intervening material were to have a zero attenuation coefficient. Then integrating the flux over the detector's sensitive area gives:

$$I(x) = I(x)|_{\Sigma=0} \exp\left[-\int_0^x \Sigma(s)\, ds\right] \tag{4.5}$$

Comparing Eq. (4.5) to Eq. (4.1) indicates that the source strength, I_0, in the latter is equal to the intensity of radiation after traveling a distance x in a medium with $\Sigma = 0$ (a condition usually satisfied for penetrating radiation in air). If the attenuation in air is not to be ignored, Eq. (4.4) would have to be written as:

$$\phi(x) = \phi(x)|_{\Sigma=\Sigma_{air}} \exp\left[-\int_0^x (\Sigma(s) - \Sigma_{air})\, ds\right] \tag{4.6}$$

where Σ_{air} is the attenuation coefficient of air (a constant value for the incident source radiation). Then:

$$I(x) = I(x)|_{\Sigma=\Sigma_{air}} \exp\left[-\int_0^x (\Sigma(s) - \Sigma_{air})\, ds\right] \tag{4.7}$$

Therefore, if the projection of Eq. (4.2) is to be obtained by normalizing the detected intensity at x to that at the same location, but in air, one would be evaluating the material's attenuation coefficient relative to that of air. The practical implication of this is that the source intensity, I_0, need not be known, and, if Σ_{air} is known, its value should be added to that obtained from the projection of Eq. (4.2) evaluated with respect to that measured in air, if a precise value of Σ is desired.

Another implication of Eq. (4.7) is that the distance of integration, x, must be known. In computed tomography, this is achieved by setting a spatial domain of fixed dimensions within which the attenuation coefficients are determined at pre-designated pixels (small picture elements). If the imaged object is smaller than the imaging domain, then the unfilled space will produce a zero attenuation coefficient. On the other hand, if the object happens to be larger than the imaging domain, the additional attenuation caused by the excess material will be added to those within the imaging domain, which will produce an image artifact with overestimated attenuation coefficients. It is, therefore, important to ensure that the entire imaged object resides within the pre-defined imaging domain.

4.2 Physical Relevance

Equations (4.1) and (4.2) provide the physical model for relating a measurement, and its projection, to the attenuation coefficient, Σ. For proper physical interpretation of the attributes of an image based on such measurement, one must understand the physical significance of Σ. As stated in Section 2.5, Σ is the total macroscopic cross section of the incident radiation for the material of the imaged object. The totality of Σ implies that it includes all interaction types that cause radiation removal from the incident beam, i.e. Σ encompasses both absorption and scattering. The macroscopic cross section, Σ, is related to the microscopic cross section per atom, σ, by the atomic density N, as Eq. (2.4) shows. Therefore, one must understand the nature of radiation interactions and how they affect σ, while keeping in mind that N is related to the density, ρ, and mass number of the material, A, by $N = \frac{\rho}{A\mathrm{u}}$, where u is the atomic mass unit, see Eq. (2.5). That is,

$$\Sigma(E) = N\sigma = \frac{\rho}{A\mathrm{u}}\sigma(E) \tag{4.8}$$

where the source energy, E, is included to emphasize that the microscopic cross section is also dependent on E. Equation (4.8) directly indicates that Σ is not only indicative of the material density, but is also affected by the nature of the material via A and σ. In other words, two materials of the same density can have two different values of Σ. To further understand the physical significance of Σ, we will consider the two types of radiation employed in imaging: photons and neutrons.

4.2.1 *Photons*

Photon interactions in the domain of x-rays and gamma-rays are governed by three main dominant interactions: (1) photoabsorption, (2) Compton scattering, and (3) pair production. Tabulated values for the photon cross sections for various reactions and different elements, compounds, and mixtures, at different energies, are provided by the XCOM: Photon Cross Sections Database `http://www.physics.nist.gov/PhysRefData/Xcom/Text/XCOM.html` (Berger et al., 1999).

In photoabsorption, also called the photoelectric effect, a photon is fully absorbed by the atom, liberating an orbital (bound) atomic electron. The microscopic cross section for this interaction, σ_a, can be approximately expressed as (Hussein, 2003):

$$\sigma_{pe}(E) \approx 12.1\frac{Z^{4.6}}{E^{3.15}} \times 10^{-24}\,\text{cm}^2 \tag{4.9}$$

where E is the photon energy in keV and Z is the atomic number of the atom absorbing the photon. It is obvious that σ_{pe} is highest at low energy and for heavy (high Z) metals and is strongly dependent on both the atomic-number and the incident energy of the photon. The photoabsorption cross section also exhibits sharp edges when the photon energy approaches the binding energy of an atomic orbit, in particular the electron rich K and L shells. When E is below the binding energy of an electron shell, σ_{pe} declines sharply, since the electron cannot be ejected from the shell, producing the so-called "absorption edges." These absorption edges, being electron-level dependent, are unique (characteristic) for each element.

The other dominant photon interaction mechanism is Compton scattering, in which the photon in a particle-like collision is scattered by a free electron in the atom, losing some energy and changing its direction in the process. The macroscopic cross section, σ_{cs}, is approximately given by (Hussein, 2003):

$$\sigma_{cs}(E) \approx 0.665Z\frac{1+\frac{2E}{m_ec^2}}{1+2\left(\frac{E}{m_ec^2}\right)^2} \times 10^{-24}\,\text{cm}^2 \tag{4.10}$$

where $m_ec^2 = 511\,\text{keV}$ is the rest-mass energy of an electron. Therefore, the microscopic cross section for photons tends to decrease with increasing energy and atomic number, i.e. is roughly proportional to $\frac{Z}{E}$.

The third common mode of photon interaction is that of pair production, in which a photon is annihilated in the presence of the strong electromagnetic field of a high atomic-number atom, leading to the generation of an electron-positron pair. This requires a photon energy of at least 1.022 MeV (= rest mass of the pair = $2m_ec^2$). The microscopic cross sections for pair production, σ_{pp}, is approximately given by (Hussein, 2003):

$$\sigma_{pp}(E) = 0.579Z^2 \times 10^{-27}\,\text{cm}^2; \qquad E > 1.022\,\text{MeV} \tag{4.11}$$

Therefore, this interaction is only relevant in industrial imaging where accelerator-generated high-energy x-rays are used to image high Z materials.

Now focusing on the photoelectric effect and Compton scattering, which are the two most likely encountered photon interactions, the total interaction cross section, Σ, using Eqs. (4.8), (4.9) and (4.10), becomes:

$$\Sigma(E) \approx \frac{\rho(\text{g/cm}^3)}{1.6606}\left(\frac{Z}{A}\right)\left[12.1\frac{Z^{3.6}}{E^{3.15}} + 0.665\frac{1+\frac{2E}{m_ec^2}}{1+2\left(\frac{E}{m_ec^2}\right)^2}\right]\text{cm}^{-1} \tag{4.12}$$

with ρ expressed in g/cm^3, and u in grams, u = 1.660565×10^{-24} g. Since $\frac{Z}{A} \approx 0.5$, for most materials except hydrogen, Eq. (4.12) shows that at not-too-low photon energy, when Compton scattering dominates, Σ becomes proportional to ρ for a given source energy, E. However, at low photon energy, the material composition, via the atomic number Z, has an effect on the value of Σ, and the cross section ceases to be entirely dependent on ρ, at a given source energy.

4.2.2 Neutrons

The total macroscopic cross section for neutrons is also dependent on material density, through the atomic density, N, in Eq. (4.8). However, it is strongly dependent on the nature of the medium via the mass number, N, and the microscopic cross section, σ. The latter cross section is more difficult to relate to material composition than that for photons. In fact, when employing neutron transmission imaging, one must look for a material with cross sections that are distinctly different from those of its surrounding. For example, one can examine the presence of hydrogen, or a hydrogen-rich material, within a metallic structure, since hydrogen has a much higher neutron removal cross section than most metals. A good source for neutron microscopic cross sections can be found in the Nuclide Table of the Nuclear Data Evaluation Laboratory, Korea Atomic Energy Research Institute (`http://atom.kaeri.re.kr/`).

4.3 Discretization

When numerically implementing any of Eqs. (4.1) to (4.7), one must deal with the integrals they contain. The image reconstruction process is a discrete process in which the value of Σ is evaluated at small pixels or voxels of fixed area or volume. It is essential, therefore, to write the above equation in a discrete form, by replacing the integrations with summations. For example, Eq. (4.1) has to be expressed as:

$$I(x) = I_0 \exp\left[-\sum_{i}^{N} \Sigma_i \Delta x_i\right] \tag{4.13}$$

where Δx_i is the distance traveled by the incident radiation in voxel i, with N being the number of voxels encountered by the incident radiation before leaving the attenuating object. The formulation of Eq. (4.13) implicitly assumes that Σ_i in voxel i has a constant value, i.e. the material within the voxel is uniform, or homogenized, to provide a constant value if more than one material exists within the voxel. This is, in effect, a process of "lumping" the contents of a voxel into a point. One then would desire to have as small a voxel as possible, so that the effect of lumping is minimized. In practice, however, there is a limit on the size of the voxel one can choose. The more voxels one has, the more measurements are required, since the content of each voxel presents an unknown parameter for which at least one measurement is to be provided. A large number of measurements is not only demanding in terms of acquiring the measurements themselves, but also increases the computational difficulty in terms of computer storage and manipulation. In addition, to resolve the contents of a small voxel from other voxels in the same projection, the contribution of that voxel to the projection must exceed the statistical variability in the recorded value of the projection. A low statistical variability requires either an intense radiation source or a long exposure time, conditions which are not always easy to achieve in practice.

Another inherent limitation of the discretization in Eq. (4.13) is that the value of Δx_i must be known. This value may be found by ray-tracing the radiation passage from the source position to the detector position for a given projection. However, in order to trace a ray, its point of origin at the source position and its end point at the detector must be well defined. A source of radiation in never a single point, neither is a detector, but each can be considered to consist of many contiguous points. This by itself is another discretization process for the source and the detector and demands that all source points be connected to all detector points. The result is that, in each voxel, i, many values of Δx_i can exist, each corresponding to a different source-point to detector-point trace. Using the average value is a sensible solution, provided that the source is isotropic (same intensity for all traced rays) and the detector efficiency is not directionally dependent (for each incident ray). These conditions are not difficult to meet in practice, but one must keep in mind that the averaging process is only meaningful if the variance of the averaged values is small in comparison to the mean value, i.e. if the intersection values for each ray crossing a certain voxel are not too different from each other. If the latter condition is not satisfied, the average value would not be representative of the distances of radiation travel within a pixel.

4.4 Nature of Radiation Source

So far, we have not mentioned the radiation type and its effect on the measurement model for transmission. The exponential relationships associated with the models of Eqs. (4.1) to (4.7) are applicable only to neutral radiation,[1] and they inherit the assumptions implicit in the application of the attenuation law, as discussed in Section 2.5. Direct application of those equations, to obtain a value of Σ, requires that the incident radiation be monoenergetic, so that Σ assumes a fixed value determined by the source's single energy. However, radiation employed in transmission imaging is not always monoenergetic, as explained below for various types of radiation.

4.4.1 Gamma Rays

Gamma radiation is employed in industrial applications because of the higher energy, hence the higher penetrability, it provides. Gamma-rays are emitted from unstable nuclei as they decay from excited nuclear levels to more stable ones. Therefore,

[1] The exponential attenuation relationship is also empirically applicable to beta-rays because of their continuous energy spectrum, which results in them stopping in matter gradually at different distances.

Table 4.1 Properties of Typical Gamma Radioisotopes Used in Transmission Imaging (Periodic Table of the Isotopes: `http://ie.lbl.gov/education/isotopes.htm` and XCOM: Photon Cross Sections Database: `http://www.physics.nist.gov/PhysRefData/Xcom/Text/XCOM.html`).

Isotope	Half-Life	Photon Energy (keV)	Intensity (%)	Attenuation coefficient in: H_2O	Al (cm^{-1})	Fe
^{169}Yb	32.03 d	63.12	44.2	1.89E-01	6.33E-01	7.76
		109.78	17.47	1.62E-01	4.11E-01	2.31
		177.21	22.16	1.41E-01	3.37E-01	1.18
		197.96	35.80	1.36E-01	3.23E-01	1.05
^{75}Se	119.78 d	121.12	17.20	1.58E-01	3.95E-01	2.04
		136.00	58.30	1.53E-01	3.75E-01	1.69
		264.66	58.90	1.24E-01	2.93E-01	0.88
		279.54	24.99	1.22E-01	2.87E-01	0.85
^{192}Ir	73.83 d	295.96	28.68	1.19E-01	2.79E-01	0.81
		308.46	30.00	1.14E-01	2.76E-01	0.79
		316.51	82.81	1.10E-02	2.73E-01	0.78
		468.07	47.83	9.89E-02	2.33E-01	0.65
		604.41	8.23	8.91E-02	2.09E-01	0.57
		612.47	5.31	8.87E-02	2.08E-01	0.57
^{137}Cs	30.07 y	661.66	85.10	8.61E-02	2.02E-01	0.54
^{60}Co	5.27 y	1173.24	99.97	6.55E-02	1.54E-01	0.42
		1332.50	99.96	6.13E-02	1.44E-01	0.39

gamma sources are characterized by discrete energy levels that vary in intensity, depending on the decay probability of each nuclear level. The common sources currently in use, in order of ascending photon energy, are: ytterbium-169, selenium-75, iridium-192, cesium-137 and cobalt-60. Table 4.1 gives their most dominant photon energy levels and their corresponding attenuation coefficients in water, aluminum and iron, with nominal densities of 1,000, 2,700, and 7,600 kg/m^3, respectively. It is obvious from this table that the attenuation coefficient for a given source can vary considerably for the same material, particularly for low-energy sources. It is, therefore, important to formulate a measurement model that accounts for the changes in source energy and the relative intensity of the radiation emitted at a given energy. Note that in Table 4.1 the relative intensities can exceed 100%, since the radiation intensity is per one distintegration, not per photon. For example, ^{60}Co emits two photons per disintegration.

Let $\Sigma(E_i, s)$ be the attenuation coefficient for the material at position s, of photons emitted from a radioisotope at a relative intensity χ_i per photon, so that $\sum_i \chi_i = 1$.

Then the attenuation relationship of (4.1) can be expressed as:

$$I(x) = I_0 \sum_i \chi_i \exp\left[-\int_0^x \Sigma(E_i, s)ds\right] = \sum_i I(E_i, x) \tag{4.14}$$

$$p(x) = -\ln\frac{I(x)}{I_0} = -\ln\left\{\sum_i \chi_i \exp\left[-\int_0^x \Sigma(E_i, s)ds\right]\right\} \tag{4.15}$$

$$p(E_i, x) = -\ln\frac{I(E_i, x)}{\chi_i I_0} = \int_0^x \Sigma(E_i, s)ds; \quad p(x) \neq \sum_i p(E_i, x) \tag{4.16}$$

where $p(E_i, x)$ is the projection along x, if the transmitted radiation is monitored at each individual energy, E_i. The variation of Σ with energy makes it difficult to reconstruct unique values of Σ from $p(x)$ without energy discrimination. On the other hand, energy discrimination offers the opportunity to extract more than one value for Σ for the same material, a feature that can be utilized for material identification. For example, if one monitors the projections of the lowest and highest source energies of ^{169}Yb, reported in Table 4.1, and reconstructs the corresponding Σ values, one would obtain lowest-energy values that are about 1.4, 2.8 and 7.4 times higher than the corresponding highest-energy values, for H_2O, Al and Fe, respectively. These ratio values can be used as indicators of the nature of the material involved. This is the essence of dual (or multiple) energy tomographic systems.

For a multienergetic source, one cannot claim that a projection $p(x)$, obtained over all energies, is representative of the mean value of the attenuation coefficient averaged over energy. In other words, $\bar{\Sigma} = \sum_i \chi_i \Sigma(E_i)$ is not directly obtainable from $p(x)$ due to the complicated nature of Eq. (4.15). Neither can one obtain the value $\Sigma(\bar{E})$ directly from $p(x)$, where $\bar{E} = \sum_i \chi_i E_i$ is the average source energy. What one obtains from $p(x)$ is a value of Σ that is representative of the radiation properties for the source used. In other words, the reconstructed Σ values would not directly correspond to tabulated values for a particular material, such as those obtainable from Berger et al. (1999), unless the source is monoenergetic, such as ^{137}Cs, or can be considered to be nearly monoenergetic, as in the case of ^{60}Co.

4.4.2 X-Rays

X-ray generators produce a continuous energy spectrum extending from an energy corresponding to the potential (voltage) applied on the x-ray tube down to zero energy, as schematically shown in Figure 4.3. This spectrum continuum is caused by the radiation emitted as bombarding electrons slow-down within the target of the x-ray tube (bremsstrahlung effect). Superimposed on the x-ray continuum are peaks corresponding to the characteristic transitions between deep electronic shells of the target. For a target of atomic-number Z, the continuum portion of the energy spectrum takes

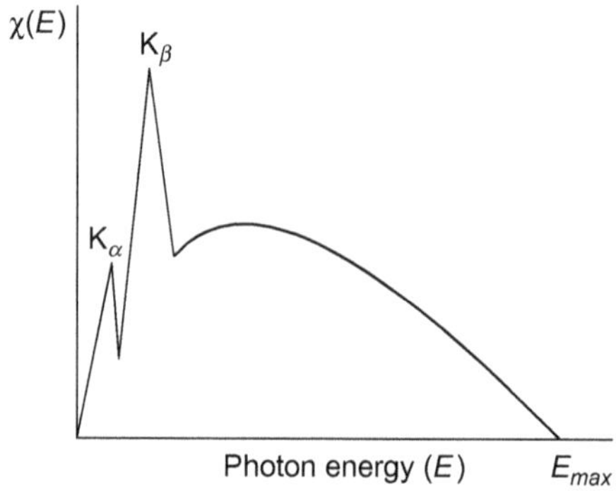

Figure 4.3 A typical x-ray spectrum, showing the bremsstrahlung continuum and characteristic peaks.

approximately the form (Lachance and Claisse, 1995):

$$\chi(E) = CZE^2\left(\frac{E_{max}}{E} - 1\right) + BZ^2E \tag{4.17}$$

where $\chi(E)$ is the fraction of photons emitted at energy E, B, and C are constants, and $E_{max} = eV_p$, with e being the electron's charge and V_p being the applied (peak) voltage. These jumps in the intensity of the characteristic peaks are given by jump ratios that also depend on the material of the target, see (Hussein, 2003).

The shape of the energy spectrum of x-rays has two implications on the measurement model of transmitted radiation. The first implication is similar to that for a gamma-ray source of multiple energy levels, i.e. the attenuation coefficient varies with energy. This effect is, however, more pronounced in x-rays, because the spectrum is a continuum, it has imposed peaks, and it continues down to the zero energy. Therefore, the summations in Eqs. (4.14) and (4.16) for gamma-rays are replaced with integrals when dealing with the continuum portion of the x-ray spectrum, so that:

$$I(x) = I_0\left\{\int_0^{E_{max}} \chi(E)\exp\left[-\int_0^x \Sigma(E,s)ds\right] + \sum_i(\xi_i - \chi(E_i))\exp\left[-\int_0^x \Sigma(E_i,s)ds\right]\right\} \tag{4.18}$$

$$p(x) = -\ln\frac{I(x)}{I_0} = -\ln\left\{\int_0^{E_{max}} \chi(E)\exp\left[-\int_0^x \Sigma(E,s)ds\right] + \sum_i(\xi\,\chi(E_i))\exp\left[-\int_0^x \Sigma(E_i,s)ds\right]\right\} \tag{4.19}$$

with $\chi(E)$ expressed by Eq. (4.17) or another suitable function, and ξ_i being the jump ratio for a characteristic peak at energy E_i, with the summation being over these peaks. Therefore, a value for Σ obtained from an x-ray projection, Eq. (4.19), is far more convoluted in relation to any particular photon energy in the spectrum than the case of gamma-rays. Nevertheless, a value of Σ extracted from an x-ray projection is still indicative of the material content, due to the dependence of all $\Sigma(E)$ values on the material density and atomic number.

The second implication of the complex form of the x-ray spectrum is far more serious. If one is to monitor the energy spectrum of a beam of x-rays after it passes through a fixed thickness of some uniform material, one would find that the energy spectrum is "hardened," i.e. has a median energy larger than of the incident radiation. This so-called *beam-hardening* effect is due to the fact that the low-energy component of the spectrum is subdued, as x-rays travel through matter. The attenuation coefficient increases with decreasing photon energy, with the rate of increase becoming much more pronounced as the energy becomes so low that the photoabsorption (photoelectric effect) dominates over Compton scattering, see Hussein (2007). Therefore, the low energy component of the x-ray spectrum is continually eliminated by photoabsorption as x-rays penetrate matter. This beam hardening effect is further enhanced by the fact that the lower energy portion of an x-ray spectrum tends to contribute more to the overall beam intensity than the higher energy part, as depicted in Figure 4.3 and indicated by the $\frac{1}{E}$ dependence of $\chi(E)$ in Eq. (4.17) (note that the second term in the right-hand-side of Eq. (4.17) is small compared to the first term). Further, radiation detectors are typically more absorbing of low-energy photons because of the same photoelectric effect. As such, low-energy photons tend to contribute more to a detected signal than a higher-energy component of the same intensity. Therefore, usually the very low energy component of the x-ray spectrum is eliminated with the aid of a filter consisting of an aluminum sheet several millimeters thick, placed in front of the emitted x-ray beam. This helps maintain a higher effective photon energy and reduces beam hardening, so that the obtained projection is more reflective of the density of the encountered material. Note that the beam hardening effect is not of much concern with gamma sources, since they do not have a continuous low-energy component, but still exists to some extend when a source has a significant low-energy level, as in the case of ^{169}Yb (see Table 4.1).

The x-ray bremsstrahlung is not isotropic, and, subsequently, the intensity, I_0, of the x-ray continuum varies with direction (angle) of emission. The angular distribution should, therefore, be measured and utilized in modeling transmission in various directions. In practice, this is not a very difficult task, since attenuation in air is negligible, and as such I_0 can be replaced by the intensity in air measured with a particular detector. Therefore, the change in x-ray intensity for detectors located at various directions from the vector normal to the target in an x-ray tube can be accounted for by normalizing the transmission measurements to a detector-specific intensity-in-air measurement.

4.4.3 Neutrons

Sources used to generate neutrons are either passive (isotopic sources) or active (electronically triggered), see (Hussein, 2003). The latter generate mainly monoenergetic neutrons. Radioisotopic sources produce, however, neutrons with a continuous energy spectrum. For example, californium-252, a common source of neutrons, produces neutrons with an energy spectrum that can be expressed by the Watt formula (Watt, 1952):

$$\chi(E) = c\exp(-aE)\sinh\sqrt{bE} = c'\sqrt{E}\exp(-aE) \tag{4.20}$$

where a and b, c and c' are constants. This spectrum extends over a wide energy range within which the attenuation coefficient (total cross section) for neutrons in a given material can vary significantly. The obtained transmission projections will still, however, be indicative of the overall change in the total cross section, but not to a particular neutron energy. The representative measurement model is then given by:

$$I(x) = I_0\left\{\int \chi(E)\exp\left[-\int_0^x \Sigma(E,s)ds\right]\right\} \tag{4.21}$$

$$p(x) = -\ln\frac{I(x)}{I_0} = -\ln\left\{\int \chi(E)\exp\left[-\int_0^x \Sigma(E,s)ds\right]\right\} \tag{4.22}$$

where the integration over energy extends over the entire energy range of the spectrum.

Neutron transmission measurements are typically conducted with thermal (slow) neutrons extracted either from a nuclear reactor or a thermalization facility driven by an isotopic source. This is because of the ease and efficiency of detecting thermal neutrons. Thermal neutrons also have an energy spectrum governed by the Maxwell-Boltzmann distribution (Glasstone and Sesonske, 1981):

$$\chi(E) = \frac{2\pi\sqrt{E}}{(\pi kT)^{\frac{3}{2}}}\exp\left(-\frac{E}{kT}\right) \tag{4.23}$$

where k is the Boltzmann constant and T is the absolute temperature of the medium in which neutrons are thermalized.

4.5 Secondary Radiation

The essence of transmission imaging is the reconstruction of an attenuation-coefficient image, with the aid of the transmission measurement model of Eq. (4.1). This model is based on the attenuation law of Eq. (2.17), which stipulates, see Section 2.5, that the attenuated radiation is removed completely (by absorption and scattering) from the incident beam. However, complete removal is not assured even in the case of absorption; though one may tend to assume that absorbed radiation leaves no trace behind it. In fact, absorbed photons produce electrons via the photoelectric effect, and the emerging electrons lose energy as they travel through matter producing secondary photons via the bremsstrahlung effect. The latter photons are subsequently absorbed via the photoelectric effect, producing secondary electrons, which further generate more bremsstrahlung, and so on. In effect, a single photon absorption creates a "shower" of secondary photons of lower energy. Some of these photons can reach detectors monitoring transmitted radiation. This in turn will affect the value of the projection derived from the affected transmission imaging, somewhat subduing the effect of radiation attenuation. Most of the secondary photons of the bremsstrahlung effect tend, however, to be of much lower energy than those of the incident radiation and can be removed from the detector by setting a low-energy window to electronically cutoff the effect of these photons. A filter consisting of a thin sheet or a metal, e.g. aluminum, can also be placed in front of the detector to remove these secondary lower-energy photons. Such a filter, however, also absorbs part of the primary beam's radiation, producing its secondary bremsstrahlung photons. The secondary photons are not only produced in the material of the examined object, and within a detector's filter if present, but also in the walls of a collimator placed around and in front of a detector to confine its field-of-view. The resulting secondary photons are typically absorbed within the material of the collimator's body, and only those produced near the internal surface of the collimator's walls are likely to leak out. Most of these photons are most likely to reach the opposite side of the collimator walls and be absorbed within these walls.

The bremsstrahlung effect is stronger in elements with high atomic number (Hussein, 2007). Therefore, it is typically neglected in medical imaging, since biological material have a relatively low atomic number. In imaging industrial objects, this effect may be present and may produce photons that can reach transmission detectors. Typically, however, the contribution of this secondary radiation is very small in comparison to the intense recorded signal of transmitted radiation. This is because secondary radiation tends to be emitted in all directions, and a small portion of which is directed toward any particular transmission detector. Moreover, the latter portion of radiation is further attenuated by the intervening material between the point of origin of the secondary radiation and the detector.

In the case of neutrons, secondary-neutron production is only likely to occur when imaging fissionable materials. The most likely form of secondary radiation production will be gamma-ray photons produced by neutron capture or inelastic scattering (Hussein, 2007), in the material of the object, collimators and/or surrounding shielding. Such photons are emitted isotropically and can reach transmission detectors. Some neutron detectors are also sensitive to photons, and the recorded transmission signal can be accordingly enhanced. Therefore, the effect of these secondary photons should be assessed and effort be made to ensure that the employed detector is not too sensitive to photons, or that their presence in the detected signal can be discriminated against.

4.6 Scattering

Like secondary radiation, scattered radiation can contaminate the idealistic formulation of the forward model for transmission, which assumes that all interacting radiation is removed from the transmission signal. In fact, a considerable amount of effort in the design of transmission-based imaging systems is devoted to eliminating the effect of scattered radiation, by collimation and/or by placing transmission detectors away from the object so that scattered radiation is widely spread and only a small portion is seen by the detector. One can also use energy discrimination to eliminate scattered radiation from the detected signal, taking advantage of the fact that the scattered radiation has an energy lower than that of the source, since it has to interact before it scatters. However, when the source is multienergetic, as in the case of x-rays and isotopic neutron sources, energy discrimination against scattered radiation from a higher source energy can also lead to the elimination of transmitted (uncollided) source radiation from a lower-energy source component.

The reason scattered radiation is more problematic than secondary radiation, discussed in Section 4.5, is that the probability of scattering tends to be higher than that of absorption. Moreover, scattered radiation is likely to have an energy higher than that of secondary radiation, and as such has a better chance of penetrating intervening material to reach transmission detectors. In addition, no matter how much effort is made to prevent scattered radiation from reaching a transmission detector, it is inevitable that some scattered radiation still reaches transmission detectors. Utilizing transmission projections contaminated with scattered radiation will lead to an overestimation of the attenuation coefficient, due to the increase in the strength of the detected signal. The detected scattered radiation tends to carry information from regions outside the field-of-view of a transmission detector, since that radiation could have suffered a collision or more somewhere outside the detector's field-of-view before reaching the detector. Therefore, scattered radiation in a transmission projection, if not properly removed, blurs the sharpness of a transmission-based image, due to the non-localized nature of its origin.

4.7 Sensitivity

Using the definition of sensitivity, S, of Eq. (3.5), along with Eq. (4.2), one obtains:

$$S = \frac{\partial p(x)}{\partial \Sigma(s)} = \Delta s \tag{4.24}$$

where Δs is the width, in the direction the incident radiation, of the voxel within which the value of Σ is evaluated. Here we defined sensitivity with respect to the projection, $p(x)$, rather than the measured intensity, $I(x)$, since it is the projection that is used to reconstruct the physical parameter, Σ. It is obvious from Eq. (4.24) that the sensitivity of a transmission tomography system is defined by its spatial resolution. In other words, as one would expect, the smaller the voxel size, the more difficult it would be to discern changes in Σ, since the influence of that Σ on the value of the projection would be quite weak. A coarser image would be more sensitive to changes in the value of Σ, but would have a poor resolution, i.e. the spatial variation in Σ would not be sharply depicted in the image. Therefore, in order to have good sensitivity to changes in Σ, and maintain a good resolution, one must increase the influence of each voxel in the measurement domain as a whole, not in a single projection. This is done by having many projections across each voxel.

4.8 Variability

The susceptibility of a transmission system to "error" in measurements is determined by the variance of a projection, σ_p, which, with the aid of Eq. (4.1) and (4.2), is as follows:

$$\sigma^2_{p(x)} = \left[\frac{\partial p(x)}{\partial I(x)}\right]^2 \sigma^2_I + \left[\frac{\partial p(x)}{\partial I_0(x)}\right]^2 \sigma^2_{I_0} \tag{4.25}$$

$$= \frac{\sigma^2_I}{I^2(x)} + \frac{\sigma^2_{I_0}}{I_0^2(x)} \tag{4.26}$$

where $\sigma^2_{I(x)}$ is the variance in the measured transmission measurement, $I(x)$, and $\sigma^2_{I_0}$ is the variance in the corresponding measurement in air, i.e. the latter is the variance in the scaling parameter. Radiation counting is governed by radiation statistics in which $\sigma^2_I = I(x)$ (Hussein, 2003). Therefore, $\sigma^2_{p(x)}$ is inversely proportional to $I(x)$, i.e. the higher the value of $I(x)$, the less its statistical variability is, as one would expect. A measurement in the absence of the object is considered a system constant, since its value depends on the source and detector configuration, in terms of intensity of source, efficiency of detector and the distance between the two. Therefore, this normalizing value of $I_0(x)$ should be recorded with care over a longer counting time, then adjusted to the same counting period used to acquire $I(x)$, so that the statistical variability of $I_0(x)$ is negligible. However, a long counting period makes the variability of $I_0(x)$ dependent on the stability of the source and whether it provides a constant mean value. The stability of an x-ray source depends on the stability of the applied voltage and that of the current bombarding its target. For a radioisotope, the counting period for $I_0(x)$ has to be much lower than its half-life, so that the source intensity does not significantly drop during the measurement time. If the normalization values for $I_0(x)$ are recorded at a time significantly in advance of using an isotopic source for imaging, the imaging measurements should be adjusted for the source decay. For radiation extracted from the core of an experimental nuclear reactor, changes in the reactor power during measurements may also affect the recorded values of $I_0(x)$ and $I(x)$, and corrections may be needed to ensure that the measurements used in image reconstruction are related to the image and not external parameters related to reactor operation.

Another time-varying effect is the object motion. Such motion could be due to inadvertent effects, and as such it is not possible to determine in advance and will appear as an artifact upon image reconstruction. In some cases, motion is inherent in the imaged object. In medical imaging, cardiac, respiratory and muscle motion are unavoidable. Motion in industrial objects can be due to fluid motion or turbulence, vibration, or because the object itself is in motion or rotation. The effect of motion can be smoothed out by acquiring multiple transmission measurements for the same transmission trajectory at different time intervals.

References

M. J. Berger, J. H. Hubbell, S. M. Seltzer, J. S. Coursey, and D. Zucker, *XCOM: Photon Cross Section Database (version 1. 2), [Online]*, National Institute of Standards and Technology, NBSIR 87-3597, Gaithersburg, 1999, (http://physics.nist.gov/xcom, accessed, May 2010).

S. Glasstone and A. Sesonske, *Nuclear Reactor Engineering*. New York: Van Nostrand Reinhold, 1981.

E. M. A. Hussein, *Handbook on Radiation Probing, Gauging, Imaging and Analysis: Basics and Techniques*. Dordrecht: Kluwer Academic Pub., 2003, vol. I.

——, *Radiation Mechanics: Principles and Practice*. Amsterdam: Elsevier, 2007.

G. R. Lachance and F. Claisse, *Quantitative X-ray Fluorescence Analysis: Theory and Application*. Chichester: John Wiley & Sons, 1995.

B. E. Watt, "Energy spectrum of neutrons from thermal fission of U^{235}," *Physical Review*, vol. 87, no. 6, pp. 1037–1041, Sep 1952.

5 Emission

We will distinguish here between two forms of radiation emission: one resulting from internal sources embedded within an object (by injection, precipitation, or nature) and another is induced by an external radiation source. When the emitted radiation is induced by an external radiation source, the intensity of the secondary source emitting the monitored radiation would depend on the nature of the external source, as well as on the macroscopic cross section of that radiation in the investigated material.

5.1 Embedded Radiation

Let us consider an internal source embedded within an object, so that its has an intensity per unit volume (concentration), $I(\vec{r})$, at a point at position, $\vec{r}$, measured from some fixed origin, as shown in Figure 5.1. This source would result in a count rate, $C(\vec{R})$, at a detector located at position $\vec{R}$, external to the object, that can be estimated by:

$$C(\vec{R}) = K \int_{V_D} \frac{I(\vec{r})}{|\vec{R} - \vec{r}|^2} \exp\left[- \int_0^{|\vec{R}-\vec{r}|} \Sigma(\vec{r'})\, dr' \right] dV_D \tag{5.1}$$

where K is a system (normalization) constant that accounts for detector efficiency and geometry, $\Sigma(\vec{r'})$ is the total cross section of the material at $\vec{r'}$ along the radiation path from the source to the detector, and the integration is over the volume of the detector's field-of-view, V_D.

In the model of Eq. (5.1), both the attenuation and divergence effects, discussed in Chapter 2, are taken into account. In Eq. (5.1), the physical parameter to be reconstructed is the source intensity, $I(\vec{r})$, within the voxels of an image frame fictitiously laid over the examined object. The cross section, Σ, of the material within the object may not be known. However, unlike in transmission imaging, the objective here is not to reconstruct the value of Σ, but to find the spatial distribution of I. It is, therefore, reasonable to assume that Σ is known everywhere in the object, at the source energy. Nominal average values of Σ can be used, assuming that the nature of the material containing the embedded source is known. However, any error in estimating Σ will lead to erroneous values of I. A separate, transmission-based imaging can be used to determine the distribution of Σ within the object, but then one must ensure that the source energy used in transmission imaging is the same as that of the internal

Computed Radiation Imaging. DOI: 10.1016/B978-0-12-387777-2.00005-7

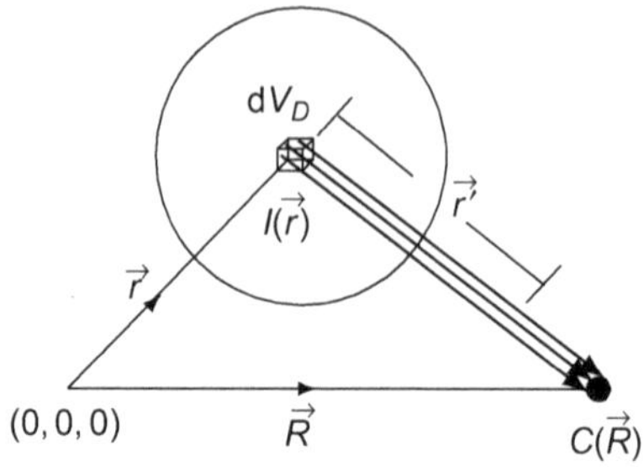

Figure 5.1 A schematic for embedded radiation emission.

source. If the latter condition is not met, the transmission-image value of Σ should be corrected to provide values at the energy of the internal source. When independent transmission imaging is used to determine Σ, care must be taken to ensure that it is conducted on the same object (in terms of material and geometry) for which emission imaging is to performed; changes due to object instability or motion while conducting either form of imaging will introduce image artifacts. One may ignore radiation attenuation altogether, by equating the exponential term in Eq. (5.1) to unity. Also, taking the average of emission measurements at two opposite sides of the object would reduce the effect of attenuation, since an internal source in the proximity of one detector would be farther away from the opposite detector. However, due to the exponential nature of the attenuation process, such averaging will not entirely eliminate the spatial dependence of attenuation, even in a uniform object.

The system constant, K, in Eq. (5.1) can be determined with the aid of a source, of the same type as the internal source, of known total intensity (activity), I_0, positioned in air at some location, $\vec{r}_0$, with the detector positioned at some distance, $\vec{R}_0$, giving a count rate, $C_{Air}(\vec{R}_0)$. Then according to Eq. (5.1):

$$C_0(\vec{R}_0) = K\frac{I_0(\vec{r})}{|\vec{R}_0 - \vec{r}_0|^2}\ \exp\left[-|\vec{R}_0 - \vec{r}_0)|\Sigma_{air}\right] \tag{5.2}$$

where Σ_{air} is the attenuation coefficient for air (usually considered to be equal to zero) and the integration over the volume is incorporated in the value of I_0.

The model of Eq. (5.1) assumes that the emitted radiation that succeeds in reaching a detector is "uncollided," i.e. has not been removed by absorption or scattering interactions. Although, generally, the absorbed radiation can be considered to have been fully removed from the system, in some cases it can produce secondary radiation that may contribute to the detector, as discussed in Section 4.5. Such radiation is likely to have an energy that is different from that of the original source, and as such can be removed by energy discrimination, if its contribution is found to be significant. Scattered radiation can be removed in a similar manner, since it also has an energy different (lower) than that of the internal source, due to the energy loss caused by collisions. Removing secondary and scattered radiation is necessary to satisfy the measurement model of Eq. (5.1), so that it can be used in the inverse problem of image reconstruction and to avoid image blurring caused by recording radiation originating from positions unrelated to the location of the internal source radiation.

5.2 Induced Emission

Images can also be reconstructed by monitoring one type of radiation induced by another type. Positron emission tomography is based on such induced emission, where the positrons of an internal source combine with the electrons of the matter, causing mutual annihilation and the production of two 511 keV photons at opposite directions. Then, the emitted photons, on which image reconstructions is based, are induced by positrons. However, since the positrons are charged particles, they do not travel very far from their point of origin before being annihilated by the electrons of the surrounding medium. Then, the point of radiation emission and the source that induced it can be considered, for all practical purposes, to coincide with each other. The source in positron emission tomography can be considered to be an internal source, and the model of Eq. (5.1) can be used, with the assumption that there is an inherent error in determining the location of the source of emitted photons equal to the range of the positrons in the imaged medium. Emission induced by charged particles can be produced by exposing an object to a beam of charged particles, e.g. protons, alpha particles, or accelerated ions (Hussein, 2003). Such particles do not penetrate deep within the object, and as such the associated radiation emission is from near the surface and its attenuation is negligible. Characteristic fluorescent x-rays, i.e. those induced by the excitation of atomic levels, can also be used to image the concentration of certain elements by induced emission imaging, see Cesareo and Mascarenhas (1989) and Takeda et al. (2000).

When one type of penetrating radiation is used to induce another type of penetrating radiation, the attenuation of both types of radiation must be accounted for. Examples of such emissions include: gamma-ray emission by neutron activation or inelastic scattering and photoneutron production. Hussein (2003) compiled the various means for producing such emissions. We will consider here a generic emission process, schematically shown in Figure 5.2, in which a primary radiation of energy, E_p,

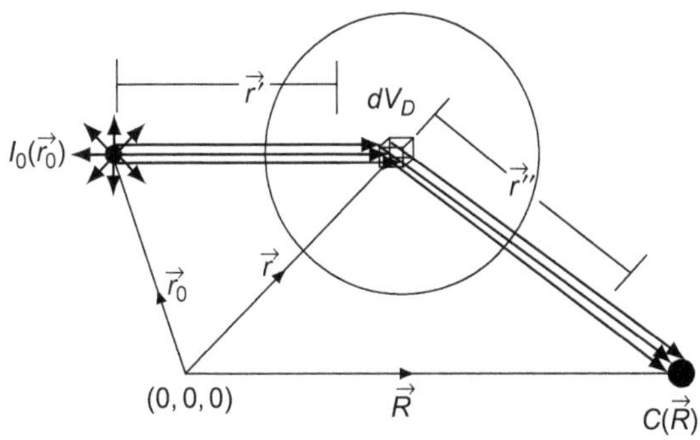

Figure 5.2 A schematic for induced radiation emission.

induces a secondary radiation with energy, E_s. For an emission to emerge from a point at $\vec{r}$ within an object, the primary radiation, produced by an external source, has first to penetrate the object, to reach $\vec{r}$. An external source at $\vec{r}_0$, with an intensity $I_0(\vec{r}_0)$, reaches $\vec{r}$ with a flux, $\phi_p(\vec{r})$, given by:

$$\phi_p(\vec{r}) = \frac{I_0(\vec{r}_0)}{4\pi\,|\vec{r}-\vec{r}_0|^2} \exp\left[- \int_0^{|\vec{r}-\vec{r}_0|} \Sigma_p(\vec{r'})dr' \right] \tag{5.3}$$

assuming that the external source is isotropic and $\Sigma_p(\vec{r'})$ is the attenuation coefficient of the primary radiation for the material present at $\vec{r'}$. Equation (5.3) takes into consideration both the attenuation and divergence of the primary radiation. The emission rate per unit volume, $\mathcal{E}$, of the secondary radiation, in accordance to Eq. (2.10), is determined by:

$$\mathcal{E}(\vec{r}) = \sigma_{p\to s}(\vec{r})N(\vec{r})\phi_p(\vec{r}) \tag{5.4}$$

where $\sigma_{p\to s}(\vec{r})$ is the microscopic cross section at $\vec{r}$ for producing emitted radiation of type s, from a primary radiation of type p, $N(\vec{r})$ is the atomic density of the element with which radiation interacts. The emitted radiation has to exit the object for detection by a detector at $\vec{R}$. The flux, $\phi_D(\vec{R})$, at the detector (assuming isotropic emission) is then:

$$\phi_D(\vec{R}) = \frac{\mathcal{E}(\vec{r})}{4\pi\,|\vec{R}-\vec{r}|^2} \exp\left[- \int_0^{|\vec{R}-\vec{r}|} \Sigma_s(\vec{r'})dr' \right] \tag{5.5}$$

Using Eqs. (5.3) to (5.5), the detector's count rate can be expressed as:

$$\begin{aligned} C(\vec{R}) = K \int_{V_D} & \frac{I_0(\vec{r}_0)}{4\pi\,|\vec{r}-\vec{r}_0|^2} \exp\left[- \int_0^{|\vec{r}-\vec{r}_0|} \Sigma_p(\vec{r'})dr' \right] \\ & \times \frac{\sigma_{p\to s}(\vec{r})N(\vec{r})}{4\pi\,|\vec{R}-\vec{r}|^2} \exp\left[- \int_0^{|\vec{R}-\vec{r}|} \Sigma_s(\vec{r''})dr'' \right] dV_D \end{aligned} \tag{5.6}$$

where again K is a system constant and V_D is the volume seen by the detector. In the measurement model of Eq. (5.6), the physical parameter that an emission imaging process will aim at determining is $N(\vec{r})$. The microscopic cross, $\sigma_{p\to s}(\vec{r})$, contains information on the elemental composition of the material. However, in emission imaging, if a particular element is identifiable by means other than the image reconstruction process, then its macroscopic cross section, $\sigma_{p\to s}$, can be determined in advance.

The presence of a particular element can usually be characterized by the energy of the radiation it emits. This is because nuclear emissions are either the result of the de-excitation from characteristic nuclear levels or are caused by a particular interaction that releases a specific energy, see Hussein (2007). Similarly, atomic emissions are the result of de-excitations of electrons from one level in the atom to a lower level, at energies characteristic of the element.

Like the case with internal sources, the attenuation of the emitted radiation through the inspected object needs to be accounted for. In induced imaging, the attenuation of the primary radiation should also be taken into account. The problem is complicated in induced imaging by the fact that it aims, directly or indirectly, at reconstructing elemental densities, which also determine the attenuation coefficients of the primary and secondary radiation. That is, if the latter microscopic cross sections are assumed to be predetermined to correct for radiation attenuation, then one may argue that $\Sigma_{p\rightarrow s} = N\sigma_{p\rightarrow s}$ can also be predetermined, and there is no need to utilize emission imaging at all. However, emission imaging is typically used to detect the presence of one or a few elements, not all elements, while the attenuation coefficient incorporates all present elements and their corresponding densities. Therefore, using estimated or independently determined attenuation coefficients will enable the determination of the atomic density of the element(s) of interest, while accounting for the overall attenuation of all present elements.

The intensity of the external source, I_0 in Eq. (5.6), can be determined using transmission measurements in air, as explained in Section 4.1. The system constant, K in Eq. (5.6), is more difficult to determine for induced emission than in the case of internal sources, since in the latter case an internal source can be placed in air and used to calibrate the system. However, a standard calibration object of known composition and density can be employed as a reference to calibrate an induced emission source/detector arrangement, to determine K, or even $KI_0(\vec{R})$ as a single system constant. The system constant is then evaluated as the ratio between a recorded measurement for the reference object and the corresponding measurement-model value determined using Eq. (5.6) with K, or $KI_0(\vec{R})$, set equal to unity.

An inherent assumption in the measurement model of Eq. (5.6) is that the emitted radiation is generated promptly, without delay, as the primary radiation interacts with matter. In some interactions, however, the emission is delayed, if the interaction product is a transmuted nucleus that continues to decay, with a certain half-life, after exposure to the external source of radiation ceases. Measuring delayed emission has the advantage of allowing the acquisition of measurements in the absence of the external source, which reduces the radiation background normally associated with the primary radiation source. In delayed emission imaging, the emitted radiation should be allowed to fully decay before further exposure of the object to the primary source, unless the intensity of a residual precursor element from a previous solution is accounted for.

When monitoring delayed emission, a number of time-dependent factors must be introduced to correct for radiation decay. While the object is being exposed to radiation, it produces the precursor element that emits radiation, but that element also decays while the irradiation process is ongoing. Therefore, the rate of production of the precursor is such that:

$$\frac{d\hat{N}(\vec{r},t)}{dt} = \sigma_{p\rightarrow s}(\vec{r})N(\vec{r})\phi_p(r) - \lambda N(\vec{r},t) \tag{5.7}$$

where $\hat{N}(\vec{r},t)$ is the atomic density of the precursor element at time, t, which will result in emission with a decay constant, λ (characteristic of the precursor). Recall that $N(\vec{r})$ is the atomic density of the element that is to be monitored by emission imaging, i.e. the element that produces the precursor as a result of being exposed to a flux of primary radiation, $\phi_p(r)$ in Eq. (5.7). Assuming constant flux of the primary source, one has the general solution:

$$\hat{N}(\vec{r},t) = \frac{\sigma_{p\rightarrow s}(\vec{r})N(\vec{r})\phi_p(r)}{\lambda}\left[1-\exp(-\lambda t)\right] + \hat{N}_0(\vec{r},t)\exp(-\lambda t) \tag{5.8}$$

where $\hat{N}_0(\vec{r},t)$ is the atomic density of the precursor element at time zero, if present from a previous exposure without being allowed to fully decay. Therefore, after exposure to the primary radiation for a time, t_p, there will be $\hat{N}(\vec{r},t_p)$ atoms of the radiation emitting element. If the emission rate is recorded after another period, t_d, the emitting element would have decayed to an atomic density of $N(\vec{r},t_p)\exp(-\lambda t_d)$. Image reconstruction aims at determining the spatial distribution of $N(\vec{r})$, not of the delayed emission measurements, $\hat{N}(\vec{r},t_p+t_d)$. However, one obtains measurements corresponding to $\hat{N}(\vec{r},t_p+t_d)$, not $\hat{N}(\vec{r})$. But, one can reconstruct an image for $\hat{N}(\vec{r})$, then restore the original value, $N(\vec{r})$, with the aid of Eq. (5.8):

$$N(\vec{r}) = \frac{\lambda\left[\hat{N}(\vec{r},t_p+t_d)\exp(\lambda t_d) - \hat{N}_0(\vec{r},t)\exp(-\lambda t_p)\right]}{\sigma_{p\rightarrow s}(\vec{r})\phi_p(r)\left[1-\exp(-\lambda t_p)\right]} \tag{5.9}$$

where $\hat{N}(\vec{r},t_p+t_d)$ is the image reconstructed value after exposing an object to radiation of a time, t_p, and performing measurements after a delay time, t_d.

As in the case of internal sources, the models for induced emission do not account for the fact that secondary emissions can be produced by radiation absorption, and that scattering can contribute to the detected signal. However, like internal emission, such additional contributions produce radiation with an energy different from that of the emission of interest, and energy discrimination can be used to remove the effect of secondary emissions.

5.3 Discretization

As with transmission modeling, image reconstruction with emission requires discretization of the forward model. This necessitates approximating the volume integrations in Eq. (5.1) and (5.6) by summations over small voxels. Within each voxel, it is assumed that each of the parameters that appear in these models has a constant value, i.e. the reconstructed image parameter has a value averaged over the voxel's volume. Consequently, the smaller the voxel's size, the better the spatial resolution of the reconstructed image. However, obtaining a high resolution image requires the gathering of a large number of measurements, with at least one measurement per voxel.

5.4 Sensitivity

With integration replaced with summation for discretization purposes, the definition of Eq. (3.5) for sensitivity applied to Eq. (5.1) gives the following sensitivity, $S_{internal}$, for embedded emission:

$$S_{internal} = \frac{\partial C(\vec{R}_j)}{\partial i(\vec{r}_i)} = K \sum_i \frac{1}{|\vec{R}_j - \vec{r}_i|^2} \exp\left[-\sum_k \Sigma(r'_k)\, \Delta r_k \right] \Delta V_i \tag{5.10}$$

where the subscripts i and k refer to voxels within the field-of-view of detector j, Δr_k is the distance traveled by the emitted radiation, and ΔV_i is the volume of voxel i. It is obvious from Eq. (5.10) that the sensitivity of an internal emission imaging system is reduced by the weakening of the recorded signal by radiation attenuation and divergence, but is enhanced with increased voxel size. The latter comes, however, at the expense of reduced spatial resolution. Detector efficiency and size also affect the sensitivity, via the factor K in Eq. (5.10).

Similarly, the sensitivity for induced emission, $S_{induced}$, is expressed with the aid of Eq. (5.6) as:

$$S_{induced} = \frac{\partial C(\vec{R}_j)}{\partial N(\vec{r}_i)} = K \sum_i \frac{I_0(\vec{r}_0)}{4\pi |\vec{r}_i - \vec{r}_0|^2} \exp\left[-\sum_k \Sigma_p(\vec{r'}_k)\Delta r'_k \right]$$
$$\times \frac{\left[\sigma_{p \to s}(\vec{r}_i)\right]}{4\pi |\vec{R}_j - \vec{r}_i|^2} \exp\left[-\sum_{k'} \Sigma_s(\vec{r'}_{k'})\Delta r'_{k'} \right] \Delta V_i \tag{5.11}$$

The sensitivity for induced emission imaging is also reduced by increased attenuation and divergence of primary and secondary radiation, but can be enhanced by increasing the voxel size, the intensity of the primary source, and/or the detector size and efficiency. Having a large interaction cross section also enhances the sensitivity.

5.5 Sources

Sources that can be used in embedded emission must be in a physical and chemical form suitable for inclusion within the medium in which it is introduced. The embedded source should have a short half-life, so that a small amount produces an easily measurable high source activity. A short-lived source also decays rapidly, returning the

Table 5.1 Some Radioisotopes Suitable for use in Embedded Emission Imaging.

Isotope	Half-Life	Energy (keV)	Comment
^{24}Na	14.96 h	1369, 2754	(a)
^{41}Ar	109.34 m	1294	(b)
^{67}Ga	3.26 d	93, 185, 300	(c)
^{82}Br	35.30 h	777	(a)
^{85m}Kr	4.48 h	151	(b)
^{99m}Tc	6.01 h	141	(c)
^{111}In	2.80 d	245	(c)
^{123}I	13.27 h	159	(c)
^{131}I	8.02 d	365	(c)
^{201}Tl	72.91 h	167	(c)

Half-lives and dominant gamma energies are obtained from: `http://ie.lbl.gov/education/isotopes.htm`, which also gives the relative intensities of emitted photons.
(a) Suitable for aqueous or organic liquids (Hussein, 2003).
(b) Available as gases (Hussein, 2003).
(c) Used in medical imaging.

Table 5.2 Some Positron Sources.

Source	^{11}C	^{13}N	^{14}O	^{15}O	^{44}Sc	^{68}Ga
Half-life	20.39 m	9.965 m	70.606 s	122.24 s	3.927 h	67.629 m

Half-lives obtained from `http://ie.lbl.gov/education/isotopes.htm`.

examined object to its original state quickly. However, a source with a very short half-life may not allow sufficient time for acquiring meaningful measurements. Table 5.1 lists some of the sources that can be used in emission imaging, along with their half-lives and gamma-energy of emission. Note that only gamma-ray sources are listed, because their penetrability makes them viable for use in emission imaging.

Positron sources can also be used in induced emission imaging to produce 511 keV photons by annihilation. Short-lived positron sources are listed in Table 5.2. Such sources are to be incorporated into a chemical and a physical form (pharmaceuticals in case of medical applications) that can be embedded within the inspected medium in a manner that labels the desired imaging feature.

In order to induce secondary emissions, a radiation source (other than positron sources) must be able to disturb the nuclear or atomic structure, causing excitation of the nucleus or the atom. Atomic excitation causes electrons to rise to a higher orbit, with subsequent prompt (in less than a picosecond) emission of x-ray photons as the electrons return to more stable orbits. However, this process, known as x-ray fluorescence (XRF), produces low energy (a few to tens of keV (Hussein, 2003)) photons, which makes them only useful for shallow imaging of materials. Higher energy, more penetrating photons, can be produced by nuclear excitation using neutron sources, or highly energetic photons. The emission can be prompt, with an excited nucleus immediately emitting its excitation energy in the form of photons, or delayed as the product nucleus decays at a slower rate to a more stable state. Prompt emission has the advantage of providing an instantaneous emission signal, but the simultaneous presence of the inducing source radiation can interfere with the detection of the emitted radiation, since most detectors are sensitive to both types of radiation and the discrimination between the emitted and the incident radiation can be difficult.

Prompt emission induced by neutrons sources is typically via an inelastic scattering process that causes excitation of the target nucleus. This excitation requires, however, a relatively high neutron energy above a threshold energy that differs from one nuclide to another. Such threshold energy is typically in the MeV range, requiring the use of accelerator-produced neutrons. Delayed emission can, however, be produced by low-energy (thermal) neutrons. Such neutrons can be extracted from a research reactor or from an isotopic neutron source embedded within a moderating (neutron slowing-down) material, since isotopic sources produce high-energy neutrons.

Nuclear radiation emission can also be induced by high-energy photons. However, such photonuclear activation is only possible at very high photon energy and requires high-intensity photon sources produced by high-voltage electron accelerators.

5.6 Interfering Effects

Radiation emission can be viewed as a radiation-transmission process from the position of the internal source of emission to the detector site, in which the intensity of the radiation source is attenuated. The modeling formulations of Sections 5.1 and 5.2 aim at accounting for the modulating effect of attenuation. However, similar to the modeling of the transmission process in Section 4.1, one must keep in mind that radiation scattering and secondary radiation emissions during the transmission process can influence the quantity and energy of the detected radiation. Therefore, the factors discussed in Sections 4.5 and 4.6, for the transmission of external radiation, should also be considered for the transmission process of internally emitted radiation. The statistical variability of the detected signal discussed in Section 4.8 is also applicable, when measuring the intensity of emitted radiation.

References

R. Cesareo and S. Mascarenhas, "A new tomographic device based on the detection of fluorescent x-rays," *Nuclear Instruments and Methods in Physics Research A*, vol. 277, pp. 669–672, 1989.

E. M. A. Hussein, *Handbook on Radiation Probing, Gauging, Imaging and Analysis: Basics and Techniques*. Dordrecht: Kluwer Academic Pub., 2003, vol. I.

——, *Radiation Mechanics: Principles and Practice*. Amsterdam: Elsevier, 2007.

T. Takeda, A. Momose, Q. Yu, T. Yuasa, F. A. Dilmanian, T. Akatsuka, and Y. Itai, "New types of x-ray computed tomography (CT) with synchrotron radiation: fluorescent x-ray CT and phase-contrast x-ray CT using interferometer," *Cellular and molecular biology*, vol. 46, pp. 1077–1088, 2000.

6 Scattering

6.1 Introduction

Radiation scattering can either be particle-like or wave-like. The latter occurs at low radiation energy, where the radiation wavelength becomes comparable in value to the spacing between atoms in a lattice structure. Then, it is possible to examine crystal structures by the optical-like process of Bragg diffraction, with either low-energy (soft) x-rays, cold neutrons (with energy below the 0.025 eV, the thermal neutron energy at room temperature), or fast electrons. Crystallography, being concerned with the arrangement of atoms in a crystallized material, is not among the imaging processes at the macroscopic level considered in this book. Similarly, other neutron scattering methods that take advantage of the wave properties of cold neutrons to probe matter are not addressed here. Instead, we focus on imaging methods that rely on the corpuscular nature of the Compton scattering of photons and the elastic scattering of neutrons, as they provide macroscopic properties similar to those of transmission and emission imaging.

Scatter imaging resembles induced-emission imaging in the sense that the radiation path within the imaged medium is not well-defined by a straight-line from the source of radiation to its detector. Like induced emission, the source's radiation travels within the imaged object for some distance until it interacts. In the case of emission, the interaction of interest produces a secondary radiation that differs in nature from the radiation of the primary source. In scattering, the radiation emerging from the interaction is of the same type as that of the source, but with an altered direction and energy. The scattered radiation can itself scatter again and again, producing signals that are often difficult to trace back. Therefore, scatter imaging limits itself to the detection of once-scattered radiation: the so called *single scattering* process, where the radiation path from the source to the detector is identifiable. Then, multi-scattered radiation is considered undesirable, interfering radiation that needs to be eliminated, which is not difficult due to its much reduced energy and its lack of directionality. The other extreme that allows imaging with scattered radiation is the use of highly scattered radiation, so that the detected scattered radiation loses completely any relationship to the source radiation. Modeling the transport of multi-scattered radiation is quite complex, making its inversion to extract imaging information quite difficult, due to the multiple folding of radiation transport information carried by multi-scattered radiation. Nevertheless, multiple scattering can provide an overall global information

Computed Radiation Imaging. DOI: 10.1016/B978-0-12-387777-2.00006-9

on the nature of the object from which it emerges. Our focus here is on once-scattered radiation.

6.2 Single-Scattering Model

Let us trace the passage of a radiation beam of intensity $I_0(\vec{r}_0)$, emanating from a source at $\vec{r}_0$ that travels some distance within an object until experiencing a collision at position $\vec{r}$ within the object, which makes it scatter toward a detector at position $\vec{R}$ outside the object, contributing to a detector count rate of $C(\vec{R})$, see Figure 6.1.

Before reaching point $\vec{r}$, the incident beam's intensity will be reduced by attenuation by the traversed material. Therefore, when it reaches point $\vec{r}$, the radiation flux will be:

$$\phi(\vec{r}, E) = \frac{I_0(\vec{r}_0)}{4\pi\,|\vec{r} - \vec{r}_0|^2} f(\vec{r}_0, \vec{r}; E) \tag{6.1}$$

$$f(\vec{r}_0, \vec{r}; E) = \exp\left[- \int_0^{|\vec{r}-\vec{r}_0|} \Sigma(\vec{r'}, E)\, dr' \right] \tag{6.2}$$

where it is assumed that the incident radiation emerges from an isotropic source at an energy E, with $\Sigma(r', E)$ being the attenuation coefficient (total cross section) at position $\vec{r'}$ along the incident beam, at the source energy, E. The factor f in Eq. (6.2) is the attenuation factor of the incident radiation.

Within a volume, ΔV, around the point at $\vec{r}$, radiation will scatter at a rate determined according to Eq. (2.10):

$$\mathcal{S} = \sigma_s(\vec{r}, E) N(\vec{r}) \phi(\vec{r}, E) \Delta V \tag{6.3}$$

where $\sigma_s(\vec{r}, E)$ is the microscopic scattering cross section at the designated indices, and $N(\vec{r})$ is the atomic density at $\vec{r}$.

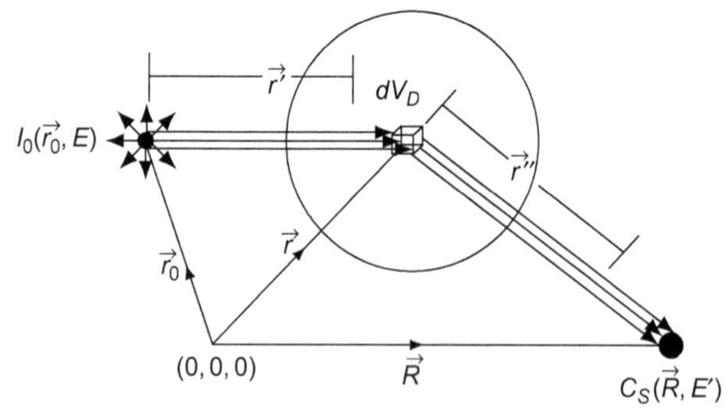

Figure 6.1 A schematic for scattered radiation emission.

Radiation scattered at $\vec{r}$ can scatter to any direction. Let $p(\mu, E \to E')$ be the probability of radiation scattering by an angle $\cos^{-1}\mu$, with $\mu = (\vec{r} - \vec{r}_0) \cdot (\vec{R} - \vec{r})$, to reach a detector at $\vec{R}$, and in the process change its energy to $E' < E$. Assuming that scattering in the azimuthal direction is isotropic, as it is usually the case for unploarized radiation, the scattering probability per solid angle is then $\frac{p(\mu, E \to E)}{2\pi}$. The scattering rate, $\mathcal{S}_\Omega$, within solid angle $\Delta\Omega$ per unit area for a detector at $\vec{r}$ is:

$$\mathcal{S}_\Omega \frac{\Delta\Omega}{\Delta A_d} = \mathcal{S}\frac{p(\mu, E \to E')}{2\pi\,|\vec{R} - \vec{r}|^2} \tag{6.4}$$

where ΔA_d is an area on the detector's surface; recall that $\Delta\Omega = \frac{\Delta A_s}{|\vec{R}-\vec{r}|^2}$. In Eq. (6.4), it is assumed that the solid angle is sufficiently small so that the probability of scattering remains constant.

After scattering, radiation will be attenuated by the intervening material before reaching the detector at $\vec{R}$. The attenuation factor of the outgoing radiation can be expressed as:

$$g(\vec{r}, \vec{R}; E; E') = \exp\left[-\int_0^{|\vec{R}-\vec{r}|} \Sigma(\vec{r''}, E')\,dr''\right] \tag{6.5}$$

where $\Sigma(r', E')$ is the attenuation coefficient at position $\vec{r'}$ along the direction of the scattered radiation, at the scattering energy, E'.

Combining Eqs. (6.1) to (6.5), the count rate at detector, positioned at $\vec{R}$ to measure scattered radiation, is:

$$C_S(\vec{R}, E') = K\int_{V_D} \frac{I_0(\vec{r}_0, E)}{4\pi\,|\vec{r} - \vec{r}_0|^2} \exp\left[-\int_0^{|\vec{r}-\vec{r}_0|} \Sigma(\vec{r'}, E)\,dr'\right] \sigma_s(\vec{r}, E) N(\vec{r})$$
$$\times \frac{p(\mu, E \to E')}{2\pi\,|\vec{R} - \vec{r}|^2} \exp\left[-\int_0^{|\vec{R}-\vec{r}|} \Sigma(\vec{r''}, E')\,dr''\right] dV_D \tag{6.6}$$

where K is a system constant that incorporates the detector's properties (size and efficiency), and V_D is the scattering volume seen by the detector.

6.3 Multiple Scattering

Obviously, there is nothing that can stop radiation from encountering more than one collision. Therefore, when applying the single-scattering equation of Eq. (6.6) to model a measurement, one must ensure that only once-scattered radiation is detected.

This can be achieved by collimating both the source and detectors to confine their path, since multiple scattering, unlike single scattering, does not have a well-defined path. The finer (narrower) the collimator, the lower the probability that multi-scattered radiation reaches the detector. Energy discrimination can also be used to eliminate the contribution of multiple scattering, since the energy of once-scattered radiation is uniquely related to the angle of scattering, as shown below for the Compton scattering of photons and the elastic scattering of neutrons. In essence, energy discrimination acts as a "soft" collimator, which can supplement the "hard" collimation process that employs physical collimators, allowing the use of wider collimators and eliminating the contribution of radiation scattered on the inner walls of such collimators. Limiting the size of the imaged object to within the range of one-mean-free path (mfp) of the incident radiation also reduces the probability of multiple scattering, since, on average, radiation interacts once within a mfp; recall that one mfp is equal to $\frac{1}{\Sigma}$, see Eq. (2.8). The contribution of multiple-scattering may also be compensated for within the image reconstruction process.

While multiple-scattered radiation may be dealt with using one or more of the above mentioned methods, radiation that suffers only two subsequent collisions, one of them at a small angle, is difficult to eliminate. This is because a small-angle scattering hardly changes the direction or energy of radiation, making it appear to be almost identical to once-scattered radiation. However, such a small-angle scattering carries information that relates to regions in the imaged object that can differ significantly from those of the monitored single scattering, particularly if the small-angle scattering takes place downstream of the path of incident or second-scattered (larger angle) radiation.

Analogous to the measurement model of induced emission, Eq. (5.6), one can deduce that the model of Eq. (6.6) maps the atomic density, $N(\vec{r})$, to the measured scattering counts, $C_S(\vec{R}, E')$. However, in induced emission, the energy of the emitted radiation is defined by the type of interaction that induces the emission, hence the value of the interaction cross section, $\sigma_{p \rightarrow s}$. In scattering, both the scattering cross section, σ_s, and the angular probability of scattering, $p(\mu, E \rightarrow E')$, depend on the nature of the scattering atoms. As such, one must know the nature of the scattering medium to determine the type of atoms to which $N(\vec{r})$ refers. In order to elaborate further, let us consider the scattering of photons and neutrons.

6.4 Compton Scattering

Photons with energy from about 300 keV to 2 MeV interact predominantly in most materials via Compton scattering, in which a photon possessing corpuscular properties is scattered by the "free" electrons of the atom. The target electrons are considered to be free and at rest, since their binding energy is much lower than that of the bombarding photons. Therefore, the atomic cross section for scattering, σ_s, in Eq. (6.6), can be expressed in terms of the electron scattering, σ_e, as:

$$\sigma_s = \sigma_e Z \tag{6.7}$$

where Z is the number of electrons per atom. Using the expression for the atomic density, N in Eq. (2.5), then:

$$\sigma_s N = \sigma_e \frac{Z\rho}{A\mathrm{u}} = \sigma_e \frac{\rho_e}{A\mathrm{u}} \approx \frac{1}{2}\sigma_e \frac{\rho}{\mathrm{u}} \tag{6.8}$$

where ρ is the mass density, A is the mass number of the scattering material, $\rho_e = ZN$ is the electron density, and the approximation is due to the fact that, for most elements, $\frac{Z}{A} \approx 0.5$.

Replacing $\sigma_s N(\vec{r})$ in Eq. (6.6) by the expression of Eq. (6.8), one can say that the measurement model maps a material's electron density, or approximately its mass density, to a scattering measurement, irrespective of the type of atom present in the scattering material. This conclusion is aided by the fact that σ_e is determined by the energy, E, of the incident photons, while the angular scattering probability, $p(\mu, E \rightarrow E')$, is determined by the angular scattering probability of the electron as given by the well-known Klein-Nishina relationship, which is determined by the incident photon energy and the energy of the scattered photon and is not material dependent, see for example Hussein (2003).

The kinematics of Compton scattering dictates a unique relationship between the energy of incident photon, E, and that of the scattered photon, E', for a given scattering angle:

$$E' = \frac{E}{1 + \frac{E}{m_e c^2}(1-\mu)} \tag{6.9}$$

where $m_e c^2 = 511$ keV is the rest-mass energy of the electron, and μ, as indicated earlier, is the cosine of the angle of scattering. Equation (6.9) explains why it is sufficient to know E and E' to determine $p(\mu, E \rightarrow E')$ using the Klein-Nishina relationship, as stated above. In practice, if a parallel (collimated) monoenergetic beam of incident photons is used, the direction of scattered photons can be determined without collimation by measuring the photon energy and applying Eq. (6.9) to find μ. Even if the incident beam is not collimated, but is still monoenergetic, the energy of the photon can be used to draw an equi-angle (isogonic) curve at which incident photons scatter by the same angle, determined from Eq. (6.9), to a detector at a given location.

Employing a monoenergetic source of photons facilitates the formulation of the single-scattering model of Eq. (6.6), since the cross sections (per unit electron density) required by the model can be easily found from cross section libraries, such as XCOM (Berger et al., 1999). Fortunately, many of gamma sources are monoenergetic, or contain distinct gamma energies that can be individually monitored, since each source energy will lead to a different scattering energy at a given scattering angle, as Eq. (6.9) indicates. When using x-rays, which have a continuum energy spectrum, the cross section has to be integrated over the energy spectrum of the incident photons for the incoming radiation and over the spectrum of the scattered photons for the scattered radiation. The latter spectrum is quite convoluted, as it involves shifting the spectrum of incident photons according to Eq. (6.9) and adjusting its magnitude by the probability of scattering. Alternatively, one can divide the energy spectrum of incident x-rays into energy bands (groups) and use the average energy of each band along with the corresponding scattering energy dictated by Eq. (6.9).

One should keep in mind, though, when applying the single-scattering model, that the attenuation coefficients in the arguments of the exponential attenuation factors are total cross sections, i.e. they include the cross section for photoabsorption, Rayleigh (coherent) scattering, Compton scattering and, when present, pair production. All these cross sections, except for Compton scattering, have a strong (nonlinear) dependence on the atomic number, Z, of the material, which is usually not known in advance in an imaging exercise. The problem is further complicated by the presence of many elements, hence different Z values. Fortunately, however, Compton scattering dominates over a wide range of energy, and, consequently, the error made in approximating the other cross sections, iterating for their values, or ignoring them altogether, is not significant. The competing reactions can, however, interfere with a measurement that is assumed to be based on Compton scattering. Rayleigh scattering does not change the photon energy and is dominant at small scattering angles and at low energy. Therefore, Rayleigh scattering can produce a scattering signal at small angles, but its effect can be removed by energy discrimination, unless the angle of Compton scattering is so small that its corresponding scatter energy is not very different from that of the source. The energy lost in photon absorption is given to an atomic electron (via the photoelectric effect). This causes atomic ionization or excitation. In the latter case, secondary photons are emitted as the excited atoms return to their original state. The resulting fluoroscopic photons are, however, low in energy (in the energy range of the electron's binding energy) and can be easily discriminated against by their low energy. In many cases, if the source energy, E, is sufficiently high, fluoroscopic photons do not even interfere with Compton scattered photons, since the minimum energy of a scattered photon, according to Eq. (6.9), $E'_{min} = \frac{E}{1+\frac{2E}{m_ec^2}}$, becomes larger than that of the secondary photons. The third competing interaction, pair production, occurs only when the energy of the incident photon is greater than, $2m_ec^2 = 1.022\,\text{MeV}$, the minimum energy required to convert the photon energy into the mass of an electron-positron pair. The resulting positron readily absorbs an electron in its proximity, and the two particles annihilate each other, producing two 511 keV photons, emanating at two opposite directions. Such photons can interfere with Compton-scattered photons of the same energy.

6.5 Neutron Elastic Scattering

The elastic scattering of neutrons results in, like in Compton scattering, an angle-dependent change in energy. Unlike Compton scattering, however, neutrons interact with the nucleus. Therefore, in Eq. (6.6), both the neutrons scattering microscopic cross section, σ_s, and the angular scattering probability, $p(\mu, E \rightarrow E')$, depend on the nature of the nuclide encountered by the incoming neutrons. Therefore, in a medium containing more than one type of nuclide, relating the detected scattering signal to the atomic number of all nuclides or that of any particular nuclide is not that straightforward. However, some information on the nature of the scattering nuclide can be retrieved from the kinematics of neutron scattering.

In the single scattering of a neutron of energy, E, by a nucleus of mass number, A, the energy of the scattered neutron is determined (considering the target nucleus to be at rest) by (Hussein, 2003):

$$E' = E\left[\frac{\mu + (\mu^2 + A^2 - 1)^{\frac{1}{2}}}{(A+1)}\right]^2 \tag{6.10}$$

with $\cos^{-1}\mu$ being the angle of scattering. For heavy nuclides, $A >> 1$, $E' \approx E$, and not much change in energy is observed at any angle. Therefore, if one detects scattered neutrons with an energy almost equal to that of the source, one can conclude that the scattering is caused by heavy nuclides. Since neutron scattering by such heavy nuclides tends to be isotropic, i.e. $p(\mu, E \rightarrow E') = \frac{1}{2}$, then one can determine from the intensity of the scattering signal, at an energy close to E', the number density, $N(\vec{r})$, of heavy nuclides. This can be useful for imaging heavy nuclides in the presence of light nuclides. One will not, however, be able to determine the nature of scattering nuclides or distinguish between them.

For hydrogen, $A = 1$, Eq. (6.10) gives $E' = E\mu^2$, showing a drastic change in energy with angle. At $\cos^{-1}\mu = \frac{\pi}{2}$, a neutron scattered by the proton nucleus of hydrogen loses its entire energy, since $E' = 0$. In addition, no backscattering, $\cos^{-1}\mu > \frac{\pi}{2}$, can take place when a neutron is scattered once by a hydrogen nucleus, and the same value of E' cannot be attained at two different angles. This makes neutron scattering a natural candidate for imaging hydrogenous materials, by monitoring the forward scattering of neutrons. Then $N(\vec{r})$ in Eq. (6.6) will refer to the number density of hydrogen nuclei. The values of σ_s of $p(\mu, E \rightarrow E')$ in the same equation can be determined in advance from the cross section of hydrogen at the considered energies.

Eq. (6.10) also indicates that the energy of the scattered photon is at minimum equal to:

$$E_{min} = E\left(\frac{A-1}{A+1}\right)^2 \tag{6.11}$$

For nuclides of intermediate mass number, and with a detector with a good energy resolution, one can utilize E_{min} to discriminate between different nuclides. For example, in a mixture containing hydrogen ($A = 1$), carbon ($A = 12$) and oxygen ($A = 16$), single scattering with hydrogen gives rise to neutrons within the entire energy range, i.e. with $E' \geq 0$, while carbon ($A = 12$) will scatter neutrons to $E' \geq \left(\frac{11}{13}\right)^2 E = 0.72E$, and oxygen produces neutrons within energy range $E' \geq \left(\frac{15}{17}\right)^2 E = 0.78E$. Therefore, neutron scattering in the energy range $0 < E' \leq 0.72E$ will only be indicative of the hydrogen content, while detecting neutrons within $0.72E < E' < 0.78E$ will provide a measure of the hydrogen plus carbon content, and at $0.78E < E' \leq E$ all three elements are detected. Therefore, the individual atomic density of each element can be determined by energy-selective detection of the scattered neutrons. However, this is complicated by the fact that the attenuation of the incident neutrons before scattering depends on all elements present, while that of the scattered neutrons is affected only by the elements that contribute to scattering in the considered energy range. This requires an iterative solution of the problem, but such an iterative process is required anyway, even in a monoatomic medium, due to the nonlinearity of the forward problem of Eq. (6.6).

Monoenergetic neutrons produced by neutron generators enable direct use of the single-scattering model of Eq. (6.6), along with the neutron kinematics of Eq. (6.10). In addition, neutron generators also produce intense beams that can give rise to a strong scattering signal. On the other hand, isotopic neutron sources, such as ^{252}Cf and ^{241}Am/Be, possess a wide energy spectrum and a relatively low intensity, unless large sources are employed. Neutron production is typically accompanied with the release of gamma rays. In addition, incident neutrons may also produce gamma rays by inelastic scattering or radiative capture. Such gamma rays may affect detectors used to measure the scattered neutrons, and precautions must be taken to electronically discriminate their contribution. The produced gamma rays can, in some cases, persist for a while, leaving residual radioactivity even after the termination of radiation exposure, which is not only a safety concern, but it can also affect subsequent measurements unless adequate gamma discrimination is applied to the detected signal. Inelastically scattered neutrons can also interfere with the desired elastic scattering signal, but the probability of inelastic scattering is much lower than that of the elastic one, and the former is isotropically produced, so that the inelastic neutron scattering component is typically negligible. In fissionable materials, the neutron generated by fission will obviously interfere with the scattering signal. However, fissile nuclides all have a high mass number, and the energy of scattered neutrons hardly changes, while the energy of fission neutrons spans a wide range.

6.6 Discretization

Numerical implementation of Eq. (6.6) requires lumping some of the parameters into corresponding equivalent average values. If the source is not monoenergetic and has a wide energy spectrum, one can divide the spectrum into a number of discrete energy groups, with each group assigned a single mean energy. A source that is not monodirectional can also be divided into a number of contiguous solid angles, each given a unique direction. The model of Eq. (6.6) can then be applied to a particular direction with a given source energy, to obtain the detector response at a given location and direction of scattering. Scattering can be assumed to occur at a point at the center of a voxel, or at a number of points, or within smaller subvoxels within the voxel in the direction of the incident beam. The scattering direction may also be divided into a number of discrete directions to cover the field-of-view of a detector of a given surface area. Applying such discretization processes implies that a single averaged value is utilized for the cross section and the scattering probability of Eq. (6.6). In addition, an average value is used for the atomic density within each voxel. Like all discretization processes, excessive discretization can lead to accumulation of numerical error, while crude discretization may lead to in accurate values. Model validation against Monte Carlo simulations can, therefore, be beneficial. Such simulations should be first performed against simulations that include only single scattering, ignoring multiple scattering and any other interfering interactions. Once the single-scattering model is verified, multiple scattering and the other interfering interactions, discussed in this chapter, can be included in the simulations, to determine their effect. Monte Carlo simulations can also incorporate the detector's response function, i.e. the change in its efficiency with radiation energy. This model of Eq. (6.6) can then be modified to include the detector efficiency, as a factor within the system constant, K.

6.7 Sensitivity

The physical parameter of interest in scatter imaging is $N(\vec{r})$, the density of the atoms that causes scattering at some point within the imaged object. Then, discretization and replacing integrals with summations, applying the definition of sensitivity, Eq. (3.5), to the single-scatter model of Eq. (6.6), leads to:

$$S_S = \frac{\partial C_S(\vec{R},E)}{\partial N(\vec{r})} = K\frac{I_0(\vec{r}_0)}{4\pi\,|\vec{r}-\vec{r}_0|^2}\exp\left[-\sum_k \Sigma(\vec{r'}_k,E)\Delta r'_k\right]$$
$$\times\ \frac{\sigma_s(\vec{r},E)p(\mu,E\to E')}{2\pi\,|\vec{R}-\vec{r}|^2}\exp\left[-\sum_{k'}\Sigma_s(\vec{r''}_{k'},E')\Delta r''_{k'}\right]\Delta V_D \tag{6.12}$$

The sensitivity for imaging with scattered radiation is governed by the scattering probability, the $\sigma_s(\vec{r},E)p(\mu,E\to E')$ term in Eq. (6.12). However, the sensitivity is hampered by the attenuation and divergence of incident and scattered radiation. The sensitivity is, however, improved by increasing the voxel size, ΔV, increasing the intensity of the radiation source, $I_0(\vec{r}_0)$, and increasing the detector size and efficiency, which affect the value of K in Eq. (6.12).

References

M. J. Berger, J. H. Hubbell, S. M. Seltzer, J. S. Coursey, and D. Zucker, *XCOM: Photon Cross Section Database (version 1. 2), [Online]*, National Institute of Standards and Technology, NBSIR 87-3597, Gaithersburg, 1999, (`http://physics.nist.gov/xcom`, accessed, May 2010).

E. M. A. Hussein, *Handbook on Radiation Probing, Gauging, Imaging and Analysis: Basics and Techniques*. Dordrecht: Kluwer Academic Pub., 2003, vol. I.

Part II

The Inverse Problem

"[A]esthetics" derives from a Greek verb meaning "to perceive with the senses." It is a scientist's prerogative to introduce–and to be governed a little by–aesthetics in his [or her] work. This implies things like restraint, nonexaggeration, nonreliance on innumerable assumptions, criteria or data banks, and so on.

[S]ometimes we tend to resort to inversion techniques too blindly, without using our judgment or "feel" about handling a given problem, which may lead to "anti-aesthetic" excesses.

*Diran Deirmendjian (1976)**

* In: Deepak, A., Ed. (1977). *Inversion methods in atmospheric remote sounding*, p. 138, Academic Press, New York.

7 Features

Before reading this and subsequent chapters, the reader may like to consult Appendix 7.A for a brief reminder of the basics of functional analysis, which are useful in many of the presented formulations.

7.1 Discretization

As the discussion of the forward problem in Part I of this book indicated, the forward problem provides an *approximate* mapping, $\boldsymbol{m}$, of a set of physical parameters, $\boldsymbol{c} \in C$, into a set of modeled measurements, $\boldsymbol{e} \in E$, with C and E being, respectively, the spaces of $\boldsymbol{c}$ and $\boldsymbol{e}$, see Section 1.5. The mapping, $\boldsymbol{m}$, is approximate due to the inevitable simplifications of the measurement modeling process, the various interfering effects that are not taken into account, and the errors associated with the discretization process required for a numerical solution. The availability of a forward mapping is necessary to solve the inverse problem. Since an image consists of a number of discrete pixels or voxels within which the physical attributes are estimated, it is necessary to discretize the forward models, as indicated in Chapters 4 to 6. The imaging process itself requires the acquisition of a finite, though usually large, number of measurements. Therefore, it is also necessary to have a discrete mapping formulation that relates discrete values of c to discrete values of e.

The discrete form of the forward model can be a matrix or a set of functions that accept discrete image parameters. In radiation imaging, measurements typically correspond to counts, and the image parameters are attenuation coefficients (cross sections), source intensities, or material densities, depending on the image modality, as explained in Part I of this book. Then, the image parameters and the measurements must be non-negative real-valued numbers, otherwise they are meaningless. The measurements and image parameters can, therefore, be presented in a discrete form as vectors, in Euclidean space.[1] The two can be then related in a matrix form:

$$\boldsymbol{e} = \boldsymbol{A}\boldsymbol{c} \tag{7.1}$$

or a functional form:

$$\boldsymbol{e} = \boldsymbol{f}(\boldsymbol{c}) \tag{7.2}$$

[1] An n-dimensional Euclidean space is the space of all n-dimensional sets of real numbers.

Computed Radiation Imaging. DOI: 10.1016/B978-0-12-387777-2.00007-0

where $\boldsymbol{e}$ is a vector whose elements are individual modeled measurements, $\boldsymbol{c}$ is a vector containing the imaging parameters, $\boldsymbol{A}$ is a matrix relating $\boldsymbol{e}$ to $\boldsymbol{c}$, and $\boldsymbol{f}$ is a vector that includes functions of $\boldsymbol{c}$. Both $\boldsymbol{A}$ and $\boldsymbol{f}$ incorporate the fact that a single measurement can be influenced by many parameters. The matrix $\boldsymbol{A}$ is called the system matrix.

The difference between the formulation of Eq. (7.2) and that of Eq. (7.1) is that in the former the image parameters vector, $\boldsymbol{c}$, is not explicitly expressed as a separate vector, but rather its elements appear in each function formulation that relates one modeled measurement to many elements of $\boldsymbol{c}$. We will use:

$$\boldsymbol{e} = \boldsymbol{m}\boldsymbol{c} \tag{7.3}$$

to indicate either the matrix form of Eq. (7.1) or the functional form of Eq. (7.2).

Recall that a Euclidean space is a space of real numbers. In this space, the translation (or a rotation) of a point results in a shifting (or turning) of every point in the same direction and by the same displacement (or angle). In the Euclidean space, one can define the following norms that measure the "length" of the physical parameters vector, $\boldsymbol{c}$, and the measurement vector, $\boldsymbol{e}$:

$$\|\boldsymbol{c}\| = \sqrt{\sum_{i=1}^{N} c_i^2}; \quad \|\boldsymbol{e}\| = \sqrt{\sum_{i=1}^{M} e_i^2} \tag{7.4}$$

where c_i designates the ith parameter in an image of N voxels, and e_i is the ith measurement in a set of M measurements. With this definition one can measure how close an image, $\hat{\boldsymbol{c}}$, is to the actual one, $\boldsymbol{c}_0$, (if known, e.g., when testing or calibrating), via the norm $\|\hat{\boldsymbol{c}} - \boldsymbol{c}\|$, and for an iterative process one can determine how the estimated vector, $\hat{\boldsymbol{c}}_k$, at iteration k differs from that at a preceding iteration $k-1$ through the norm $\|\hat{\boldsymbol{c}}_k - \hat{\boldsymbol{c}}_{k-1}\|$. Similarly, one can measure how a modeled measurement, $\hat{\boldsymbol{e}} = \boldsymbol{m}\hat{\boldsymbol{c}}$, corresponding to an estimated image, $\hat{\boldsymbol{c}}$, differs from the actual measurement, $\boldsymbol{e}$ using the norm $\|\hat{\boldsymbol{e}} - \boldsymbol{e}\|$.

The Euclidean norm for the mapping, $\boldsymbol{m}$, that relates $\boldsymbol{c}$ to $\boldsymbol{e}$, is defined as:

$$\|\boldsymbol{m}\| = \sup \frac{\|\boldsymbol{m}\boldsymbol{c}\|}{\|\boldsymbol{c}\|}; \quad \boldsymbol{c} \neq \boldsymbol{0} \tag{7.5}$$

where sup (supremum, least upper bound) is used here so that a value for $\|\boldsymbol{m}\|$ can still be found even if the maximum is not attainable with a non-zero vector $\boldsymbol{c}$. The ratio $\frac{\|\boldsymbol{m}\boldsymbol{c}\|}{\|\boldsymbol{c}\|}$ can be seen as a factor that measures the extent with which the mapping, $\boldsymbol{m}$ "stretches" $\boldsymbol{c}$. The norm $\|\boldsymbol{m}\|$ is then the largest stretching factor of all non-zero vectors, $\boldsymbol{c}$. For the inverse mapping, $\boldsymbol{m}^{-1}$, the Euclidean norm is:

$$\|\boldsymbol{m}^{-1}\| = \sup \frac{\|\boldsymbol{m}^{-1}\boldsymbol{e}\|}{\|\boldsymbol{e}\|}; \quad \boldsymbol{e} \neq \boldsymbol{0} \tag{7.6}$$

with $\boldsymbol{m}^{-1}\boldsymbol{e}$ being equal to the sought solution, $\boldsymbol{c}$.

Images are usually discretized into square pixels or cubic voxels, with a constant image attribute value inside each pixel/voxel. In other words, with each voxel, associated is a basis rectangular function that has a value of unity inside the voxel and zero outside it. Lewitt (1992) suggested a spherically symmetric volume element, called a "blob," with a radially symmetric basis function that smoothly and monotonically decrease from unity at the center to zero at the edge of the blob. Lewitt (1992) proposed a modified from of the Kaiser-Bessel window, $b(r)$, used in digital signal processing:

$$b(r) = \begin{cases} \left(1-\left(\frac{r}{a}\right)^2\right)^{\frac{m}{2}} \frac{I_m\left(\alpha\sqrt{1-\left(\frac{r}{a}\right)^2}\right)}{I_m(\alpha)} & \text{for } r \geq 0 \\ 0 & \text{for } r \geq 0 \end{cases} \tag{7.7}$$

where I_m is a modified Bessel function of the first kind[2] of order m, a is the radius of the blob, r is a radial distance from the center of the blob, and α is tapering parameter that determines the overall shape of the blob (a small α value gives a wide blob and a large value produces a narrowly peaked blob with long tails). While a step basis function averages image features within a voxel, a radially symmetric basis function preserves peaks and valley features within a voxel, and as such is better at depicting the characteristics of an image, which would have otherwise been smoothed out in the usual step basis function (Matej and Lewitt, 1996). However, the use of blobs is computationally demanding, and needs to be optimized for the problem at hand, see Garduño and Herman (2004).

7.2 Well-Posed Problem

The inverse problem of the forward mapping $\boldsymbol{e} = \boldsymbol{mc}$ is well-posed[3] (according to Hadamard, 1923) if:

1. There *exists* a solution for any measurement subset in the possible set of measurements, E, i.e., $\hat{\boldsymbol{e}} \in E$.
2. The obtained solution, $\hat{\boldsymbol{c}}$, is in the permitted set of image parameters, i.e. $\hat{\boldsymbol{c}} \in C$, and is *unique*.
3. The inverse mapping, $\hat{\boldsymbol{e}} \xrightarrow{\boldsymbol{m}^{-1}} \hat{\boldsymbol{c}}$, is *continuous*.

where $\hat{\boldsymbol{e}}$ is a measurement with some uncertainty, and $\hat{\boldsymbol{c}}$ is the corresponding reconstructed image parameter. The first two conditions require $\boldsymbol{m}$ to have an inverse that produces realistic image parameters, while the third condition indicates that there has to be a continuous change in the image parameters with the continuous change in

[2] $I_m(r) = \frac{1}{2\pi i}\oint \exp\left[\left(\frac{r}{2}\right)\left(t+\frac{1}{t}\right)\right] t^{-n-1} dt$; with the contour traversed in a counterclockwise direction enclosing the origin (Weisstein, Eric W., "Modified Bessel Function of the First Kind", From MathWorld–A Wolfram Web Resource. `http://mathworld.wolfram.com/ModifiedBesselFunctionoftheFirstKind.html`).

[3] Discussion in this section is guided in part by that given by Bertero and Boccacci (1998).

measurements. The latter condition, obviously, controls the amount of error propagation from measurements to system parameters. Continuity is, therefore, indicative of solution *stability*.[4]

An ill-posed inverse problem is accordingly a problem in which the inverse of $\boldsymbol{m}$ produces a solution that is not in C ($\hat{\boldsymbol{c}} \notin C$), or is not unique (i.e. more than one $\hat{\boldsymbol{c}}$ correspond to same $\hat{\boldsymbol{e}}$), or when a small change in the value of $\hat{\boldsymbol{e}}$ results in a large change in the reconstructed image parameters, $\hat{\boldsymbol{c}}$. Note that since in radiation imaging one usually seeks a nonnegative solution (a solution $\boldsymbol{c} \in C$ of nonnegative real numbers), while having a set of measurements $\boldsymbol{e} \in E$ contaminated by random uncertainties, there is no assurance that the solution of $\hat{\boldsymbol{c}} = \boldsymbol{m}^{-1}\hat{\boldsymbol{e}}$ will not produce negative members that do not belong to C. This nonnegative constraint can by itself make the problem an ill-posed one.

7.3 Existence

The discrete forms of Eqs. (7.1) and (7.2) should be definite, and repeatable in their structure for the same measurement configuration. As such, their inverse is likely to exist. This is easier to ascertain for the matrix form, Eq. (7.1), as long as its inverse can be numerically constructed. Although a functional form for the inverse of Eq. (7.2) is not usually explicitly possible, one can argue that the inverse function exists as long as the functional form of the forward mapping is definite. Therefore, the existence condition of well-posedness is usually satisfied. If not, the system's physical layout and design should be altered to ensure the existence of solution, otherwise the entire exercise of solving the inverse problem is not worth pursuing.

7.4 Uniqueness

For a *linear* system, the uniqueness of solution can be ascertained by the argument that if the system has two solutions, $\boldsymbol{c}_1$ and $\boldsymbol{c}_2$, corresponding to the same measurement, $\boldsymbol{e}$, then both must satisfy Eq. (7.3), leading to:

$$\boldsymbol{e} = \boldsymbol{m}\boldsymbol{c}_1 = \boldsymbol{m}\boldsymbol{c}_2 \tag{7.8}$$

$$\boldsymbol{m}(\boldsymbol{c}_1 - \boldsymbol{c}_2) = \boldsymbol{0} \tag{7.9}$$

which necessitates that $\boldsymbol{c}_1 = \boldsymbol{c}_2$ and the solution must be unique. Also, Eq. (7.9) indicates that $\boldsymbol{c}_{1,2} = \boldsymbol{c}_1 - \boldsymbol{c}_2$ is a non-trivial solution of the homogeneous equation: $\boldsymbol{m}\boldsymbol{c} = \boldsymbol{0}$. If such solution exists, then for a linear problem, the linear combination $\boldsymbol{c}_2 = \boldsymbol{c}_{1,2} + \boldsymbol{c}_1$ must also be a solution of Eq. (7.3), but Eq. (7.8) indicates that $\boldsymbol{m}\boldsymbol{c}_1 = \boldsymbol{m}\boldsymbol{c}_2$. Therefore, one has:

$$\boldsymbol{m}\boldsymbol{c}_2 = \boldsymbol{m}(\boldsymbol{c}_{1,2} + \boldsymbol{c}_1) = \boldsymbol{m}\boldsymbol{c}_{1,2} + \boldsymbol{m}\boldsymbol{c}_1 = \boldsymbol{m}\boldsymbol{c}_1$$

[4] Stability refers to the insensitivity of solution to small random errors in data (measurements). Robustness, on the other hand, is the solution's insensitivity to a small number of large errors (outliers) in the measurement set.

Then $\boldsymbol{c}_1 = \boldsymbol{c}_2$, and again Eq. (7.3) has a unique solution. Therefore, Eq. (7.3) has a unique solution if and only if the only solution for the homogeneous equation $\boldsymbol{mc} = \mathbf{0}$ is $\boldsymbol{c} = \mathbf{0}$. The discrete form of Eq. (7.3) satisfies this condition, since if $\boldsymbol{c} = \mathbf{0}$ it will produce $\boldsymbol{e} = \mathbf{0}$.

The above discussion indicates that the uniqueness of solution can be assessed for a linear problem by examining the homogeneous problem: $\boldsymbol{mc} = \mathbf{0}$. If the homogeneous problem has a solution other than the trivial solution, $\boldsymbol{c} = \mathbf{0}$, then the problem does not have a unique solution. If $\boldsymbol{mc} = \mathbf{0}$ has a non-travail solution, then there are "invisible" images, $\boldsymbol{c} \neq \mathbf{0}$, that cannot be observed because they do not produce any corresponding measurements. Such invisible images exist in the so-called null space, and because they have nil corresponding measurements they can be added to a real (visible) image without affecting the recorded measurements; hence one cannot be assured of the presence of a unique solution corresponding to a certain set of measurements.

For nonlinear problems, the homogeneous problem can have more than one observable image (solution) corresponding to the same set of measurements. A nonlinear problem, by definition, does not have a single unique solution, but may have a set of unique solutions. Nevertheless, under some restrictions, a unique solution can exist within some bounds, as shown in Section 10.7.

7.5 Continuity

The third condition for the well-posedness of a problem, continuity of solution, follows naturally from the fact that the forward mapping, $\boldsymbol{e} = \boldsymbol{mc}$, and its discretized form, represent an observable natural process in which a change in $\boldsymbol{c}$ produces a corresponding change in $\boldsymbol{e}$. If such change is not observable, it is usually an indication of a poor measurement arrangement, in terms of design, sensitivity of observation or physical layout, or poor counting statistics that needs improvement by increasing the source strength, counting period and/or detector efficiency. A continuous parameter is differentiable, i.e. $\frac{\partial e_i}{\partial c_j}$ exists, at least in the neighborhood of the value of c_j, where the subscripts refer to individual values.

7.6 Ill-Posed Problem

From the above discussion, one may conclude that the discrete form of the forward problem of Eq. (7.3), is well-posed. This conclusion is, however, based on a strict mathematical analysis. In practice, the discrete formulation is an approximation of the continuous general form of $\boldsymbol{e} = \boldsymbol{mc}$, which is generally ill-posed. The measurements used in the solution of the inverse problem are always contaminated with uncertainties, and the forward mapping itself is an approximation, as shown in Part I of this book. That is, if the true image is represented by $\boldsymbol{c}_0$, then the recorded measurement, $\hat{\boldsymbol{e}}$, corresponding to $\boldsymbol{c}_0$, can be expressed as:

$$\hat{\boldsymbol{e}} = \boldsymbol{mc}_0 + \boldsymbol{\delta}_e \tag{7.10}$$

where $\boldsymbol{\delta}_e$ is the difference between the measured $\boldsymbol{e}$ and the computed modeled measurement corresponding to the true solution, if known. Then, the inverse problem produces a solution:

$$\hat{\boldsymbol{c}} = \boldsymbol{m}^{-1}\hat{\boldsymbol{e}} = \boldsymbol{c}_0 + \boldsymbol{m}^{-1}\boldsymbol{\delta}_e \tag{7.11}$$

That is, the obtained solution differs from the true solution by $\boldsymbol{m}^{-1}\boldsymbol{\delta}_e$. Then $\boldsymbol{m}^{-1}$ $\boldsymbol{\delta}_e \neq \mathbf{0}$, even for the trivial image, $\boldsymbol{c}_0 = \mathbf{0}$. The homogeneous problem, $\boldsymbol{m}\boldsymbol{c}_0 = \mathbf{0}$, does not exist even if $\boldsymbol{c}_0 = \mathbf{0}$, because of the ever present measurement error, $\boldsymbol{\delta}_e \neq \mathbf{0}$. One cannot then address the uniqueness question. However, if there exists a trivial solution $\hat{\boldsymbol{c}} = \mathbf{0}$ for which $\boldsymbol{m}\hat{\boldsymbol{c}} = \mathbf{0}$, then according to Eq. (7.11) $\boldsymbol{c}_0 = -\boldsymbol{m}^{-1}\boldsymbol{\delta}_e$, and the true image corresponding to the trivial solution $\hat{\boldsymbol{c}} = 0$ will vanish only if $\boldsymbol{m}^{-1}\boldsymbol{\delta}_e = \mathbf{0}$. With the measurement error being a definite measurable quantity, the condition $\boldsymbol{m}^{-1}\boldsymbol{\delta}_e = 0$ will require an indefinite inverse mapping, and accordingly the solution of Eq. (7.11) will not exist, while the uniqueness condition $\boldsymbol{m}\hat{\boldsymbol{c}} = \mathbf{0}$ with $\hat{\boldsymbol{c}} = \mathbf{0}$ is still satisfied.

Moreover, since $\boldsymbol{\delta}_e$ can be independent of the solution and usually contains a random component (due to statistical fluctuations), the solution $\hat{\boldsymbol{c}}$ may no longer continually depend on the measurement $\hat{\boldsymbol{e}}$. In other words, $\hat{\boldsymbol{e}}$ can change due to measurement uncertainties and not because of genuine changes in the imaged object. A true image, $\boldsymbol{c}_0$, can, therefore, have different measurements, $\hat{\boldsymbol{e}}$, producing a solution (when it exists) that is not exactly unique; though may approximately resemble the true image. Therefore, the image reconstruction problem is in effect an ill-posed problem.

In the discrete form of Eq. (7.3), if the inverse mapping $\boldsymbol{m}^{-1}$ exists, a solution $\boldsymbol{m}^{-1}\boldsymbol{e}$ will exist. The homogeneous equation, $\boldsymbol{m}\boldsymbol{c} = \mathbf{0}$, has only a trivial solution, $\boldsymbol{c} = \mathbf{0}$, since $\boldsymbol{m}$ is definite. Therefore, the uniqueness of a linear problem is also assured. Any solution $\hat{\boldsymbol{c}} = \boldsymbol{m}^{-1}\hat{\boldsymbol{e}}$ will also continuously depends on $\hat{\boldsymbol{e}}$, irrespective of the measurement error. Therefore, one arrives at the interesting conclusion that although the continuous inverse problem is, in general, an ill-posed problem, its discrete form satisfies the conditions of a well-posed problem. However, the solution provided by the discrete form is *approximate* because: (1) discretization itself is an approximation with its own truncation error, (2) the forward model it applies is an approximation of reality (as indicated in Part I of this book), (3) the employed measurements carry their own uncertainties, and (4) the inversion process introduces its own numerical round-off error. The next logical question is then how good is the obtained approximate solution, or in other words, how much error is propagated through the solution process.

7.7 Ill-Conditioning

Let us first consider the propagation of measurement error through the inversion process. For measurements with an error $\boldsymbol{\delta}_e$, the inverse mapping produces an error in the image parameter, $\boldsymbol{\delta}_c$ such that:

$$\boldsymbol{\delta}_c = \boldsymbol{m}^{-1}\boldsymbol{\delta}_e \tag{7.12}$$

Therefore,

$$\|\boldsymbol{\delta}_c\| \leq \|\boldsymbol{m}^{-1}\| \; \|\boldsymbol{\delta}_e\| \tag{7.13}$$

but we have,

$$\|\boldsymbol{e}\| \leq \|\boldsymbol{m}\| \; \|\boldsymbol{c}\| \tag{7.14}$$

If the mapping is linear, one can estimate $\|\boldsymbol{m}^{-1}\|$ as:

$$\|\boldsymbol{m}^{-1}\| = \frac{1}{\inf \frac{\|\boldsymbol{mc}\|}{\|\boldsymbol{c}\|}}; \quad \boldsymbol{c} \neq \boldsymbol{0} \tag{7.15}$$

where inf is the greatest lower bound (infimum) of the forward mapping, $\boldsymbol{m}$. Using the above relationships, one can state that:

$$\frac{\|\boldsymbol{\delta}_c\|}{\|\boldsymbol{c}\|} \leq \mathcal{N} \frac{\|\boldsymbol{\delta}_e\|}{\|\boldsymbol{e}\|} \tag{7.16}$$

with

$$\mathcal{N} = \frac{\sup \frac{\|\boldsymbol{mc}\|}{\|\boldsymbol{c}\|}}{\inf \frac{\|\boldsymbol{mc}\|}{\|\boldsymbol{c}\|}}; \quad \boldsymbol{c} \neq \boldsymbol{0} \tag{7.17}$$

The quantity $\mathcal{N}$ is known as the condition number. As evident from Eq. (7.16), $\mathcal{N}$ controls error propagation from the measurements to the solution (reconstructed image parameters). It is obvious that a low condition number is desired for control of error propagation. A *well-conditioned* problem has a condition number of nearly unity, i.e. with no significant magnification of measurement error through the inversion process. An ill-conditioned problem, on the other hand, has a large condition number, leading to magnification of error propagation during the solution of the inverse problem. The condition number is a measure of the "elasticity" of the problem, with a fully elastic problem having no error growth.

Based on the condition number, Demmel (1987, 1988) redefined the ill-posed problem as an infinitely-conditioned problem. This is a more practical definition than that of Hadamard, as it provides a measure of the difficulty in the numerical solution of an inverse problem.

In practice, discretization produces a high condition number; the finer the discretization, the higher the condition number. The resulting discrete inverse problem then exhibits many of the features of the continuous inverse problem. Therefore, one can speak, according to Hansen (1998), of a discrete ill-posed problem. The discrete problem may also be rank deficient. Recall that the rank of a matrix is the maximum number of its linearly independent rows or columns. A matrix that has a full rank will have its rows and columns all linearly independent; if not the matrix is rank deficient.

The number of distinct eigenvalues of a full rank matrix is equal to its rank, since linear dependence produces repeated eigenvalues. Therefore, computed measurements in the forward mapping of a system that has a full-rank matrix are all independent. However, the presence of uncertainties in a practical discrete inverse problem interferes with measurement independence. That is, two measurements which are too close in value that they are not practically distinguishable (given measurement uncertainties) are in effect not independent measurements. Therefore, while an idealized (error-free) discrete inverse problem is well-posed, its practical counterpart is ill-posed. One other way of viewing a discrete inverse problem is to consider the fact that a zero value is in practice hardly achievable. For example, the difference between two conceptually identical measurements recorded at different times is not likely in practice to be equal to zero, but is rather a small definite value within the level of measurement uncertainty. In other words, for any two parameters to be distinguishable from each other, the difference between them must be larger than the lowest level of uncertainty. Recalling that a matrix is singular if it has a zero eigenvalue, then while an ideal problem may have a non-zero eigenvalue, a "practical" zero value could exist due to problem uncertainties.

7.A Basics of Functional Analysis

The forward problem maps a set of physical attributes (parameters) into the measurement space. These parameters depend on the position in the object, and in some cases the radiation energy, and are compiled together as an N-dimensional vector, where N is the number of parameters. In the limit, $N \longrightarrow \infty$, the space of the physical parameters is a continuous space. The parameters space consists of all possible mutations of vectors representing physical attributes. The inverse problem maps a set of measurements into the parameters space. The measurements are also formulated into an M-dimensional vector, where M is the number of measurements, whose values are functions of the position, direction and/or energy at which the measurements are acquired. The measurement functions on the limit can approach an infinite-dimensional space. The measurement space encompasses all possible measurement vectors. Both the physical parameters and measurements are real numbers, and the mapping process are operators on the functions. The branch of mathematics that deals with these functions and their mapping is called functional analysis. The objective of this appendix is to summarize the basics of functional analysis necessary for Part II of this book. For more details, the reader can consult many of the available textbooks on functional analysis. For definitions, the Wolfram MathWorld (`http://mathworld.wolfram.com`) is quite useful.

7.A.1 Within a Space

We will start by considering a vector space (a set of vectors), representing either the parameters space or the measurements space.

Inner-Product

In a vector space, one can take any two vectors $\boldsymbol{u}$ and $\boldsymbol{v}$, and multiply them to obtain a scalar quantity, $\langle \boldsymbol{u}, \boldsymbol{v} \rangle$, called the inner product, (a generalization of the dot product), and the space then is called an inner-product space. The inner product has the following properties:

$$\langle \boldsymbol{u}, \boldsymbol{v} \rangle \geq 0 \tag{7.18a}$$

$$\langle \boldsymbol{u}, \boldsymbol{v} \rangle = \mathbf{0} \text{ if and only if } \boldsymbol{u} = 0 \text{ or } \boldsymbol{v} \text{ (empty vector)} \tag{7.18b}$$

$$\langle \boldsymbol{u}, \boldsymbol{v} \rangle = \langle \boldsymbol{v}, \boldsymbol{u} \rangle \quad \text{(commutive)} \tag{7.18c}$$

$$\langle \boldsymbol{u}, \boldsymbol{v} + \boldsymbol{w} \rangle = \langle \boldsymbol{u}, \boldsymbol{v} \rangle + \langle \boldsymbol{u}, \boldsymbol{w} \rangle \tag{7.18d}$$

$$\langle a\boldsymbol{u}, \boldsymbol{v} \rangle = a \langle \boldsymbol{u}, \boldsymbol{v} \rangle \tag{7.18e}$$

$$\langle \boldsymbol{u}, \boldsymbol{v} \rangle \leq \langle \boldsymbol{u}, \boldsymbol{u} \rangle \ \langle \boldsymbol{v}, \boldsymbol{v} \rangle \quad \text{(Schwarz inequality)} \tag{7.18f}$$

where $\boldsymbol{w}$ is also a vector and a is a scalar.

Euclidean Space

This is an inner-product space in which the inner product is the same as the dot product so that:

$$\langle \boldsymbol{u}, \boldsymbol{v} \rangle = \boldsymbol{u} \cdot \boldsymbol{v} = \sum_i u_i v_i \tag{7.19}$$

where u_i and v_i are the ith elements in the vectors.

Vector Norm

The length of a vector $\boldsymbol{u}$ is defined by the norm, $\|\boldsymbol{u}\|$ in a number of ways:

$$\|\boldsymbol{u}\|_1 = \sum_i |u_i| \qquad L_1 \text{ norm} \tag{7.20a}$$

$$\|\boldsymbol{u}\|_2 = \left(\sum_i |u_i|^2 \right)^{\frac{1}{2}} \qquad L_2 \text{ norm (Euclidean)} \tag{7.20b}$$

$$\|\boldsymbol{u}\|_p = \left(\sum_i |u_i|^p \right)^{\frac{1}{p}} \qquad L_p \text{ norm} \tag{7.20c}$$

$$\|\boldsymbol{u}\|_\infty = \lim_{p \to \infty} \left(\sum_i |u_i|^p \right)^{\frac{1}{p}} = \max_i |u_i| \qquad L_\infty \text{ norm} \tag{7.20d}$$

where $p \geq 1$ is an integer. Each norm is such that:

$$\|\boldsymbol{u}\| \geq 0 \tag{7.21a}$$
$$\|\boldsymbol{u}\| = 0 \text{ if and only if } \boldsymbol{u} = 0 \tag{7.21b}$$
$$\|a\boldsymbol{u}\| = a\|\boldsymbol{u}\| \tag{7.21c}$$
$$\|\boldsymbol{u}+\boldsymbol{v}\| \leq \|\boldsymbol{u}\| + \|\boldsymbol{v}\| \quad \text{(triangle inequality)} \tag{7.21d}$$

With a norm, one can formulate mathematical structures (topology) to study convergence, continuity, etc., of sequences and functions.

Metric Space

This is an inner-product space in which a metric is constructed to measure the distance, $d(\boldsymbol{u}-\boldsymbol{v})$, between any two vectors, $\boldsymbol{u}$ and $\boldsymbol{v}$:

$$d(\boldsymbol{u}-\boldsymbol{v}) = \langle \boldsymbol{u}-\boldsymbol{v}, \boldsymbol{u}-\boldsymbol{v}\rangle^{\frac{1}{2}} = \|\boldsymbol{u}-\boldsymbol{v}\| \tag{7.22}$$

This metric has the following properties:

$$\|\boldsymbol{u}-\boldsymbol{v}\| \geq 0 \tag{7.23a}$$
$$\|\boldsymbol{u}-\boldsymbol{v}\| = 0 \text{ if and only if } \boldsymbol{u} = \boldsymbol{v} \tag{7.23b}$$
$$\|\boldsymbol{u}-\boldsymbol{v}\| = \|\boldsymbol{v}-\boldsymbol{u}\| \tag{7.23c}$$
$$\|\boldsymbol{u}-\boldsymbol{v}\| \leq \|\boldsymbol{u}-\boldsymbol{w}\| + \|\boldsymbol{w}-\boldsymbol{v}\| \text{ for any } \boldsymbol{w} \quad \text{(triangle inequality)} \tag{7.23d}$$

When the distance is defined in terms of the L_1 norm, one has the largest distance possible between two vectors, or the so-called taxicab or Manhattan distance. The distance defined by the Euclidean norm, L_2 norm, is a generalization of the geometric shortest distance between two points. The L_∞ norm defines a distance known as the Chebyshev distance (largest differences among any two elements of the vectors), which in a plane is the distance a king travels between two squares on a chessboard. The distance L_1 reflects the worst deviation between the two entities it measures. The L_2 is the most probable deviation, since random deviations for the same physical phenomenon tend to obey normal (Gaussian) statistics (if the deviations were independent from each other).

Sequences

In an iterative process, one obtains a sequence, $\boldsymbol{u}_0, \boldsymbol{u}_2, \ldots, \boldsymbol{u}_n, \ldots$, of vectors that should be all within the space. If this process results in a complete metric space (see below), it is called a Hilbert space. This sequence forms a *Cauchy sequence* if:

$$\lim_{m,n\rightarrow\infty} \|\boldsymbol{u}_n - \boldsymbol{u}_m\| = 0 \tag{7.24}$$

One then hopes that the iterative process converges to a solution, $\boldsymbol{u}$. The condition for such convergence is:

$$\lim_{n\rightarrow\infty} \|\boldsymbol{u}_n - \boldsymbol{u}\| = 0 \tag{7.25}$$

From the triangle inequality Eq. (7.23d): $\|\boldsymbol{u}_n - \boldsymbol{u}\| \leq \|\boldsymbol{u}_n - \boldsymbol{u}_m\| + \|\boldsymbol{u}_m - \boldsymbol{u}\|$. Therefore, a convergent sequence of vectors is also a Cauchy sequence.

Complete Spaces

A metric space in which every Cauchy sequence is a convergent sequence is a complete space. If a complete metric space has a norm defined by an inner product (such as in a Euclidean space), it is called a Hilbert space. Recall that in a Euclidean space the scalar product is defined by Eq. (7.19). Therefore, with the L_2-norm of Eq. (7.20b), the distance metric is given as:

$$d(\boldsymbol{u} - \boldsymbol{v}) = \|\boldsymbol{u} - \boldsymbol{v}\| = \left(\sum_i |u_i - v_i|^2\right)^{\frac{1}{2}} = \langle \boldsymbol{u} - \boldsymbol{v} \rangle^{\frac{1}{2}} \tag{7.26}$$

Note that a complete vector space with a norm is called a Banach space. A Hilbert space is, therefore, a Banach space with a norm defined by the inner product.

7.A.2 Mapping between Spaces

In the discussion below, it is assumed that the parameters space, C, and the measurements space, E, are both Banach spaces.

Mapping Operator: Domain and Range

The forward problem, Eq. (7.3), in a vector form, is:

$$\boldsymbol{e} = \boldsymbol{mc} \tag{7.27}$$

where $\boldsymbol{m}$ designates an operator which maps a subset in the parameters space, C, represented by the vector $\boldsymbol{c}$, to a subset in the measurement space, E, expressed by the vector $\boldsymbol{e}$. The subset, $\boldsymbol{c}$, is the domain of m, while $\boldsymbol{e}$ is its range.

Boundedness

The operator $\boldsymbol{m}$ is bounded if:

$$\|\boldsymbol{mc}\| \leq B\boldsymbol{c} \tag{7.28}$$

where B is a constant irrespective of $\boldsymbol{c}$. The least upper bound (smallest B) defines the operator's norm:

$$\|\boldsymbol{m}\| = \sup_{c \in C} \frac{\|\boldsymbol{mc}\|}{\boldsymbol{c}} \tag{7.29}$$

Continuity

If a sequence $\boldsymbol{c}_n, n = 1, 2, 3, \ldots$ converges to $\boldsymbol{c}$ as $n \to \infty$ in C, and the corresponding mapped sequence $\boldsymbol{mc}_n, n = 1, 2, 3, \ldots$ also converges to $\boldsymbol{mc}$, then $\boldsymbol{m}$ is a continuous operator. A continuous operator also maps a null element, $\boldsymbol{c} = 0$, in C, to a null vector, $\boldsymbol{mc} = 0$, in E, and as such is bounded. All bound operators map a null vector to a null vector, since then the inequality of (7.28) is always satisfied, and are therefore continuous.

References

M. Bertero and P. Boccacci, *Introduction to Inverse Problems in Imaging*. Bristol: Institute of Physics, 1998.

J. W. Demmel, "On condition numbers and the distance to the nearest ill-posed problem," *Numerische Mathematik*, vol. 51, pp. 251–289, 1987.

——, "The probability that a numerical analysis problem is difficult," *Mathematics of Computation*, vol. 50, pp. 449–480, 1988.

E. Garduño and G. T. Herman, "Optimization of basis functions for both reconstruction and visualization," *Discrete Applied Mathematics*, vol. 139, pp. 95–111, 2004.

P. C. Hansen, *Rank-Deficient and Discrete Ill-Posed Problems: Numerical Aspects of Linear Inversion*. Philadelphia: Society for Industrial and Applied Mathematics, 1998.

R. Lewitt, "Alternatives to voxels for image representation in iterative reconstruction algorithms," *Physics in Medicine and Biology*, vol. 37, pp. 705–716, 1992.

S. Matej and R. Lewitt, "Practical considerations for 3-D image reconstruction using spherically symmetric volume elements," *IEEE Transactions on Medical Imaging*, vol. 15, pp. 68–78, 1996.

8 Formulation

The solution of an inverse problem can be tackled via the matrix or functional formulations of the forward problem, Eqs. (7.1) and (7.2), or by analytic methods. The forms of solution are discussed broadly in this chapter and in more details in subsequent chapters.

8.1 Matrix

The direct inverse of the forward problem of the matrix formulation of Eq. (7.1) is:

$$\boldsymbol{c} = \boldsymbol{A}^{-1}\hat{\boldsymbol{e}} \tag{8.1}$$

We used here $\hat{\boldsymbol{e}}$, instead of $\boldsymbol{e}$, to emphasize that actual measurements, with their associated errors, are employed. Direct inversion of Eq. (8.1) may not be practical, due to the large size of the matrix. Nevertheless, matrix analysis of the inverse problem provides insight into its nature. If the number of measurements, M, hence the length of the vector $\hat{\boldsymbol{e}}$, is not equal to the number of image parameters, N (length of vector, $\boldsymbol{c}$),[1] one has a non-square matrix that cannot be directly inverted. However, a matrix solution can be found if one solves the equation[2]: $\boldsymbol{A}^{\mathrm{T}}\boldsymbol{A}\boldsymbol{c} = \boldsymbol{A}^{\mathrm{T}}\hat{\boldsymbol{e}}$, instead of $\boldsymbol{A}\boldsymbol{c} = \hat{\boldsymbol{e}}$, and inverts the square matrix $\boldsymbol{A}^{\mathrm{T}}\boldsymbol{A}$ to obtain the solution: $\boldsymbol{c} = (\boldsymbol{A}^{\mathrm{T}}\boldsymbol{A})^{-1}\boldsymbol{A}^{\mathrm{T}}\hat{\boldsymbol{e}}$, where $\boldsymbol{A}^{\mathrm{T}}$ is the transpose matrix of $\boldsymbol{A}$. This solution happens also to be an optimal solution that minimizes the least-squares objective function:

$$\chi^2 = [\hat{\boldsymbol{e}} - \boldsymbol{A}\boldsymbol{c}]^{\mathrm{T}}[\hat{\boldsymbol{e}} - \boldsymbol{A}\boldsymbol{c})] \tag{8.2}$$

which defines the square of the difference (residual) between the given measurements and those modeled using the forward model. The use of the squares signifies that the differences between measured and modeled values are likely to follow a normal

[1] The image parameters are all included into a single vector, $\boldsymbol{c}$, though a section image is formed by two-dimensional pixels, and a volume image is composed of three-dimensional voxels. Therefore, elements of $\boldsymbol{c}$ are allocated to the corresponding pixels and voxels, in accordance to a pre-designated order determined by the spatial grid over which the image is to be reconstructed.

[2] One can also solve: $\boldsymbol{A}\boldsymbol{c} = (\boldsymbol{A}\boldsymbol{A}^{\mathrm{T}})(\boldsymbol{A}\boldsymbol{A}^{\mathrm{T}})^{-1}\hat{\boldsymbol{e}}$, to obtain: $\boldsymbol{c} = \boldsymbol{A}^{\mathrm{T}}(\boldsymbol{A}\boldsymbol{A}^{\mathrm{T}})^{-1}\hat{\boldsymbol{e}}$.

Computed Radiation Imaging. DOI: 10.1016/B978-0-12-387777-2.00008-2

distribution with a zero mean, see Appendix 8.A. Therefore, the minimization of χ^2 of Eq. (8.2) leads to a physically meaningful estimate of $\boldsymbol{c}$ given by:

$$\hat{\boldsymbol{c}} = [\boldsymbol{A}^{\mathrm{T}}\boldsymbol{W}\mathbf{A}]^{-1}\boldsymbol{A}^{\mathrm{T}}\boldsymbol{W}\hat{\boldsymbol{e}} \tag{8.3}$$

where $\boldsymbol{W}$ is an $M \times M$ weighting diagonal matrix, introduced here to favor measurements with lower uncertainties, as explained in Appendix 8.A. Given the uncertainties associated with $\hat{\boldsymbol{e}}$, and the approximation of the forward model that determines $\boldsymbol{A}$, the inversion of Eq. (8.1) produces an estimate, $\hat{\boldsymbol{c}}$, of the image parameters. Note that the solution of Eq. (8.3) is equally applicable to an overdetermined problem, $M > N$, an underdetermined one, $M < N$, and a well-determined problem, $M = N$, since in all cases the matrix $[\boldsymbol{A}^{\mathrm{T}}\boldsymbol{W}\boldsymbol{A}]$ is a square $N \times N$ matrix. In imaging problems, one strives to have an overdetermined problem, in order to better accommodate measurement uncertainties.

If an imaging system is properly designed, the matrix $[\boldsymbol{A}^{\mathrm{T}}\boldsymbol{W}\boldsymbol{A}]$ will not be singular, and its inverse will exist, enabling the least-square solution of Eq. (8.3). However, in some cases the matrix $[\boldsymbol{A}^{\mathrm{T}}\boldsymbol{W}\boldsymbol{A}]$ can be nearly singular, i.e. has an eigenvalue that is about equal to zero. The condition number of the matrix will then be quite large, and a very small error in the measurement vector, $\hat{\boldsymbol{e}}$, can greatly affect the estimated value of $\hat{\boldsymbol{c}}$. One must then resort to some form of so-called regularization to allow a solution that gives a near-minimum value of χ^2, while avoiding the singularity problem, see Section 10.4.

When the problem is linear, as is the case in transmission imaging, the matrix inversion process of Eq. (8.3) can directly lead to a solution of the inverse problem. Such solution is not necessarily unique due to the influence of measurement uncertainties. However, if the condition number is not very high, error propagation can be controlled, and the obtained solutions (for different measurement noise levels) will be close to each other, within the range of error propagation. For the nonlinear problem of scatter imaging, one can construct an estimate of the matrix $\boldsymbol{A}$ based on some guess of the solution, and use Eq. (8.3) to obtain a new approximate estimate of the solution, which is used to update the matrix $\boldsymbol{A}$, and so on. Such successive approximation process can lead to an acceptable solution under certain conditions, discussed in Section 10.7.

The matrix formulation of Eq. (7.1) is easily accommodated in transmission imaging, using measurement projections instead of the measurements themselves, by discretizing Eq. (4.2). However, in a realistic imaging system, the size of matrix $\boldsymbol{A}$ is too large to readily manipulate or directly invert. For scatter imaging, the matrix $\boldsymbol{A}$ will contain the exponential attenuation factor that appears in the measurement model of Eq. (6.6). This produces a nonlinear system matrix that inhibits direct solution of the inverse problem, since the system matrix is not fully known.

8.2 Functional

The function formulation of Eq. (7.2) is a convenient way to express each single measurement as a function of the corresponding image parameters. Such formulation also avoids the need to explicitly segregate the image parameter vector, $\boldsymbol{c}$, from the forward mapping operation. For a linear problem, a unique solution will exist if and only if $\boldsymbol{c} = \boldsymbol{c}_0$ is the only solution for $\boldsymbol{f}(\boldsymbol{c}) = 0$, since the difference between two solutions (if they exist) corresponding to the same $\boldsymbol{e}$ has to be zero, i.e. the two solutions have to be identical. For nonlinear problems, uniqueness of solutions is only assured under certain conditions.

For the function formulation: $f(\boldsymbol{c}) = \hat{\boldsymbol{e}}$, the least-squares objective function becomes:

$$\chi^2 = [\hat{\boldsymbol{e}} - \boldsymbol{f}(\boldsymbol{c})]^{\mathrm{T}}[\hat{\boldsymbol{e}} - \boldsymbol{f}(\boldsymbol{c})] \tag{8.4}$$

One cannot directly obtain an explicit solution for $\boldsymbol{c}$ by minimizing χ^2 of Eq. (8.4), in the manner the solution of Eq. (8.3) was obtained by minimizing Eq. (8.2). One must then resort to optimization methods that aim at finding an optimal value for $\boldsymbol{c}$ that minimizes the objective function of Eq. (8.4). Chapter 11 presents some of the optimization methods that can be used in radiation imaging, for both linear and nonlinear inverse problems.

The functional formulation of Eq. (7.2) and Eq. (8.4) do not lead to a matrix, the condition number of which can be determined to assess the system's susceptibility to error caused by measurement uncertainties and model approximations. Therefore, one may need to rely on numerical experimentation to examine the extent of error propagation. Regularization can also be applied to control error propagation, see Section 10.4.

8.3 Analytic

The analytic approach solves an integral analytical form of the forward problem. Fourier series expansions and integral forms amenable to backprojection are two of such analytic formulations.

8.3.1 Fourier Transform

Instead of attempting to obtain a direct solution for the image parameter, $\hat{c}$, at each voxel, let us assume that the parameters of an image constitute some analytical function, $\hat{c}(\vec{r})$, where $\vec{r}$ is the vector that determines the position of a voxel in the image domain. One can attempt to compose this function from its modal components (harmonics), or its spatial frequencies. This can be done with the aid of the Fourier integral:

$$\hat{c}(\vec{r}) = \int_{-\infty}^{\infty} C(\vec{k}) \exp[2\pi i(\vec{k}\cdot\vec{r})]\, d\vec{k} = \int_{-\infty}^{\infty} C(\vec{k})[\cos(2\pi i(\vec{k}\cdot\vec{r})) + i\sin(2\pi i(\vec{k}\cdot\vec{r}))]\, d\vec{k} \tag{8.5}$$

where $\vec{k}$ is a spatial frequency (cycles per unit length), with dimensions of inverse distance, and $C(k)$ is the coefficient of expansion (or the amplitude of the partial wave) at frequency $\vec{k}$, with i being the imaginary value $\sqrt{-1}$. The infinite limits in Eq. (8.5) allow for all frequencies.

In order to determine the function $\hat{c}(\vec{r})$, the values of each $\mathcal{C}(k)$, at different frequencies, are to be found. These expansion coefficients can be determined by the Fourier transform:

$$\mathcal{C}(\vec{k}) = \int_{-\infty}^{\infty} c(\vec{r}) \exp[-2\pi i(\vec{k} \cdot \vec{r})] \, d\vec{r} \tag{8.6}$$

However, $c(\vec{r})$ is not known; it is the function one is trying to find. But the forward model can be used to relate $c(\vec{r})$ to the measurement function, $\hat{e}(\vec{r})$. The integral over $d\vec{r}$ is equivalent to integrating over volume, dV. If that volume is the volume seen by a detector recording a measurement, e, then the forward model that maps $c(\vec{r}) \longrightarrow e$ will contain the integral: $\int c \, dV$. The forward model for transmission, Eq. (4.2), directly contains an integral, $\int c \, ds$, along the direction of the incident radiation, with $c = \Sigma$, $e = p$, $dV = dAds$, and dA being the detector area, which is a constant value. Therefore, for transmission image reconstruction, the inverse transform of Eq. (8.6) can be rewritten as:

$$\begin{aligned}
\mathcal{C}(\vec{k}) &= \int_{-\infty}^{\infty} \int_{-\infty}^{\infty} c(s) \exp[-2\pi i(\vec{k} \cdot \vec{r})] \, ds \, dA \\
&= \int_{-\infty}^{\infty} \left\{ \int_{-\infty}^{\infty} c(s) ds \right\} \exp[-2\pi i(\vec{k} \cdot \vec{r})] \, dA \\
&= \int_{-\infty}^{\infty} e(x) \exp[-2\pi i(\vec{k} \cdot \vec{r})] \, dA \\
&= \mathcal{E}(\vec{k})
\end{aligned} \tag{8.7}$$

where $\mathcal{E}(\vec{k})$ is a Fourier coefficient of the measurement. Equation (8.7) indicates that a Fourier transform of a function of the image parameter, $\mathcal{C}(\vec{k})$, is equal to a Fourier coefficient of a corresponding measurement, $\mathcal{E}(\vec{k})$. This is the essence of the Fourier-based solution of the inverse problem of image reconstruction: utilize the Fourier coefficients of a function representing measurements to obtain those of the image parameters, and in turn reconstruct the parameter's function via Eq. (8.5). The main advantage of the use of the modal solution via the Fourier transform is its computational efficiency, via the use of readily available digital fast-Fourier transforms (FFT's), which enable approximate but quick computation of discrete Fourier transform and their inverses.

More detailed analysis of this method for transmission imaging is given in Section 12.2. The procedure of Eq. (8.7) can also be applied to the inverse problem of emission imaging, using their corresponding forward models given by Eqs. (5.1) and (5.6), when the response of a detector is viewed as the summation of the source activity within its field-of-view. For scatter imaging, applying the Fourier transforms to the model of Eq. (6.6) is not as straightforward, due to the presence of the nonlinear (exponential) terms that cannot be taken out of the integral. However, if such terms are evaluated based on some guessed value of c, to render a linear problem, the Fourier inverse transforms of the linear terms can be used to determine a new estimate of c, which in turn can be utilized to update the value of the nonlinear terms, and the solution process can be repeated in an iterative manner.

8.3.2 Backprojection

Each measurement, e, is an integral of many values of c, as all measurement models discussed in Part I of this book show. One can then express a function representing the measurements, $e(\vec{R})$, by the generalized integral equation:

$$e(\vec{R}) = \int_{-\infty}^{\infty} G(\vec{R};\vec{r})\, c(\vec{r})\, d\vec{r} \tag{8.8}$$

where $G(\vec{R};\vec{r})$ is some integration kernel that represents the response of the measurement recorded at $\vec{R}$ to the physical attribute at the point at $\vec{r}$. This kernel can be deduced from the forward model of the problem. Now taking the opposite view that each voxel in an image claims a share in each measurement it contributes to, one can estimate an image parameter as:

$$\hat{c}(\vec{r}) = \int_{0}^{4\pi} F(\vec{r};\vec{\Omega})\, e(\vec{\Omega})\, d\vec{\Omega} \tag{8.9}$$

where $\vec{\Omega}$ is a direction emanating from a voxel at $\vec{r}$ toward a measurement $e(\vec{\Omega})$, and $F(\vec{r};\vec{R})$ is some "filtering" function that relates a measurement, $e(\vec{R})$, to a physical parameter, $c(\vec{r})$. The integration in Eq. (8.9) is over all directions at which measurements stimulated by $c(\vec{r})$ are acquired. The operation within the integral of Eq. (8.9) performed over a single direction, $\vec{\Omega}$, takes a measurement in that direction, filtered with $F(\vec{r};\vec{R})$, and spreads it (project it back) over the direction $\vec{\Omega}$. Therefore, the operation of Eq. (8.9) defines the so-called filtered backprojection. In the inverse-problem sense, the function $F(\vec{r};\vec{R})$ is an inverse filter, i.e. a filter that leads to a solution of the inverse problem. At a first glance, the filtering function $F(\vec{r};\vec{R})e(\vec{R})$ appears to be the inverse, or the adjoint, of the kernel $G(\vec{R};\vec{r})$, but the inversion process is not always that straightforward. Therefore, some reasonable estimates are utilized. Note that if $F(\vec{r};\vec{R})$ is taken as a constant value, the operation of Eq. (8.9) results in a uniform backprojection, which is the simplest form of backprojection. Backprojection methods are discussed in Section 12.3.

Comparing Eq. (8.8) to the forward model for transmission imaging, Eq. (4.2), one can deduce that the Kernel function $G(\vec{R};\vec{r}) = 1$; when $\vec{R} - \vec{r} = (|R| - |r|)\hat{r}$ and $|R| > |r|$, with $\hat{r}$ being a unit vector in the direction $\vec{r}$, and $G(\vec{R};\vec{r}) = 0$ otherwise. That is, when the vectors $\vec{R}$ and $\vec{r}$ overlap in the same direction, with the former surpassing the latter, $G(\vec{R};\vec{r})$ exists and becomes equal to unity. In this case, the integral of Eq. (8.8) is the classical Radon transform (Radon, 1907) of c on the line $(|R| - |r|)\hat{r}$. Inverse radon transform methods can then be used to solve the inverse problem of transmission imaging, as discussed in Section 12.3. The observation that, in the forward model of transmission imaging, the existence of the kernel $G(\vec{R};\vec{r})$ depends on $\vec{R}-\vec{r}$ makes it possible to express this kernel as $G(\vec{R}-\vec{r})$, then the integral of Eq. (8.8) represents a convolution process. Equation (8.9) can then be resolved through deconvolution, see Section 12.3.

The forward models of emission, Eqs. (5.1) and (5.6), can lend themselves directly to the integral formulation of Eq. (8.8), if an emission measurement is viewed as a summation of the source activity within its field-of-view. In the case of scatter imaging, the forward model of Eq. (6.6) is nonlinear. However, if the nonlinear terms are ignored, assumed to be known in advance, or iteratively dealt with, the associated forward models become equivalent to linear integrals, and the backprotection methods can be applied.

The backprojection approach of Eq. (8.9) does not allow direct assessment of the overall susceptibility of the inverse problem to error. However, this can be accomplished by the application of some level of uncertainty to the measurements and finding numerically how the error is propagated to the solution. As Eq. (8.9) indicates, an error, $\delta_e(\vec{R})$, is directly backprojected as an error, $\delta_c(\vec{r})$, in the solution, such that:

$$\delta_c(\vec{r}) = \int_0^{4\pi} F(\vec{r};\vec{\Omega})\, \delta_e(\vec{\Omega})\, d\vec{\Omega} \tag{8.10}$$

Therefore, the obtained solution contains the same error features as the measurements, unless of course the filter function, $F(\vec{r};\vec{\Omega})$, is chosen to dampen the effect of error, while still maintaining its backprojection capability. Additional separate filtering may be applied to the measurements prior to subjecting them to backprojection, see Chapter 9. However, no constraints can be imposed on the solution via the backprojection process, making it difficult to control error propagation.

Like the Fourier method of Section 8.3.1, the backprojection method is an analytical method and enjoys the advantage of computational efficiency. The backprojection process can also be directly applied once a measurement is acquired, without awaiting for the recording of subsequent measurements. It can, therefore, be applied online while subsequent measurements are being acquired.

8.4 Probabilistic

All the above methods aim at reconstructing an image that is consistent with the provided measurements, often supplemented with constrains based on some *a priori* knowledge about the nature of the sought solution. The probabilistic approach to the solution of image reconstructions incorporates probability distributions into the solution of the inverse problem. This, in turn, introduces into the supplementary solution features that are not provided explicitly by the measurements. In such approach, an estimate of the to-be-reconstructed image parameters vector, $\boldsymbol{c}$, is viewed as a "hypothesis" that is to be retrieved from a "datum" given by the measurement vector, $\boldsymbol{e}$. The conditional probability theorem of Bayes (1763) relates the hypothesis and datum by:

$$P(\boldsymbol{c}|\boldsymbol{e}) = \frac{P(\boldsymbol{e}|\boldsymbol{c})P(\boldsymbol{c})}{P(\boldsymbol{e})} \tag{8.11}$$

where both $\boldsymbol{c}$ and $\boldsymbol{e}$ are assumed to have probability distributions, $P(\boldsymbol{c})$ and $P(\boldsymbol{e})$, respectively. When the hypothesis probability is not known in advance, an estimate of this *a priori* probability, $P(\boldsymbol{c})$, is made, and Eq. (8.11) is referred to as the Bayes' postulate. The forward mapping $\boldsymbol{c} \rightarrow \boldsymbol{e}$, when properly normalized, represents a conditional probability, $P(\boldsymbol{e}|\boldsymbol{c})$; also called the "likelihood," as it indicates how likely $\boldsymbol{e}$ would have been arrived at if the value of $\boldsymbol{c}$ was given. The inverse mapping then corresponds to *a posteriori* probability, $P(\boldsymbol{c}|\boldsymbol{e})$.

If $P(\boldsymbol{c})$ and $P(\boldsymbol{e})$ are known, designated some presumed distributions, or proclaimed to be equi-probable (uniform) distributions (which amounts to complete ignorance of their statistical nature), then the *a posteriori* probability, $P(\boldsymbol{c}|\boldsymbol{e})$, can be determined from $P(\boldsymbol{e}|\boldsymbol{c})$ using the Bayesian relationship of Eq. (8.11). The likelihood, $P(\boldsymbol{e}|\boldsymbol{c})$, can be deduced from the forward mapping, since both relate $\boldsymbol{e}$ to given values of $\boldsymbol{c}$. The application of the Bayes' hypothesis provides a means to estimate the inverse mapping, which is directly related to the *a posteriori* probability, $P(\boldsymbol{c}|\boldsymbol{e})$.

Another approach is to maximize the likelihood, $P(\boldsymbol{e}|\boldsymbol{c})$. This produces the most likely solution. By relating $\boldsymbol{e}$ to $\boldsymbol{c}$ using the forward mapping, i.e. $\hat{\boldsymbol{e}} = \boldsymbol{m}\hat{\boldsymbol{c}}$, one can redefine the likelihood, $P(\boldsymbol{e}|\boldsymbol{m}\hat{\boldsymbol{c}})$, which relates a measurement, $\boldsymbol{e}$, to a modeled measurement, $\boldsymbol{m}\hat{\boldsymbol{c}}$. This makes it possible to define a distribution for $P(\boldsymbol{e}|\hat{\boldsymbol{e}})$, since both $\boldsymbol{e}$ and $\boldsymbol{m}\hat{\boldsymbol{c}}$ are members of the same population of measurements. With $\hat{\boldsymbol{e}}$ being a random variable with a corresponding mean, $\boldsymbol{e}$, the likelihood, $P(\boldsymbol{e}|\hat{\boldsymbol{e}})$, becomes in effect a probability distribution of the departure of $\hat{\boldsymbol{e}}$ from the mean value. Minimizing this difference (variance), or equivalently maximizing $P(\boldsymbol{e}|\hat{\boldsymbol{e}})$, leads to the most likely solution (called in statistics the "mode"). If a normal (Gaussian) distribution of the departure from the mean is assumed, then the solution is equivalent to the matrix-based least-squares method, as shown in Appendix 8.A.

One other approach is that of the maximum *a posteriori* method (MAP). As the name indicates, it maximizes $P(\boldsymbol{c}|\boldsymbol{e})$. Maximizing this probability leads to the mean, or expected, solution. Using the Bayes' postulate of Eq. (8.11) and considering $P(\boldsymbol{e})$ to be a constant probability distribution, one can focus on:

$$\max P(\boldsymbol{c}|\boldsymbol{e}) = \max[P(\boldsymbol{e}|\boldsymbol{c})P(\boldsymbol{c})] \tag{8.12}$$

As mentioned above, maximizing $P(\boldsymbol{e}|\boldsymbol{c})$ leads to the most likely solution. Therefore, maximizing $P(\boldsymbol{c}|\boldsymbol{e})$ involves the maximization of the likelihood plus the solution probability, $P(\boldsymbol{c})$. In effect, the maximum *a posteriori* amounts to a regularized solution of the most maximum-likelihood solution, with $P(\boldsymbol{c})$ acting as the regularization term, see Section 10.4. This becomes more obvious if the logarithm of $P(\boldsymbol{c}|\boldsymbol{e})$ is minimized, which is equivalent to maximizing $P(\boldsymbol{c}|\boldsymbol{e})$:

$$\min \ln P(\boldsymbol{c}|\boldsymbol{e}) = \min[\ln P(\boldsymbol{e}|\boldsymbol{c}) + \ln P(\boldsymbol{c})] \tag{8.13}$$

where $\ln P(\boldsymbol{c})$ acts as a regularization term.

The Markov chain Monte Carlo method is the ultimate utilization of these probabilistic estimates in a stochastic sampling process. The optimization method of simulated annealing, discussed in Section 11.7, is a realization of this probabilistic approach. The optimization method of genetic evaluation, Section 11.6, also contains an element of randomness. Other methods that rely on random sampling to image reconstruction are discussed in Section 13.5.

8.A Probabilistic Basis of Maximum-Likelihood and Least-Squares Methods

Consider an experiment in which a quantity, e, is measured n times, each time with a different instrument. Each measurement gives a value, $\hat{e}_i$, with an error, σ_i, with $i = 1, 2, \ldots N$. Assuming that the errors are normally distributed around a zero mean value, so that each measurement is a realization of sampling a value from a normal distribution of a mean, e_i, and a variance σ_i^2. Therefore, the frequency (probability) of obtaining a value, $\hat{e}_i$, in the ith measurement can be estimated according to its normal distribution as:

$$P(\hat{e}_i) = \frac{1}{\sigma_i\sqrt{2\pi}} \exp\left[\frac{(\hat{e}_i - \bar{e}_i)^2}{2\sigma_i^2}\right] \tag{8.14}$$

where $\bar{e}_i$ is the expected (mean) value of e_i.

The frequency of obtaining the N measured values can then be estimated as:

$$P(\hat{e}_i, i = 1, 2, \ldots N) = \frac{1}{\sqrt{2\pi}} \prod_{i=1}^{N} \frac{1}{\sigma_i} \exp\left[\frac{(\hat{e}_i - \bar{e}_i)^2}{2\sigma_i^2}\right] \tag{8.15}$$

The logarithmic frequency of obtaining the N values together is:

$$\ln P(\hat{e}_i, i = 1, 2, \ldots N) = -\ln\sqrt{2\pi} - \sum_{i=1}^{N} \ln\sigma_i + \sum_{i=1}^{N} \frac{(\hat{e}_i - \bar{e}_i)^2}{2\sigma_i^2} \tag{8.16}$$

The value of $\bar{e}_i$ that maximizes $\ln\ P(\hat{e}_i, i = 1, 2, \ldots N)$ is the best estimate of the true value of e_i. This is the essence of the Maximum Likelihood method.

The value that minimizes:

$$\chi^2 = \sum_{i=1}^{N} \frac{(\hat{e}_i - \bar{e}_i)^2}{\sigma_i^2} \tag{8.17}$$

also maximizes $\ln P(\hat{e}_i, i = 1, 2, \ldots N)$. Equation (8.17) provides the cost (objective) function for the least-squares method.

In imaging, one attempts to match the modeled value of e_i with the measured value. The value that provides the best match becomes then the best estimate, or the "true" (as true as one can get) value of e_i. Then $\bar{e}_i$ in Eq. (8.17) can be replaced with the modeled value of e_i in each measurement, leading to:

$$\chi^2 = \sum_{i=1}^{N} \frac{(\hat{e}_i - e_i)^2}{\sigma_i^2} \tag{8.18}$$

which measures the difference between the measured and modeled values of e, weighted by the error in each measurement σ_i, see Section 15.4.3. Minimizing χ^2 of Eq. (8.18) gives then the best estimate of the true value of e. Equation (8.18), when it replaces the last term in the right-hand-side of Eq. (8.16), provides the maximum-likelihood formulation.

Reference

T. Bayes, "Essay towards solving a problem in the doctrine of chances," *The Philosophical Transaction*, vol. 53, pp. 370–418, 1763, published posthumously. Available on: `http://www.stat.ucla.edu/history/essay.pdf`, accessed October, 2008.

9 Preprocessing of Measurements

The purpose of imaging is to determine the value of some physical parameter, c, within designated voxels. The larger the number of voxels within an image the better is its depiction of the distribution of c, but practicality dictates the imposition of a finite number of voxels. No matter how small the voxels are, their number is still finite. Moreover, the smaller a voxel, the less it contributes to measurements, the more difficult to decipher its presence and determine its content. What is required then is to sample a sufficient number measurements so that the reconstructed parameters, $\boldsymbol{c}$, represent as faithfully as possible the distribution of c, among the image voxels. The discretization into voxels, as discussed in Section 7.1, imposes a limit on the spatial resolution of the image.

Image parameters are reconstructed from measurements, which are recorded at discrete points with detectors that have a finite size and produce noisy (statistically fluctuating) readings. The number and measurements and the manner with which they are sampled affect the quality of the reconstructed image. However, measurements can be preprocessed to reduce the effect of the above mentioned factors. In imaging, measurements are sampled in groups along a line, an arc, or a surface. A measurement and its neighbors tend typically to exhibit a certain trend of change with the spatial location of the detectors. Measurement filters discussed in this chapter take advantage of these spatial trends to smooth the measurements, in an attempt to extract, as much as possible, genuine information from measurements contaminated by noise and sampled at discrete locations.

The aspects associated with sampling with finite-size detectors that produce noisy measurements are discussed in the following sections, followed by a discussion of the methods that can be used to preprocess (filter) measurement data before utilizing them in image reconstruction. It should be cautioned though that filtering measurements ahead of using them in image reconstruction can smooth out, along with noise, fine details in the structure of the image profile, which can in turn affect the quality of the reconstructed images. It is often desirable to deal with noise and measurement fluctuations during the image reconstruction process, as discussed for example in Section 10.4, and/or after reconstructing the image by post-processing, as discussed in Chapter 16. Nevertheless, methods for preprocessing measurements are included here as they may be useful in some cases. Filtering methods in both the frequency and the spatial domains are presented, in Section 9.2 and 9.3, respectively. Filtering comes, however, at the expense of altering the magnitudes of measurements, which is

Computed Radiation Imaging. DOI: 10.1016/B978-0-12-387777-2.00009-4

undesirable in image reconstruction. Methods to overcome this problem are discussed in Section 9.4. We begin, however, by addressing a simple but a fundamental practical question: if one needs to reconstruct an image consisting of N voxels, how many measurements, M, should one acquire?

9.1 Number of Measurements

The number of measurements, M, required to reconstruct an image, should be sufficiently large to produce an image at the desired spatial resolution; with the latter dictated by the number, N, of image voxels/pixels. When collecting measurements, one must ensure that each voxel contributes at least to one measurement, so that each voxel has a "say" in the image reconstruction process. One also should devise a sampling strategy so that each measurement differs from all other measurements by covering at least one voxel not covered by the others, or being affected by at least one voxel in manner different from that of other measurements, so that each measurement provides some new information about the interrogated object. Since statistical fluctuations (noise) are almost always associated with measurements, and given that the forward problem may not accurately model physical reality, as discussed in Part I of this book, one should aim at overdetermining the inverse problem, by acquiring a total number of measurements, M, greater than the number of image voxels (i.e. unknown parameters), N. In the presence of noise, a number of measurements exactly equal to the number of unknowns, $M = N$, can in effect render an underdetermined problem, as some measurements may not differ from others by statistically significant margin to be recognized as distinct measurements. In some cases, it is impractical to overdetermine the problem. When the problem is underdetermined, the problem is incomplete, and one must rely on some supplementary *a priori* information to solve the inverse problem, as discussed in Chapter 14. However, when it is possible to overdetermine the problem, a practical question arises: by how much should the number of measurements, M, exceed the number of unknowns, N. We attempt below to provide some guidance based on two perspectives: (1) the desire to obtain a set of measurements that is responsive as much as possible to the changes in the physical attributes of the imaged object, and (2) to obtain upon image reconstruction an as accurate image as possible. The first aspect is addressed by the sampling theorem, while a framework for the second point is introduced based on minimizing the error in image reconstruction.

9.1.1 Sampling

Typically in radiation imaging, a number of contiguous measurements are acquired to formulate a measurement profile. The measurements are along a line, an arc, or a surface, at some equally spaced intervals. Let r be a distance along the acquisition direction, $\vec{r}$, which we will consider here to be a line, but the discussion can be easily extended to a curve or a surface. We will assume that measurements are acquired at detector positions separated by a distance, Δr. Though each detector has a finite volume, we will initially attribute a detector measurement to a point at the middle of the intersection of the detector's sensitive surface with the line along $\vec{r}$. Then one speaks of sampling M measurements at a spatial frequency of $\frac{1}{\Delta r}$. Measurements then lend themselves to frequency analysis, discussed in Section 9.2, but the interest here is in determining the optimal sampling frequency.

An obvious sampling frequency is $\frac{1}{\Delta r}$, which is accomplished by choosing Δr such that it is equal to the smallest voxel width projected on the direction of acquisition of measurements along $\vec{r}$. Such arrangement amounts to discretization of the measurement profile, in a manner similar to the discretization of image pixels/voxels. However, the optimal situation is to obtain a measurement profile that is continuous and is independent of the discretization process that created the voxels. Assuming for the moment that the measurements are noise-free, or that statistical fluctuations are negligible, then the sampling theorem provides some guidance on selecting the sampling frequency.

The well-known Nyquist-Shannon (1928–1949) sampling theorem states that a continuous function containing no frequencies higher than a certain maximum frequency, f_{max}, is completely determined (i.e. fully reconstructed as a continuous function) from function points sampled at a frequency of $2f_{max}$. The latter frequency is known as the Nyquist frequency. The opposite is also true: if measurements are sampled at the Nyquist frequency, or higher, the underlying continuous function that describes the measurement trend can be reconstructed. This is articulated by Whittaker-Shannon (1935–1949) sampling theorem, which states that when the sampling frequency is greater than or equal to the Nyquist frequency, one can estimate an unmeasured value of $e(\vec{r})$ between the sampling points using:

$$e(r) = \sum_{m=0}^{M-1} e(m\Delta r)\,\text{sinc}\left(\frac{r - m\Delta r}{\Delta r}\right) \tag{9.1}$$

where

$$\text{sinc}(x) = \frac{\sin(\pi x)}{\pi x} \tag{9.2}$$

with $m = 0, 1, \ldots M-1$. In other words, in theory, from a measurement sequence sampled at the Nyquist frequency, one rebuilds a corresponding continuous function.

By constructing an image with a voxel/pixel width of Δr, one is implicitly assuming that the reconstructed images would reflect changes in the image features at a maximum rate (frequency): $f_{max} = \frac{1}{\Delta r}$. Then by sampling at a frequency of $2f_{max}$, one acquires a measurement trend that reflects continuous changes in the image features. Abrupt and edge changes which occur at higher frequencies would not be retrieved, but edge preserving regularization or image enhancement methods can be employed for this purpose, as discussed in Section 10.4 and Chapter 16, respectively.

As an example, let us consider a square tomograph of $\sqrt{N} \times \sqrt{N} = N$ square pixels. To fully-determine the problem, one must acquire $M = N$ error-free independent measurements. If at each line projection $\sqrt{N}$ measurements are acquired, then one needs $\sqrt{N}$ projections to fully determine the problem. Let us retain the number of projections as $\sqrt{N}$, but sample measurements at the Nyquist frequency along each projection. A sampling line will see at most $2\sqrt{2N}$ pixels (along the diagonal of the tomograph) and at least $2\sqrt{N}$ (along one of the tomograph's sides). Therefore, if the number of measurements along each projection line, M_p, is such that $M_p \geq 2\sqrt{2N} = 2.83\sqrt{N}$, the conditions for the sampling theorem will be satisfied at each projection. With $\sqrt{N}$ projections sampled at or above the Nyquist frequency, the total number of measurements M would be: $M \geq \sqrt{N}M_p = 2.83N$. In other words, the problem needs to be overdetermined by a factor or 2.83. A modern transmission tomography system reconstructing a $N = 512 \times 512 = 262,144$ pixels acquires 1000 projections, each with 750 measurements, so that $M = 750,100$, providing $\frac{M}{N} = 2.86$ degree of overdetermination (Goldman, 2007), which is in close agreement with the above analysis in terms of the overall degree of overdetermination.

It should be noted though that even if the Nyquist condition is theoretically satisfied, in practice its use to regenerate a continuous profile is hampered by the approximation of the sinc function, for digital implementation, as it introduces interpolation errors. Moreover, noise associated with measured values add uncertainties to the measurement profile reconstructed from sampled measurements. In imaging, the objective is not to reproduce exact continuous measurement profiles, but to acquire measurements that reflect changes in the imaged object. Therefore, the conditions of the sampling theorem function are only a guide on how to overcome the effect of the discretization process. However, the Nyquist condition overdetermines the inverse problem by at least a factor of two, which helps with controlling error propagation in the image reconstruction process, as discussed below.

9.1.2 Error Minimization

The ultimate goal of imaging is to obtain values for the image parameters, represented by $\boldsymbol{c}$, from measurements, $\boldsymbol{e}$, by inverting the forward mapping, which when represented by a matrix, $\boldsymbol{A}$, gives: $\boldsymbol{Ac} = \boldsymbol{e}$, where $\boldsymbol{c}$ is an $N \times 1$ vector, $\boldsymbol{e}$ is an $M \times 1$ vector, and $\boldsymbol{A}$ is an $M \times N$ matrix. The error, $\boldsymbol{\delta}_c$, associated with $\boldsymbol{c}$ depends on the error, $\boldsymbol{\delta}_e$, in $\boldsymbol{e}$ and the error, $\boldsymbol{\Delta}$, in $\boldsymbol{A}$, with the latter being present due to the modeling approximations discussed in Part I of this book. Let $\boldsymbol{c}_0$ and $\boldsymbol{e}_0$ and $\boldsymbol{A}_0$ be the exact (but not known) image parameters, measurements and forward mapping, respectively. Then the associated errors can be expressed as:

$$\boldsymbol{\delta}_c = \boldsymbol{c} - \boldsymbol{c}_0 \tag{9.3}$$

$$\boldsymbol{\delta}_e = \boldsymbol{e} - \boldsymbol{e}_0 \tag{9.4}$$

$$\boldsymbol{\Delta} = \boldsymbol{A} - \boldsymbol{A}_0 \tag{9.5}$$

Given that $\boldsymbol{\delta}_c$ is determined by $\boldsymbol{\delta}_e$ and $\boldsymbol{\Delta}$, let us introduce the error:

$$\boldsymbol{\delta} = \boldsymbol{\delta}_e - \boldsymbol{\Delta}\boldsymbol{c}_0 \tag{9.6}$$

Image reconstruction aims at matching $\boldsymbol{e}$ with $\boldsymbol{Ac}$, or equivalently minimizing $\boldsymbol{e}-\boldsymbol{Ac}$. In terms of the above errors:

$$\begin{aligned}\boldsymbol{e}-\boldsymbol{Ac} &= \boldsymbol{e}_0+\boldsymbol{\delta}_e-\boldsymbol{A}(\boldsymbol{c}_0+\boldsymbol{\delta}_c)=\boldsymbol{e}_0+\boldsymbol{\delta}_e-\boldsymbol{A}_0\boldsymbol{c}_0-\boldsymbol{\Delta c}_0-\boldsymbol{A\delta}_c\\ &= (\boldsymbol{\delta}_e-\boldsymbol{\Delta c}_0)-\boldsymbol{A\delta}_c=\boldsymbol{\delta}-\boldsymbol{A\delta}_c\end{aligned}$$

where use is made of $\boldsymbol{A}_0\boldsymbol{c}_0=\boldsymbol{e}_0$. Therefore, minimizing $\boldsymbol{e}-\boldsymbol{Ac}$ is equivalent to minimizing $\boldsymbol{\delta}-\boldsymbol{A\delta}_c$:

$$\min_{\boldsymbol{c}}\|\boldsymbol{e}-\boldsymbol{Ac}\|=\min_{\boldsymbol{\delta}_c}\|\boldsymbol{\delta}-\boldsymbol{A\delta}_c\| \tag{9.7}$$

With the norm defined as a Euclidean norm, the problem of Eq. (9.7) is a least-squares minimization problem, discussed in Sections 8.1 and 10.3, whose solution for $\boldsymbol{\delta}_c$ is $\boldsymbol{\delta}_c=\left(\boldsymbol{A}^{\mathrm{T}}\boldsymbol{A}\right)^{-1}\boldsymbol{A}^{\mathrm{T}}\boldsymbol{\delta}$. Therefore,

$$\|\boldsymbol{\delta}_c\|=\left\|\left(\boldsymbol{A}^{\mathrm{T}}\boldsymbol{A}\right)^{-1}\boldsymbol{A}^{\mathrm{T}}\boldsymbol{\delta}\right\|\geq\left\|\left(\boldsymbol{A}^{\mathrm{T}}\boldsymbol{A}\right)^{-1}\right\|\,\left\|\boldsymbol{A}^{\mathrm{T}}\boldsymbol{\delta}\right\| \tag{9.8}$$

Given that the Euclidean norm of a matrix is equal to its largest singular value (square root of eigenvalue) and that the smallest singular value of a matrix is equal to the largest singular value of its inverse[1], $\boldsymbol{\delta}_c$ depends on the singularity of the system matrix, which in turn depends on the structure of the forward model. Effort should be made, therefore, to decrease the singularity of the forward model, by using regularization (see Section 10.4), to reduce error propagation. Once the matrix $\boldsymbol{A}$ is formulated, increasing its size to accommodate additional measurements should be done in such a way that its does not make the expanded $\boldsymbol{A}^{\mathrm{T}}\boldsymbol{A}$ matrix more singular, by ensuring that the added measurements are independent from already available measurements. The error $\boldsymbol{\Delta}$ in $\boldsymbol{A}$, can be estimated by comparing the model with experiments that produce low measuring uncertainties (e.g. with long counting periods) to overcome statistical fluctuations. Even if $\boldsymbol{\Delta}$ is not known, one can assume it to be equal to zero, and determine the term in the right-hand side of the inequality of (9.8) to ensure that the added measurements do not increase $\boldsymbol{\delta}_c$, indeed they should reduce it.

The above analysis does not provide a specific value for the desired degree of overdetermination. This is because of the nature of the Euclidean norm, which provides the most probable solution, as discussed in Appendix 8.A. The most probable solution will naturally change as more measurements are added, and will depend on the quality of such measurements. A more specific answer, yet an upper-bound, can be obtained using the maximum absolute column sum norm, $\|.\|_1$; nicknamed the taxicab norm because it is a measure of the worst error. If overdetermination reduces the worst error, it will also reduce the most probable error. The analysis given here is a modified version of that given by Ben Rosen et al. (2000).

[1] Weisstein, E. W. Matrix Norm. From MathWorld–A Wolfram Web Resource. `http://mathworld.wolfram.com/MatrixNorm.html`.

By combining the matrix, $\boldsymbol{A}$, and the measurement vector, $\boldsymbol{e}$, into a single $M \times (N+1)$ matrix, $[\boldsymbol{A}:\boldsymbol{e}]$, one can assess the modeling error and measurement error together, using the $\boldsymbol{\delta}$ error defined by Eq. (9.6). In practice, only some rows of $[\boldsymbol{A}:\boldsymbol{e}]$ are likely to contain large errors. If all rows have a *large* error, then the design of the imaging system is poor, and should be improved by better forward modeling and/or more accurate measurements. Consider $K < M$ rows of $[\boldsymbol{A}:\boldsymbol{e}]$ that have large absolute errors, with a maximum value of $\boldsymbol{\delta}_{max}$. Other rows of $[\boldsymbol{A}:\boldsymbol{e}]$ would then have a small absolute error. Let, $\boldsymbol{\delta}_{min} > 0$ be the smallest error, in the rows of $[\boldsymbol{A}:\boldsymbol{e}]$, of practical relevance, i.e. error less than δ_{min} is considered to be insignificant.

From the $M \times N$ matrix, $\boldsymbol{A}$, one can select a basis (non-singular) $N \times N$ matrix, $\boldsymbol{B}$. The remaining portion of $\boldsymbol{A}$ forms a supplementary $(M-N) \times N$ matrix, $\hat{\boldsymbol{A}}$. Similarly, one can partition $\boldsymbol{\delta}$ and $\boldsymbol{\delta}_e$ so that:

$$\boldsymbol{A} = \begin{bmatrix} \boldsymbol{B} \\ \hat{\boldsymbol{A}} \end{bmatrix}, \quad \boldsymbol{\delta} = \begin{bmatrix} \boldsymbol{\delta}_B \\ \boldsymbol{\delta}_{NB} \end{bmatrix}, \quad \boldsymbol{\delta}_e = \begin{bmatrix} \boldsymbol{\delta}_{eB} \\ \boldsymbol{\delta}_{eNB} \end{bmatrix} \tag{9.9}$$

A possible minimizer of Eq. (9.7) is $\boldsymbol{\delta} = \boldsymbol{A}\boldsymbol{\delta}_c$, which gives: $\boldsymbol{\delta}_c = \boldsymbol{B}^{-1}\boldsymbol{\delta}_B$, along with: $\hat{\boldsymbol{A}}\boldsymbol{\delta}_c = \boldsymbol{\delta}_{NB}$. Hopefully, there is more than one basis $N \times N$ matrix that one can select from the rows of $\boldsymbol{A}$. Let $\mathcal{B}$ be the set of these basis matrices. Therefore, Eq. (9.7) can now be expressed using the L_1 norm as:

$$\min_{\boldsymbol{\delta}_c} \|\boldsymbol{\delta} - \boldsymbol{A}\boldsymbol{\delta}_c\|_1 = \min_{\hat{\boldsymbol{A}} \in \boldsymbol{\mathcal{B}}} \left\| \boldsymbol{\delta}_{NB} - \hat{\boldsymbol{A}}\boldsymbol{B}^{-1}\boldsymbol{\delta}_B \right\|_1 \tag{9.10}$$

Let q be a positive integer that defines the required degree of overdetermination, such that $q = \frac{M-N}{N}$, or $M = (q+1)N$. Then, one has q rows, $\hat{\boldsymbol{A}}_1, \hat{\boldsymbol{A}}_2, \ldots \hat{\boldsymbol{A}}_q$, of the supplementary matrix, $\hat{\boldsymbol{A}}$. Each row multiplied by the inverse of the basis matrix, $\boldsymbol{B}^{-1}$, gives an $N \times N$ matrix, $\boldsymbol{F}_j = \hat{\boldsymbol{A}}_j\boldsymbol{B}^{-1}$.

Let us now assert the following norm bounds:

$$\|\boldsymbol{\delta}\|_1 \leq K\delta_{max} \tag{9.11}$$

$$\|\boldsymbol{\delta}_B\|_2 \geq \sqrt{N}\delta_{min} \tag{9.12}$$

$$\|\boldsymbol{\delta}_{NB}\|_1 \leq (K-1)\delta_{max} \tag{9.13}$$

$$\|\boldsymbol{F}_j\boldsymbol{\delta}_B\|_1 \geq \|\boldsymbol{F}_j\boldsymbol{\delta}_B\|_2 \geq s_j\sqrt{N}\delta_{min} \geq s_{max}\frac{\delta_{min}}{\delta_{max}}\sqrt{N}\delta_{min} \tag{9.14}$$

with s_j being the minimum singular value (square of eigenvalue) of $\boldsymbol{F}_j$, and s_{min} is the minimum singular value of all $\boldsymbol{F}_j$'s. The inequality of (9.11) gives an upper-bound for $\|\boldsymbol{\delta}\|_1$, defined by the maximum value of the K large errors. The inequalities of (9.12) and (9.13) indicate that at least one of the K measurements with large error is included in $\boldsymbol{\delta}_B$, otherwise no large errors are involved in the solution. In arriving at (9.14), lower bounds of the norms are employed with $\|\boldsymbol{F}_j\|_2 \leq s_i$.

Let $\boldsymbol{\delta}_L$ be the error, as defined by Eq. (9.6), associated with the large errors, partitioned as in Eq. (9.9) to $\boldsymbol{\delta}_{L_B}$ and $\boldsymbol{\delta}_{L_{NB}}$ components. Then, for large errors, one wants to minimize $\left\|\boldsymbol{\delta}_{eNB} - \hat{\boldsymbol{A}}\boldsymbol{B}^{-1}\boldsymbol{\delta}_{LB}\right\|_1$ with $\boldsymbol{\delta}_{LB} = 0$, and accordingly $\|\boldsymbol{\delta_L}\|_1 = \|\boldsymbol{\delta_{L_{NB}}}\|_1$ and $\|\boldsymbol{\delta}_c\| = 0$. When this minimization is not attained, i.e. for $\boldsymbol{\delta}_{L_B} \neq 0$, a subset, $\mathcal{B}_L \in \mathcal{B}$ would exist, within the minimum of (9.10), such that:

$$\beta = \min_{\boldsymbol{B}\in\mathcal{B}_L} \left\|\boldsymbol{\delta}_{NB} - \hat{\boldsymbol{A}}\boldsymbol{B}^{-1}\boldsymbol{\delta}_B\right\|_1 \tag{9.15}$$

Then β must be greater than $\|\boldsymbol{\delta}\|_1$, since the latter is the minimum of (9.10), with $\boldsymbol{\delta}_c = 0$, over the entire set, $\mathcal{B}$, not only the subset, $\mathcal{B}_L$. Therefore, a sufficient condition to obtain a solution not affected by large errors, i.e. with $\boldsymbol{\delta}_B = 0$, and accordingly, $\boldsymbol{\delta}_c = 0$, is $\beta > \|\boldsymbol{\delta}\|_1$. However, a necessary condition is $\beta \geq \|\boldsymbol{\delta}|_1$, because if $\beta = \|\boldsymbol{\delta}|_1$ minimizes (9.10) within the subset $\mathcal{B}_L$, one must ensure that the same value minimizes it outside the subset, to ensure that $\boldsymbol{\delta} = 0$.

One can estimate a lower-bound for β as:

$$\begin{aligned}
\beta &= \min_{\boldsymbol{B}\in\mathcal{B}_L} \left\|\boldsymbol{\delta}_{NB} - \hat{\boldsymbol{A}}\boldsymbol{B}^{-1}\boldsymbol{\delta}_B\right\|_1 = \min_{\boldsymbol{B}\in\mathcal{B}_L} \left\|\boldsymbol{\delta}_{NB} - \boldsymbol{F}\boldsymbol{\delta}_{LB}\right\|_1 \\
&= \min_{\boldsymbol{B}\in\mathcal{B}_L} \left\|\boldsymbol{F}\boldsymbol{\delta}_{LB} - \boldsymbol{\delta}_{NB}\right\|_1 \geq \min_{\boldsymbol{B}\in\mathcal{B}_L} \|\boldsymbol{F}\boldsymbol{\delta}_{LB}\|_1 - \max_{\boldsymbol{B}\in\mathcal{B}_L} \|\boldsymbol{\delta}_{NB}\|_1 \\
&\geq \sum_{j=1}^{q} \|\boldsymbol{F}_j\boldsymbol{\delta}_B\|_1 - (K-1)\delta_{max} > q s_{min}\frac{\delta_{min}}{\delta_{max}}\sqrt{N}\delta_{min} - K\delta_{max}
\end{aligned}$$

$$\beta > q\frac{\delta_{min}^2}{\delta_{max}} s_{min}\sqrt{N} - K\delta_{max} \tag{9.16}$$

where use is made of (9.12), and $\boldsymbol{F} = \hat{\boldsymbol{A}}\boldsymbol{B}^{-1}$, with $\boldsymbol{F}^{\mathrm{T}} = \left[\boldsymbol{F}_1^{\mathrm{T}}\ \boldsymbol{F}_2{}^{\mathrm{T}} \cdots \boldsymbol{F}_q^{\mathrm{T}}\right]^{\mathrm{T}}$. Given that $\|\boldsymbol{\delta}\|_1 \leq K\delta_{max}$, from (9.11), to ensure satisfaction of the sufficient condition $\beta > \|\boldsymbol{\delta}\|_1$, one must have:

$$q\frac{\delta_{min}^2}{\delta_{max}} s_{min}\sqrt{N} \geq 2K\delta_{max}$$

$$q \geq \frac{2K}{s_{min}}\left(\frac{\delta_{max}}{\delta_{min}}\right)^2 \tag{9.17}$$

Keeping in mind that the requirement of (9.17) is derived for significant errors, such that $\delta_{min} > 0$, it automatically accommodates smaller errors less than δ_{min}. The condition of (9.17) is an upper-bound, and requires some knowledge of the margin of both measurement and modeling errors, and the minimum singular value of the matrix $\hat{\boldsymbol{A}}\boldsymbol{B}^{-1}$. The existence of a non-singular basis matrix, $\boldsymbol{B}$, ensures that $s_{min} \neq 0$. However, an $M \times N$ supplementary matrix, $\hat{\boldsymbol{A}}$, can only be determined if $q = \frac{M-N}{N}$ is known.

The inequality of Eq. (9.17) can be used to assess how far a given level of determination is from this ideal condition, which assures a zero error in the reconstructed value of $\boldsymbol{c}$. As one would expect, the closer s_{min} to zero, the closer $\hat{\boldsymbol{A}}\boldsymbol{B}^{-1}$ is to singularity, the higher the required degree of overdetermination, as (9.17) indicates. It is interesting though to notice that the degree of overdetermination is affected by the relative margin of error, ratio of maximum error to minimum error, not by the value of the error itself. As inequality (9.17) indicates, the higher the maximum-to-minimum error ratio, the higher the degree of overdetermination. Obviously, the higher the number, K, of measurements with larger error, the more measurements are needed. Note though for $K = 0$, the condition (9.17) simply states that $q \geq 0$, or equivalently, $M \geq N$, which is not very informative. However, no matter how small the error is, one should be able to identify a number of measurements as having a relatively large error, and define a K value. One can also rely on the inequalities of (9.12) and (9.13) to, respectively, define δ_{min} and δ_{max}.

9.2 Frequency Analysis

If a measurement set, $e(m\Delta r)$, $m = 0, 1, \ldots M-1$, is subjected to a discrete Fourier transform (DFT),[2] one gets:

$$\mathscr{F}_k\{e(r)\} = \sum_{m=0}^{M-1} e(m\Delta r)\exp\left[-2\pi ik\frac{m}{M}\right], \quad -\frac{M}{2} \leq k \leq \frac{M}{2}, \quad \mathscr{F}_{-k} = \mathscr{F}_k^{\star} \tag{9.18}$$

where k is the wavenumber and $\mathscr{F}_k^{\star}$ is the complex conjugate of $\mathscr{F}_k$. The Fourier coefficient, $\mathscr{F}_k$, consists of real and imaginary components. The amplitude (modulus) of this coefficient determines the strength of the frequency component, while little attention is typically given to its phase shift because it does not provide useful information. Owing to the second condition associated with Eq. (9.18), $-k$ indicates a spatial revolution in a direction opposite to that of k, i.e. a phase shift of π. Therefore, negative frequencies are not considered, and attention is given to the zero and the $\frac{M}{2}$ positive frequencies. The first condition of Eq. (9.18) indicates a maximum frequency that is half the sampling frequency. Both these conditions arise from the fact that one cannot deduce variations in data at a frequency higher than the rate of sampling of two successive measurements (to observe change). The Fourier transform at $k = 0$, as Eq. (9.18) indicates, gives the summation of all measurements, and does not reflect any change, as one would expect. Therefore, scaling $\mathscr{F}_0$ by $\frac{M}{2}$ produces the datum of the measurements, while scaling a higher order coefficient with the same scaling factor gives the amplitude of the considered frequency component. The fundamental frequency, at $k = 1$, is equal to $\frac{1}{M\Delta r}$, i.e. one cycle for the entire spatial range of the acquired measurements. This is the lowest frequency for observing change in measurement data.

Frequency analysis provides some useful information about the measurements, and indirectly the imaged object. First, noise in the measurements will typical appear as an uncorrelated low-amplitude background component. The signal-to-noise ratio of

[2] Discrete Fourier numerical algorithms requires that $M = 2^n$, where n is an integer; if it is not, it can be trimmed back to the nearest value or expanded by adding zeros to the end of a data set to satisfy this requirement. For a free software see: `http://www.fftw.org/`

measurements can then be assessed. Though radiation counting fluctuations can be estimated by Poisson statistics (see Section 15.2), frequency analysis can reveal other sources of noise, such as those produced by electronics, and external vibrations or electromagnetic fields.

Low amplitudes at the high frequency components of the Fourier transform of measurements indicate that the measurements are monotonic, i.e. do not contain strong periodic variations. On the other hand, steep measurement variation are reflected by strong high-frequency components. However, such strong changes may not be captured at the rate at which the measurements are sampled. Then, changes that occur at a rate higher than half the sampling rate, will appear at lower frequencies, producing the phenomenon of aliasing. The maximum spatial frequency corresponding to a particular imaging system can be determined by its impulse response, see Section 9.2.2. However, once the voxel size is preset, one would have had decided to measure change in an image at a frequency not higher than $\frac{1}{\Delta r_v}$, where Δr_v is the voxel's width projected on the measurement acquisition line along $\vec{r}$. Therefore, to avoid aliasing one should set the resolution frequency equal to or greater than the Nyquist frequency, so that the sampling frequency is such that: $\frac{1}{\Delta r} \geq \frac{2}{\Delta r_v}$. Changes occurring at a frequency greater than $\frac{1}{2\Delta r}$ will appear as aliasing that strengthens the low frequency components of the measurements, and in turn dampens the effect of high frequencies. Such finer changes may, however, be recovered in image reconstruction with the aid of *a priori* information, constraints, or regularization in image reconstruction (see Section 10.4), or by image post-processing to enhance edges (as discussed in Section 16.3).

Another cause of aliasing is due to the fact that measurements for image reconstruction are always recorded over a finite support (a width) that typically corresponds to the width of the imaged object. The outside edges of an image inevitability produce an abrupt (high frequency) change in corresponding measurements, which in turn produces aliasing. The finite sampling width, since finite number of measurements, is equivalent to imposing on an infinite set of measurements a window that sets measurements outside the support to zero. This results in a spreading or leakage of the frequency components beyond the Nyquist frequency.

When examining the frequency spectrum generated by the discrete Fourier transform of Eq. (9.18), the effect of discretization should be kept in mind. The frequencies (harmonics) produced by the DFT are multiples of the fundamental frequency, $\frac{1}{M\Delta r}$. If the measurements happen to contain a frequency in between these discrete frequencies, its amplitude will be distributed between the adjacent harmonics, giving the illusion of two frequencies of dampened amplitudes. This effect is sometimes referred to as the "picket-fence" effect, but can be overcome by artificially changing the period over which DFT is performed by padding the original set of measurements with zeros until the number of measurements reaches the next 2^n power. This will produce finer spectral lines without affecting the form of the original spectrum.

9.2.1 Frequency Filtering

Aside from the frequency information deduced from the Fourier transform of a measurement profile, one can also filter out undesirable frequency components and use inverse DFT to regenerate the filtered measurements, for subsequent use in image reconstruction. For instance, one may want to filter out high frequencies if they appear to contain low-amplitude noise. For this purpose, one may apply a low-pass digital filter in the form of a step function with a unity amplitude below a cut-off frequency, k_c, and a zero amplitude above this frequency. However, the inverse Fourier transform of a step function is a sinc function, Eq. (9.2). The sinc function has an endless rippling effect that extends indefinitely into the spatial domain. Since one always has a finite set of measurements, these ripples are in effect terminated at both ends of the sampling interval. Therefore, the sharpness of the step function becomes less steep and rippling appears at both ends of the altered step function, which in turn allows the higher frequency one intended to eliminate to reappear, see Smith (2003, Chap. 16). This problem is overcome by the use of a windowed-sinc filter in the spatial domain that produces, upon transformation, a good approximation of the step function in the frequency domain. Two common filters can be used for this purpose (Smith, 2003):

$$\text{Hamming window}: w(l) = \frac{1}{2}\left[1.08 - 0.92\cos\left(2\pi\frac{l}{L}\right)\right] \tag{9.19}$$

$$\text{Blackman window}: w(l) = \frac{1}{2}\left[0.84 - 2\cos\left(2\pi\frac{l}{L}\right) + 0.16\cos\left(4\pi\frac{l}{L}\right)\right] \tag{9.20}$$

where L is the number of measurements covered by the window, an even number. These windows, when multiplied by the truncated-sinc filter, reduce the ripple effect on both the spatial and frequency domains, and produce a reasonably sharp passband in the frequency domain. The two filters of Eqs. (9.19) and (9.20) are quite similar, but the Blackman window provides better damping of the filtered-out frequencies, while the Hamming window generates a sharper cut-off. Other windows are also available, but the Blackman and Hamming filters have better filtering characteristics. The Blackman windows is usually preferred because of its superior damping characteristics (Smith, 2003).

A low-pass filter with a window of length L (an even number) can, therefore, be applied directly to measurements (in the spatial domain) by the convolution[3]:

$$\tilde{e}(m\Delta r) = \sum_{l=0}^{L-1} e(m\Delta r - l\Delta r) \times \frac{\sin\left[2\pi f_c\left(l - \frac{L}{2}\right)\right]}{\left(l - \frac{L}{2}\right)} \times \frac{w(l)}{K}$$

$$\text{with} \quad \frac{\sin\left[2\pi f_c\left(l - \frac{L}{2}\right)\right]}{\left(l - \frac{L}{2}\right)} = 2\pi f_c, \quad \text{when } l = \frac{L}{2}$$

$$\text{and} \quad K = \sum_{l=0}^{L-1} \frac{\sin\left[2\pi f_c\left(l - \frac{L}{2}\right)\right]}{\left(l - \frac{L}{2}\right)} \tag{9.21}$$

[3] Convolution involves reversing a function, shifting it, and multiplying it by another function, then integrating.

where $\tilde{e}(r)$ designates a filtered measurement at $r = m\Delta r$, $\frac{\sin 2\pi f_c\left(l-\frac{L}{2}\right)}{2\pi f_c\left(l-\frac{L}{2}\right)}$ is the inverse DFT of a rectangular function in the frequency domain that extends from zero to a cut-off frequency, k_c, defined by f_c which is a fraction (less than half) of the sampling rate, and $w(l)$ is one of the windows of Eqs. (9.19) or (9.20). The restriction on f_c is due to the limitations on the DFT frequency dictated by the sampling theorem, see Eq. (9.18). While f_c is determined by the upper frequency one needs to filter out, the length of the window, L, determines the sharpness of the cut-off (how close to a step function); the larger L, the steeper the cut-off.

A more sophisticated filter in the frequency domain is the Wiener filter, which is based on the signal-to-noise ratio at each frequency. Its frequency response, $H(k)$, is expressed as (Smith, 2003, Chap. 17):

$$H(k) = \frac{[F_k e(m\Delta r)]^2}{[F_k e(m\Delta r)]^2 + [\mathcal{N}(k)]^2} \tag{9.22}$$

where $\mathcal{N}$ is the Fourier transform of the noise. The squaring of the components of the kernel of Eq. (9.22), indicates that the filter relies on the powers of signal and noise. The filter is implemented by multiplying the kernel by $\mathscr{F}_k$, adding the contribution of all frequencies, and applying inverse Fourier transform to obtain the filtered values. The noise level can be assumed to be constant at all frequencies, in the case of white noise, or be estimated based on some *a priori* knowledge of the nature of the noise. The Wiener filter is quite effective in producing a high signal-to-noise power (amplitude squared) ratio, and as such is a good low-pass filter in the presence of high-frequency noise. In the spatial domain, the filter is equivalent to minimizing the mean-squared-difference between the filtered and measured values, and can be implemented as a least-squares regression (curve fitting). One disadvantage of such approach is that the least-squares methods are not robust, i.e. they are not tolerant of large errors (Proakis and Manolakis, 2007).

The above discussion considered a low-pass filter, but a high-pass filter can be composed by inverting the sign of the kernel of the low-pass filter, and adding one to its center (Smith, 2003, Chap. 14 & 16). This has the effect of flipping the step function of the low-pass filter in the frequency domain, so that it becomes a high-pass filter. A band-pass filter can be created by cascading. i.e. applying a low-pass filter then a high-pass filter, or by forming a combined filter by the convolution of low-pass and high-pass filters. Adding a low-pass filter to a high-pass filter creates a band-reject filter.

9.2.2 Impulse Response

One can gain insight into the response function of an imaging system by examining its impulse response in the frequency domain. An impulse response is defined here as the measurement profile produced when a Dirac delta function is introduced into the imaged object. This can be physically done by introducing a tiny intrusion, such as a thin wire, of physical attributes that are considerably different from those of the contents of the imaged object. The corresponding measurements provide the point spread function (PSF) for the system. Note that the PSF concept is also used to measure the spread in the reconstructed image to determine the image quality (see Section 15.5), and also in image enhancement (Section 16.1) to examine the degree of image degradation. Here, we focus on examining the response of measurements to an impulse change in the physical attributes. Such response in the frequency domain will provide a direct indication of the system's highest frequency, since an impulse change is the most abrupt change one can introduce into a system. The magnitude of the resulting frequency response compared to that of the surrounding noise will give direct indication of the signal-to-noise ratio of the imaging system. Shifting the position of the introduced impulse will result in a phase shift in the frequency response. If spatial shifting produces a linear phase-shift response, then one would be able to conclude that the point spread function is symmetric (Smith, 2003, Chap. 10). If the PSF is spatially invariant, i.e. independent of position, then the magnitude of the resulting frequency response would not change with the position of the introduced impulse change. The nature of the PSF characterizes, therefore, the nature of measurements. When applying DFT for analyzing the impulse response, one should keep in mind that the frequency resolution depends on the length of the measurement sample, but the latter can be expanded by padding the sampled measurements with zeros before performing the DFT, as discussed at the beginning of this section.

9.2.3 Detector Aperture

The frequency analysis presented is Section 9.1.1 considered point-localized measurements. A physical detector has a size, hence an aperture of a finite area. One must, therefore, account for the detector size. A simple solution is to assign a detector measurement to a point representative of the detector, say at the center of its aperture. This, however, does not reflect the spatial width of the detector and its effect on the spatial profile of recorded measurements. Continuing with the analysis for a measurement acquisition line along $\vec{r}$, one could then attempt to deconvolve the effect of the detector aperture along $\vec{r}$, while compacting the detector size in the direction normal to $\vec{r}$ into points on $\vec{r}$. Let the width of the detector aperture be D, and assume that all detectors are of the same size, and that the detector efficiency is constant along D. The observed detector measurement, e_m, for the mth measurement, with $m = 1$ to M, is then the convolution of the unknown spatial resolution of $e(r)$ along D with a rectangular function representing the detector aperture:

$$e_m = \frac{1}{D}\int_0^D e(\vec{r}-\vec{r}')\,\mathrm{rect}\left(\frac{r'}{D}\right)dr' = e(r) * \mathrm{rect}\left(\frac{r}{D}\right); \quad m = 0, 1, \ldots, M-1 \tag{9.23}$$

where $\mathrm{rect}\left(\frac{r}{D}\right) = 1$ when $\left|\frac{r}{D}\right| \le \frac{1}{2}$, and is zero otherwise, and "$*$" designates convolution. Equation (9.23) represents a linear filtering of measurements by the detector aperture. A convolution in the spatial domain becomes a summation in the Fourier frequency domain, and $\mathscr{F}_k\left\{\mathrm{rect}\left(\frac{r}{D}\right)\right\} = D\,\mathrm{sinc}(kD)$ for frequency k, where $\mathrm{sinc}(kD) = \frac{\sin \pi kD}{\pi kD}$ and $\mathrm{sinc}(0) = 1$. Therefore, transforming Eq. (9.23) and re-arranging the results leads to:

$$\mathscr{F}_k\{e(r)\} = \frac{\mathscr{F}_k\{e_m)\}}{\mathrm{sinc}(kD)} \tag{9.24}$$

This indicates that the Fourier coefficients of continuous measurements over a detector can be found from the Fourier transform of discrete measurements. These Fourier transforms can then be analyzed and processed, as discussed in Sections 9.1.1 and 9.3. Note that the singularity, caused by the zero value of the sinc function in the denominator of Eq. (9.24) when $kD = 1$, is avoided by the fact that $|k| \leq \frac{1}{2\Delta r}$, i.e. is less than the Nyquist frequency (half of the sampling frequency), as indicated in Section 9.2, where Δr is the distance between two adjacent measurements. Also, $D \leq \Delta r$, unless detector measurements overlap each other. Therefore, $kD \leq \frac{1}{2}$, and $\mathrm{sinc}(kD) \leq \frac{\sin(\frac{\pi}{2})}{\frac{\pi}{2}} = 0.6366 < 1$, and the singularity at $kD = 1$ and the multiple integer values of kD are never reached. With the view that $\frac{1}{\mathrm{sinc}(kD)}$ acts as a filter to the acquired discrete measurements, e_m, any noise in measurements is amplified by less than $1/0.6366 = 1.571$, which is not an appreciable magnification factor (Barrett, 1981).

9.3 Spatial Filtering of Noise

Statistical variations and electronic noise inevitably contaminate measurements; see Section 15.4.3 for discussion on the sources of measurement fluctuations. A high signal-to-noise ratio can be attained in radiation systems by increasing the strength of the radiation source, prolonging the measurement time, using high-efficiency detectors and having good quality electronic components. However, a measurement profile recorded along a line, a curved segment, or a detection plane, is likely to exhibit some fluctuations. Such fluctuations should be smoothed out or eliminated to avoid their proliferation into the image reconstruction process.

As indicated in Section 9.2.1, one can filter out, in the frequency domain, high frequencies if they only contain low-amplitude noise. Noise would still, however, exist at lower frequencies. One can take advantage of the fact that measurement profiles tend generally to be continuous in the spatial domain, and even abrupt changes do not tend to produce steep changes in measurements because of the finite volume of a detector. Oscillations in measurements that cause deviation from this continuity are most likely caused by noise, and can be smoothed out by a filter that restores continuity. Some of these filters are discussed below.

9.3.1 Moving-Average Filter

A simple and powerful filter that can be applied directly to a measured profile, along some distance r, is the moving average filter. In its simplest form, a moving average of order L replaces a measurement, e_m, by the average of this measurement and its $L-1$ neighbors:

$$\bar{e}_m = \frac{1}{L} \sum_{-\frac{L-1}{2}}^{\frac{L-1}{2}} e_m \tag{9.25}$$

$$\bar{e}_m = \bar{e}_{m-1} + e_{m+\frac{L-1}{2}} - e_{m-1-\frac{L-1}{2}} \tag{9.26}$$

where $\bar{e}_m$ is the filtered value, and L is the number of measurements over which the moving average is conducted. For measurements at the edge, i.e. for $m = 0$ and $M-1$, where M is the total number of measurements, averaging should be taken over the available measurements only. The recursive form of the filter, Eq. (9.26), greatly simplifies its implementation. This basic moving-average process reduces the fluctuations introduced by noise by a factor of $\frac{1}{L}$ (Smith, 2003, Chap. 15). The filter also retains abrupt changes, though with reduced sharpness. Obviously, this filter treats all measurements within its range equally, which is the reason behind its effectiveness in reducing noise, but it can also significantly alter the value of individual measurements, and may lead to distorted reconstructed images. A moving-average filter with a short range can overcome this problem, but then it is not quite effective in removing noise.

The moving average of Eq. (9.25) is equivalent to convolving measurements with a rectangular window filter of length, L, and amplitude, $\frac{1}{L}$. Since the Fourier transform or a rectangular function is a sinc function $\left(= \frac{\sin(\pi f M)}{\pi f}, \text{ where } 0 \leq f < \frac{1}{2}\right)$, the moving average filter is a low-pass filter in the frequency domain, but with no clear cut-off frequency. However, repeated application of the moving average, i.e. averaging the moving average, can improve the performance of this filter as a low-pass filter. A double application of this filter (two passes), i.e. the convolution of a rectangular filter with itself, and consequently the multiplication of the frequency response by itself, is equivalent to the application of a triangle filter. The double filter is preferable to a single rectangular filter, as it gives a measurement a higher weight than its neighbor. Multiple applications of the moving average, four times or more, becomes equivalent to the application of a Gaussian filter (Smith, 2003, Chap. 15).

9.3.2 Gaussian Filter

A Gaussian filter has the advantage that its Fourier transform is also a Gaussian distribution centered around the zero frequency (with positive and negative frequencies at both sides). One can then control the effectiveness of the low-pass nature of the filter by adjusting its width. Also, the attenuation of higher frequency components, hence their relative removal, is more effective with a Gaussian filter than with moving-average filters. The Gaussian filter may also reflect the inherent statistical nature of fluctuations in many acquired measurement distributions. The Blackman window of Eq. (9.20) also provides a shape that resembles that of a Gaussian distribution, and as such can be utilized as a filter, in place of the Gaussian filter, for computational convenience.

9.3.3 Recursive Weighted Moving Average Filter

A recursive moving filter that weights a certain measurement with the moving average of a neighboring measurement enables not only the suppression of noise but also the smoothing and some shaping of a measurement profile. A simple form of such weighted filter is given by:

$$\bar{e}^u_m = we_m + (1-w)\bar{e}^u_{m-1}, \quad m = 0, 1, \ldots M-1; \ \bar{e}^u_0 = e_0 \tag{9.27}$$

$$\bar{e}^d_m = we_m + (1-w)\bar{e}^d_{m+1}, \quad m = M-1, M-2, \ldots 0; \ \bar{e}^d_{M-1} = e_{M-1} \tag{9.28}$$

$$\bar{e}_m = \frac{1}{2}\left[\bar{e}^u_m + \bar{e}^d_m\right] \tag{9.29}$$

where $0 < w < 1$. This filter is performed into steps: one upstream of the mth measurement and the other downstream of it, to ensure that the overall trend of the profile is incorporated into the filtering process (in time-series filtering, only earlier measurements are considered). To gain insight into the nature of this recursive filter, let us reformulate $\bar{e}^u_m$ as (Smith, 2003, Chap. 19):

$$w\left(e_m - \bar{e}^u_m\right) = (1-w)[\bar{e}_m - \bar{e}_{m-1}]$$

$$e_m - \bar{e}^u_m = \Delta r\left(\frac{1-w}{w}\right)\left(\frac{\bar{e}_m - \bar{e}_{m-1}}{\Delta r}\right)$$

$$e(r) - \bar{e}^u(r) = \tau\frac{d\bar{e}}{dr}, \quad \tau = \Delta r\left(\frac{1-w}{w}\right) \tag{9.30}$$

where Δr is the distance between two successive measurements. Equation (9.30) is analogous to an RC electric circuit with resistance, R, capacitance, C, and time constant, $\tau = RC \equiv \Delta r\frac{1-w}{w}$, with an input voltage equivalent to e_m and an output voltage[4] $\bar{e}^u(r)$. Here instead of time, the change is with respect to the distance from the point of measurement to a measurement point upstream or downstream of it. Therefore, a recursive moving average of the form of Eq. (9.30) assigns a full weight to a measurement at e_m and an exponentially decreasing weight to measurements upstream and downstream of it. The degree and extent (duration) of the weighting is determined by w for a given Δr. When $w \to 0$, $\tau = \!\!\to \infty$ and $\bar{e}^u_m \to \bar{e}^u_{m-1} \to \bar{e}^u_0 = e_0$, the length of the moving average covers the entire sampling range while giving a zero weight for the measurement for which a moving average is to be calculated. On the other hand, when $w \to 1$, $\tau \to 0$ and $\bar{e}^u_m \to e_m$, and very little moving averaging (if any) is accomplished. With $w = 0.5$. $\tau = \Delta r$, and the moving average weight attenuates the signal by a factor of $\mathrm{e}^{-1} = 0.368$ every Δr. Therefore, one can state that the filter of Eq. (9.29) is equivalent to a weight of $\exp\left[-\frac{|r-r'|}{\tau}\right]$ applied to both sides of r, with r' extending from r to the end of the measurement profile at either ends.

[4] Recall that the drop in voltage, $V_{in} - V_{out}$ in an RC circuit is equal iR, where i is the current and $i = C\frac{V_{out}}{dt}$ with V_{out} and V_{in} being, respectively, the output (across C) voltage and input (across R and C) voltage. Then $V_{out} = V_{in}\left[1 - \exp\left(-\frac{t}{\tau}\right)\right]$.

The moving average filter of Eq. (9.29) is a low-pass (first-order) filter of a cutoff frequency, $k_c = \frac{1}{2\pi\tau}$. The analog equivalent of the filter of Eq. (9.29) is an electric resistance-capacitance (*RC*) circuit with a time constant $\tau = RC$. The capacitor, C, has a reactance of $\frac{1}{2\pi kC}$, which decreases with frequency, reducing the output voltage across R. Note that when radiation measurements are acquired in the current mode, i.e. using a count-rate meter, they would have been already subjected to an *RC* filtering by the electric circuit of counting. The cutoff frequency, k_c is not a sharp cutoff, but it is the frequency at which the filter attenuates the amplitude by half and consequently the power of the signal by a factor of one quarter[5] (−6 dB).

Higher order recursive digital filters can also be created by incorporating weighted values of neighboring measurements, as well as weighted values of their own moving averages. For upstream filtering, one can formulate the following general recursive filter:

$$\begin{aligned}\bar{e}^u_m = w_m e_m &+ w_{m-1} e_{m-1} + \cdots\cdots\cdots + w_1 e_1 + w_0 e_0 \\ &+ \bar{w}_{m-1}\bar{e}^u_{m-1} + \bar{w}_{m-2}\bar{e}^u_{m-2} + \cdots + \bar{w}_1 \bar{m}_1\end{aligned} \quad (9.31)$$

where the w's and $\bar{w}$'s are filtering coefficients. There is no $\bar{w}_m$ value in Eq. (9.31), since $\bar{e}^u_m$ is being calculated, and there is no $\bar{w}_0$ coefficient since $\bar{e}^u_0 = e_0$ has been already given a coefficient w_0.

9.3.4 Matched Filter

The above filters introduce certain weighting functions, either explicitly or implicitly. None of these functions directly reflect the actual distribution of the considered measurement profile. A filter that employs the distribution of the measurement profile is called a matched filter, and can be implemented via a correlation operation. Convolution involves shifting a function and multiplying it by another function, then integrating it; recall that convolution requires first reversing the function before the shifting and performing of other subsequent operations. In matched filtering, one is correlating a function with itself, i.e. performing an autocorrelation. This filter requires, therefore, a functional formulation of the measurements (which can be obtained via function fitting). Correlation can also be accomplished via convolution, by convolving a function with its mirror-image flipped function. Convolution in the frequency domain becomes multiplication, which facilitates the application of the matched filter. This filter is quite effective in attenuating high-frequency noise, and in turn in accentuating the presence of discontinuities in a measurement profile to levels well above the noise level. However, the matched filter does not necessarily preserve the shape of the profile, and can cause shifting in the position of discontinuities (Smith, 2003, Chap. 17). It is best used to detect the presence of subtle continuities in a measurement profile.

[5] The power reduction in decibel (dB) is calculated using the 10 log rule: $10 \times \log$ (power reduction factor).

9.3.5 Multipole Filters

The recursive filter of Eq. (9.29) is also known as a single-pole filter. This is because its kernel, $\exp[-\frac{-r}{\tau}]$, as obtained from its analog form, Eq. (9.30), has a Laplace transform[6]: $\mathcal{L}_s\{\exp[-\frac{-r}{\tau}]\} = \frac{1}{s+\tau}$, and consequently a single pole located at: $s = \frac{1}{\tau}$. A multipole filter with poles lying on a circle is the Butterworth filter (a low-pass filter with an analog form that contains capacitances, resistances, and inductances). Recursive multipole filters can also be designed using Chebyshev polynomials, see (Smith, 2003, Chap. 20) for an algorithm. A Chebyshev filter has poles that lie on an ellipse, and provides a sharp frequency cutoff at the expense of some ripple in the pass or stop bands. A Butterworth filter, on the other hand, has a more gradual frequency cutoff but exhibits no ripples.

9.4 Consistency and Smoothing

The filtering processes of Sections 9.2.1 and 9.3 aim at removing noise, essentially by smoothing measurements with a kernel that relates one measurement to its neighbors. This can come, however, at the expense of changing the magnitude of measurements, which in turn affects the magnitude of the image parameters produced in image reconstruction. Curve fitting can preserve the magnitude of measurements while smoothing their profile. Since measurement profiles and their slope tend generally to be continuous in the spatial domain, common curve fitting methods, such as polynomial regression or least squares, can be applied. This also has the added effect of estimating measurement values between sampled points, to fill missing data. The danger of this approach is that it can conceal any fine (small scale) genuine changes in a measured profile. On the other hand, any outliers (measurements with unusually high noise content) can also bias considerably the curve fitting process, as indicated for the Wiener filter of Eq. (9.22), which is a least-square filter. However, one can place some restrictions, from previous experience or knowledge of the expected experimental trend, on the curve fitting process.

One can combine the tasks of both preserving the signal magnitude and smoothing the measurement noise via an optimization process that aims at minimizing the cost function:

$$\chi^2 = \sum_{m=0}^{M-1} w_m^2 (e_m - \bar{e}_m)^2 + \alpha^2 \|\boldsymbol{G}\bar{e}^2\|_2 \tag{9.32}$$

where w_m^2 is a weight factor and α^2 is a smoothing (regularizing) parameter[7], $\boldsymbol{G}$ is a smoothing operator, and $\|.\|_2$ denotes a Euclidean norm. The weight factor can be taken as the inverse variance of e_m, so that measurements with higher variance are given a weight lower than measurements with low variance. In radiation counting,

[6] The Laplace transform, $\mathcal{L}$, of a function, $f(x)$ is: $\mathcal{L}_s = \int_0^\infty f(x) \exp[-sx]\, dx$.
[7] A squared parameter, α^2, is used to ensure a positive number.

the variance is determined from Poisson statistics, see Section 15.2, and is equal to the measurement itself. The first term in Eq. (9.32) is a consistency metric that measures the difference between filtered and actual measurements. The second term is the regularization term, introduced to relax the minimization of the first term and allow smoothing (stability) of measurements, so that a small error in the measurements will not result in a large change in the filtered value. The smoothing operator, $\boldsymbol{G}$, can be simply taken as the identity operator, $\boldsymbol{e}$, a first spatial derivative operator, the second spatial derivative (Laplacian) operator, or some other regularization function, see Section 10.4 for more regularization functions. With $\boldsymbol{G}=\boldsymbol{I}$, one is using Tikhonov regularization, which imposes minimum constraint on the smoothing process. With the first derivative as the regularization operator, one aims at a nearly uniform smoothed measurement profile, while the Laplacian imposes minimum curvature in the distribution of the smoothed measurements. Higher-order derivatives can also be utilized. The degree such smoothing constraints impose on the final filtered profile depends on the value of the regularization parameter, α^2, which can be optimized using the cross validation method discussed in Section 10.5.5. The optimization problem of Eq. (9.32) can be solved by one of the methods given in the ensuing sections.

9.4.1 Regularized Least-Squares

The optimization problem of Eq. (9.32) can be made equivalent to an ordinary least-squares problem via the following substitutions (RSICC, 1978):

$$\chi^2 = \sum_{i=1}^{2M} v_i^2 \left(y_i - \sum_{j=1}^{M} A_{ij}\bar{e}_j \right)^2 \tag{9.33}$$

$$\boldsymbol{y} = \begin{bmatrix} e_0 \\ \cdot \\ \cdot \\ e_{M-1} \\ 0 \\ \cdot \\ \cdot \\ 0 \end{bmatrix}, \quad \boldsymbol{v} = \begin{bmatrix} w_0 \\ \cdot \\ \cdot \\ w_{M-1} \\ G_{00} \\ \cdot \\ \cdot \\ G_{M-1,M-1} \end{bmatrix}, \quad \boldsymbol{A} = \begin{bmatrix} 1\ 0\ 0\ 0 \cdots \\ 0\ 1\ 0\ 0 \cdots \\ \cdot\ \cdot\ \cdot\ \cdot\ \cdots \\ \cdot\ \cdot\ \cdot\ \cdot\ \ 1 \\ \alpha\ 0\ 0\ 0 \cdots \\ 0\ \alpha\ 0\ 0 \cdots \\ \cdot\ \cdot\ \cdot\ \cdot\ \cdots \\ 0\ 0\ 0\ 0\ \ \alpha \end{bmatrix} \tag{9.34}$$

where $\boldsymbol{y}$ and $\boldsymbol{v}$ are $2M \times 1$ vectors, $\boldsymbol{A}$ is an $2M \times M$ matrix, $\bar{\boldsymbol{e}}$ is an $M \times 1$ vector of the sought measurements, and G_{ii} represents the smoothing operation performed over $\bar{e}_i$. The minimization of Eq. (9.33) results in the solution:

$$\bar{\boldsymbol{e}} = \left[\boldsymbol{A}^{\mathrm{T}}\boldsymbol{W}\boldsymbol{A}\right]^{-1}\boldsymbol{A}^{\mathrm{T}}\boldsymbol{W}\boldsymbol{y} \tag{9.35}$$

where T denotes matrix transpose and $\boldsymbol{W}$ is a diagonal $2M \times 2M$ matrix whose diagonal elements are the elements of $\boldsymbol{v}$.

9.4.2 Dynamic Programming

When the number of measurements is large, the matrix method of Eq. (9.35) becomes computationally demanding, in terms of computer storage. Dynamic programming enables the minimization of the cost function of Eq. (9.32) through a set of sequential suboptimized problems, without performing matrix inversion. The concept of dynamic programming is explained in optimization textbooks, such as Rao (1996), but the solution given below is guided by that reported in Trujillo and Busby (1997).

Dynamic programming requires, as its name indicates, the establishment of a dynamic relationship that relates the to-be-optimized (state) variables to each other. For the problem at hand, the state variables are $\bar{e}_m, m = 0, \ldots, M-1$. The smoothing operator in the second term of Eq. (9.32) can be used to establish this dynamic relationship if the smoothing operator, $\boldsymbol{G}$, is taken as a derivative. For the first derivative, one has:

$$\begin{aligned} g_m &= \bar{e}_m - \bar{e}_{m-1} \\ \bar{e}_{m+1} &= \bar{e}_m + g_m, \quad m = 0, \ldots, M-1 \end{aligned} \tag{9.36}$$

This is a first-order dynamic function, in which g_m is the mth element of the vector, $\boldsymbol{G}\bar{\boldsymbol{e}}$ is a "forcing" variable that moves state i to state $i+1$. In dynamic programming, g_i is called the decision or design variable. For now, we will develop the optimization problem using Eq. (9.36), then expand the concept to higher order dynamic functions.

Dynamic programming starts at one of the ends of the set of state variables, say the far end point, $M-1$, by optimizing the force function, g_{M-1}. This is done by treating state $M-1$ as an isolated subproblem. Once this subproblem is optimized, the problem is expanded to include two points: $M-1$ and $M-2$, and so on, until point $m = 0$ is reached. A forward suboptimization is then performed over the state parameters starting from $m = 0$ and ending at $m = M-1$.

Backward suboptimization starts at some initial state $h = \bar{e}_{M-1}$. This initial state value is simply to facilitate derivation of the dynamic programming process, and need not be designated a specific value. Considering this subproblem in isolation of all other points, the minimum of the cost function of Eq. (9.32), assuming equal unit weight for all measurements, reduces to:

$$f_{M-1}(h) = \min_{g_{M-1}} \left[(e_{M-1} - h)^2 + \alpha^2 g_{M-1}^2 \right] = (e_{M-1} - h)^2 \tag{9.37}$$

$$= h^2 - 2he_{M-1} + e_{M-1}^2 = a_{M-1}h^2 + b_{M-1}h + q_{M-1}$$

$$a_{M-1} = 1, \quad b_{M-1} = -2e_{M-1}, \quad q_{M-1} = e_{M-1}^2 \tag{9.38}$$

with $g_{M-1} = 0$ being the value that minimizes the state subobjective function. In other words, if the system consisted of one state variable, $\bar{e}_{M-1}$, the only inconsistency would have been related to the one measurement, e_{M-1}. The quadratic form of Eq. (9.38) is to facilitate deriving a recursive relationship.

If we now consider a subproblem consisting of states $M-1$ and $M-2$, and start with the initial state h assigned to $\bar{e}_{M-2}$, then one has the suboptimization problem:

$$f_{M-2}(h) = \min_{g_{M-2}} \left[(e_{M-2} - h)^2 + \alpha^2 g_{M-2}^2 + f_{M-1}(h + g_{M-2}) \right]$$

$$= \min_{g_{M-2}} \left[(e_{M-2} - h)^2 + \alpha^2 g_{M-2}^2 + (\bar{e}_{M-1} - [(h + g_{M-2})]^2 \right] \tag{9.39}$$

$$= (e_{M-2} - h)^2 \left[1 + \frac{\alpha^2}{(\alpha+1)^2} + \left(1 - \frac{1}{\alpha^2+1}\right)^2 \right] \tag{9.40}$$

$$= a_{M-2}h^2 + b_{M-2}h + q_{M-2} \tag{9.41}$$

where the cost function is expanded to include the cost function associated with state $M-2$ and the optimized cost function of the previous state $M-1$. For the latter state: $f_{M-1}(\bar{e}_{M-1} = \bar{e}_{M-2} + g_{M-2} = h + g_{M-2}) = \bar{e}_{M-1} - (h + g_{M-1})$, where use is made of Eqs. (9.36) and (9.37). Equation (9.40) is arrived at by finding the value of g_{M-2} that minimizes the cost function of Eq. (9.39), which can be easily found to be: $g_{M-2} = \frac{e_{M-1}-h}{\alpha^2+1}$.

The above process continues for $m = M-3$, $m = M-4$, and so on until reaching state $m = 0$. At each step, one determines f_m and g_m. However, the quadratic formulations of Eq. (9.38) and (9.41) enables the establishment of the recursive relationships:

$$a_{m-1} = \left[1 + \frac{\alpha^2 a_m}{\alpha^2 + a_m} \right] \tag{9.42}$$

$$b_{m-1} = -2e_{m-1} + \frac{\alpha^2 b_m}{\alpha^2 + a_m} \tag{9.43}$$

$$q_{m-1} = q_m + e_{m-1}^2 - \frac{b_m^2}{4(\alpha^2 + a_m)} \tag{9.44}$$

$$g_{m-1} = -\frac{2a_m h + b_m}{2(\alpha^2 + u_m)} \tag{9.45}$$

These relationships enable backward suboptimization, starting with state $m = M-1$ for which Eq. (9.38) applies. Note that $a_M = b_M = q_M = 0$.

Forward suboptimization is performed with respect to h, staring with state $m = 0$, for which the cost function is:

$$f_0(h) = a_0 h^2 + b_0 h + q_0 \tag{9.46}$$

where the parameters a_0, b_0 and q_0 are determined from the relationships of Eqs. (9.42) to (9.44). The optimal initial state which minimizes $f_0(h)$ with respect to h is: $h_1 = -\frac{b_1}{2a_1} = \bar{e}_0$. The optimal forcing term, g_0, can be updated using Eq. (9.45), and the dynamic Eq. (9.36) is used to determine $\bar{e}_1 = \bar{e}_0 + g_0$. Since all the recursive coefficients can now be determined, calculations can progress using Eq. (9.36) to obtain the rest of the state variables: $\bar{e}_1, \ldots, \bar{e}_{M-1}$, with the recursive relationship:

$$\bar{e}_{m+1} = \bar{e}_m - \frac{2a_{m+1}\bar{e}_m + b_{m+1}}{2(\alpha^2 + a_{m+1})} \tag{9.47}$$

where use is made of Eq. (9.36) and (9.45), with h replaced in the latter equation with $\bar{e}_m$. Note that in calculating $\bar{e}_m$ using Eq. (9.47), there is no need to know the value of q_m, and as such there is no need to determine it.

Dynamic programing can be considered as a multistage two-sweep (backward and forward) filter. To facilitate examination of the frequency response, one needs to formulate an equivalent filter with a certain kernel. The only parameter in the recursive relationships that affects the minimized staged cost function, f_m, is a_m. Therefore, to arrive at an overall equivalent filter, one should find a unified value for a (in practice, a_m tends to approach a stable value). This can be done by setting $a_{m-1} = a_m$ in Eq. (9.42) and solving for, a, which gives:

$$a^2 - a - \alpha^2 = 0 \tag{9.48}$$

with the positive value of a giving the desired solution, since $f \geq 0$. Now, let us represent the kth frequency component of a unit measurement e_m as $\exp(ikm\Delta r)$, with Δr being the sampling interval. The corresponding frequency component for b_m is $b \exp(ikm\Delta r)$ and that for $\bar{e}_m$ is $\bar{e} \exp(ikm\Delta r)$. From Eq. (9.43), one has:

$$b = \frac{2(\alpha^2 + a)}{\alpha^2 \exp(ik\Delta r) - \alpha^2 - a} = \frac{2a^2}{\alpha^2[\exp(ik\Delta r) - 1] - a} \tag{9.49}$$

where use is made of Eq. (9.49) in arriving at the last step in the equation. From Eq. (9.47), and with the aid of Eqs. (9.48) and (9.49), one gets:

$$\bar{e}\exp(ik\Delta r) = \bar{e} - \frac{b\exp(ik\Delta r)}{2(\alpha^2 + a)} - \frac{a\bar{e}}{\alpha^2 + a} = \bar{e} - \frac{\exp(ik\Delta r)}{\alpha^2[\exp(ik\Delta r) - 1] - a} - \frac{\bar{e}}{a}$$

Therefore,

$$\bar{e}\left([\exp(ik\Delta r) - 1] + \frac{1}{a}\right)\left(\alpha^2[\exp(ik\Delta r) - 1] - a\right) = -\exp(ik\Delta r)$$

and

$$\bar{e}\left[\alpha^2[\exp(ik\Delta r) - 1]^2 - 1 - [\exp(ik\Delta r) - 1]\left(\frac{a^2 - \alpha^2}{a}\right)\right] = -\exp(ik\Delta r)$$

which leads to:

$$\bar{e} = \frac{1}{1 - \alpha^2 \frac{[\exp(ik\Delta r) - 1]^2}{\exp(ik\Delta r)}} = \frac{1}{1 + 2\alpha^2(1 - \cos k\Delta r)} \tag{9.50}$$

Since we assumed a unit measurement, then Eq. (9.50) represents the amplification of the kth frequency component of the measurements by the filtering process. It is obvious from Eq. (9.50) that when $\alpha^2 = 0$, there is no smoothing and all frequencies will be permitted. The same happens for the zero-frequency (mean value) of a measurement profile. The presence of a frequency, k, decreases as α^2 increases. Therefore, the value of the regularization parameters, α^2, can be used to control the cutoff frequency. Dynamic programming is, therefore, equivalent to a low-pass filter.

Laplacian smoothing can be introduced into dynamic programming as a second-order system by expecting $\bar{e}$ to assume the form:

$$\bar{e}_{m+1} = \bar{e}_m + \bar{e'}(r - m\Delta r) + \frac{1}{2}\bar{e''}(r - m\Delta r)^2; \quad \bar{e'} = \frac{d\bar{e}_m}{dr}, \quad \bar{e''} = \frac{d\bar{e'}}{dr} \tag{9.51}$$

which gives:

$$\begin{Bmatrix} \bar{e}_{m+1} \\ \bar{e'}_{m+1} \end{Bmatrix} = \begin{bmatrix} 1 & \Delta r \\ 0 & 1 \end{bmatrix} \begin{Bmatrix} \bar{e}_m \\ \bar{e'}_m \end{Bmatrix} + \begin{Bmatrix} \frac{(\Delta r)^2}{2} \\ \Delta r \end{Bmatrix} g_m \tag{9.52}$$

$$\bar{\boldsymbol{e}}_{m+1} = \boldsymbol{H}\bar{\boldsymbol{e}}_m + \boldsymbol{P}g_m \tag{9.53}$$

where $\bar{\boldsymbol{e}}_{m+1}$ and $\bar{\boldsymbol{e}}_m$ are 2×1 vectors associated with $\bar{e}_{m+1}$ and $\bar{e}_m$, respectively, $\boldsymbol{H}$ is a 2×2 matrix for the dynamic system, $\boldsymbol{P}$ is a 2×1 vector relating the sought variable, $\bar{e}_m$ to the controlling Laplacian parameter, g_m. The cost function of Eq. (9.32) is then expressed as:

$$\chi^2 = \sum_{m=0}^{M-1} [\bar{\boldsymbol{e}}_m - \boldsymbol{e}_m, (\bar{\boldsymbol{e}}_m - \boldsymbol{e}_m)]^2 + \alpha^2 g_m \tag{9.54}$$

where (.,.) denotes inner product[8], $\boldsymbol{e}_m$ is a 2×1 vector whose first element is e_m and its second element is e'_m (evaluated as a difference), and an equal unit weighting of measurements is assumed. In analogy with Eq. (9.41), one expresses the step-optimized functional, $f_m(\boldsymbol{h})$, as:

$$f_m(\boldsymbol{h}) = (\boldsymbol{h}, \boldsymbol{a}_m\boldsymbol{c}) + (\boldsymbol{h}, \boldsymbol{b}_m) + q_m \tag{9.55}$$

where $\boldsymbol{h}$ is an initial estimate of $\bar{\boldsymbol{e}}_m$, $\boldsymbol{a}_m$ is a 2×2 matrix, $\boldsymbol{b}_m$ is an 2×1 vector, and q_m is a scalar. Following a process similar to the first-order dynamic system, the following recursive relationships are arrived at (Trujillo and Busby, 1997):

$$\boldsymbol{a}_{m-1} = \boldsymbol{I} + \boldsymbol{H}^{\mathrm{T}}\left(\boldsymbol{a}_m - \frac{\boldsymbol{a}_m\boldsymbol{P}\boldsymbol{P}^{\mathrm{T}}\boldsymbol{a}_m}{\alpha^2 + \boldsymbol{P}^{\mathrm{T}}\boldsymbol{a}_m\boldsymbol{P}}\right)\boldsymbol{H} \tag{9.56}$$

[8] The inner product, $(\boldsymbol{x}, \boldsymbol{y})$, of two vectors $\boldsymbol{x}$ and $\boldsymbol{y}$ is generally defined as: $(\boldsymbol{x}, \boldsymbol{y}) = \sum x_i y_i^*$, where y_i^* is the complex conjugate of y_i, see also Appendix 7.A.

$$\boldsymbol{b}_{m-1} = -2\boldsymbol{e}_{m-1} + \boldsymbol{H}^{\mathrm{T}}\left(\boldsymbol{I} - \frac{\boldsymbol{a}_m\boldsymbol{P}^{\mathrm{T}}}{\alpha^2 + \boldsymbol{P}^{\mathrm{T}}\boldsymbol{a}_m\boldsymbol{P}}\right)\boldsymbol{b}_m \tag{9.57}$$

$$g_{m-1} = -\frac{\boldsymbol{P}^{\mathrm{T}}\boldsymbol{b}_m + 2\boldsymbol{P}^{\mathrm{T}}\boldsymbol{a}_m\boldsymbol{H}\boldsymbol{h}}{\alpha^2 + \boldsymbol{P}^{\mathrm{T}}\boldsymbol{a}_m\boldsymbol{P}} \tag{9.58}$$

where $\boldsymbol{I}$ is a 2×2 identity matrix. Note that q_m is not included here since it is not needed in the calculations. The matrices involved are small in size and can be easily numerically manipulated. The backward sweep starts at measurement $M-1$ with the coefficients: $\boldsymbol{a}_{M-1} = \boldsymbol{I}$, and $\boldsymbol{b}_{M-1} = -2\bar{e}_{M-1}$, which enables calculation of coefficients $\boldsymbol{a}_m$ and $\boldsymbol{b}_m$ for $m = M-2, M-3, \ldots, 0$. The forward sweep then starts by calculating an optimal estimate $\boldsymbol{h}$ by minimizing: $f_1(\boldsymbol{h}) = (\boldsymbol{h}, \boldsymbol{a}_0\boldsymbol{h}) + (\boldsymbol{h}, \boldsymbol{b}_0) + q_0$, which gives $\boldsymbol{h} = -\frac{1}{2}\boldsymbol{a}_0^{-1}\boldsymbol{b}_1 = \bar{\boldsymbol{e}}_0$. The value of g_0 is calculated using Eq. (9.58), and $\boldsymbol{e}_1 = \boldsymbol{H}\bar{\boldsymbol{e}}_0 + \boldsymbol{P}_0 g_0$, and so on for $\bar{\boldsymbol{e}}_2, \ldots \bar{\boldsymbol{e}}_{M-1}$.

9.4.3 Spline Smoothing

The first and second order dynamic programming procedures can also be expanded to third and higher order formulations, to incorporate higher order derivatives in the smoothing term or Eq. (9.32). Moreover, the minimized functional, $f_m(c)$, can be expanded in the form of splines, to take advantage of the piecewise nature of splines as an effective data smoothing tool (Craven and Wahba, 1979, Wahba, 1990). A spline is defined by an nth order polynomial that passes through k control points (knots), so that in the smoothing term, g_m, in the cost function of Eq. (9.54), can be replaced by:

$$g_m = \sum_{j=0}^{n-1}\theta_j t^j + \sum_{i=1}^{k}\eta_i(t-k)^{n-1}, \quad t = \frac{m\Delta r}{R} \tag{9.59}$$

where R is the total length of the measurement profile, ΔR is the interval between measurements, and θ's and η's are the spline coefficients. Equation (9.59) reflects the definition of a spline as pieces of polynomials, each starting with a zero value at a knot, combined linearly so that the derivatives at each knot are continuous up to order $n-2$. With the spline formation, the cost function of Eq. (9.32) can be expressed as:

$$\chi^2 = \sum_{m=0}^{M-1}\left[\bar{e}_m - \boldsymbol{e}_m, (\bar{e}_m - \boldsymbol{e}_m)\right]^2 + \alpha^2\left(\int_0^1 \frac{d^m}{dt}\right)^2 dt \tag{9.60}$$

The dynamic step-optimized functional $f(h)$ can now be expressed as a combination of $n+2$ basis function (Trujillo and Busby, 1997):

$$f(r) = \sum_{i=1}^{n+2} f_i a_i(r) \tag{9.61}$$

with $r = m\Delta r$, $m = 0, 1, \ldots m-1$. The coefficients f_i are determined by the backward sweep process of dynamic programming, and for a cubic spline are given by:

$$a_1(r) = 1$$
$$a_2(r) = (r - r_0)$$
$$a_3(h) = \frac{(r - r_1)^2}{2}$$
$$a_i(r) = \begin{cases} 0 \\ \frac{(r-r_{i-1})^3}{6} \\ \frac{(\Delta r)^3}{6} + \frac{(\Delta r)^2(r-r_i)}{2} + \frac{\Delta r(r-r_i)^2}{r} \end{cases} \quad \text{for} \begin{cases} r < r_{i-1} \\ r_{i-1} \leq r < r_i\,,\ i > 3 \\ r \geq r_i \end{cases}$$

Matrix algebra can also be used to solve the above problem, see Buja et al. (1989) and de Hoog and Hutchinson (1987).

9.4.4 Cross Validation

The value of the smoothing parameter, α^2, obviously affects the value of $\bar{\boldsymbol{e}}$ obtained by minimizing the cost function of Eq. (9.32). One can find the optimum value of α^2 empirically by attempting different values and observing the trend of the results, either directly or after using them in image reconstruction. However, one can employ the cross validation method, originally introduced by Allen (1974) and Wahba and Wold (1975). The essence of this method is that if a random element of a measurement set, $\boldsymbol{e}$, is missing, then the regularized (smoothed) solution should be able to predict its value. Therefore, for a certain value of α^2, if one leaves a measurement e_m and smooth the remaining $M-1$ measurements, one can find an estimate $\hat{e}_m$ of $\bar{e}_m$ using the dynamic programming equations, Eq. (9.36) or Eq. (9.52). Repeating the above process for M different missing elements, one at a time, then one has the cross validation function:

$$V(\alpha^2) = \frac{1}{M}\sum_{k-0}^{M-1}(\hat{e}_k - e_k)^2 \tag{9.62}$$

The value of α^2 that minimizes $V(\alpha^2)$ is the optimum value. The function of Eq. (9.62) represents ordinary cross validation, in contrast with the generalized cross validation method discussed in Section 10.5.5.

9.4.5 Mollification

The optimization of Eq. (9.32) can also be viewed as the act of finding a mollifier, I_α, (an approximation to the identity) that transforms a given measurement profile, $e(r)$, into a profile, $\bar{e}(r) = I_\alpha e(r)$, which minimizes a certain cost function (Murio, 1993). The subscript α in the mollifier I_α signifies the introduction of smoothing by a parameter, α^2. Keeping in mind that the objective of filtering a measurement profile is to find the exact, but unknown, value, $e^{(0)}(r)$, from measurements, $e(r) = e^{(\delta)}(r)$

contaminated with some error signified by δ, then one should be aiming at minimizing a cost function of the from:

$$\chi^2 = \left\| e^{(0)}(r) - I_\alpha e^{(\delta)}(r) \right\| \leq \left\| e^{(0)}(r) - I_\alpha e^{(0)}(r) \right\| + \left\| I_\alpha e^{(0)}(r) - I_\alpha e^{(\delta)}(r) \right\| \tag{9.63}$$

where use is made of the triangle inequality, with $\|.\|$ indicating a Euclidean norm.

Murio (2003) proposed mollification by a convolution operation with a Gaussian kernel:

$$I_\alpha e(r) = (g_{\alpha,p} * e)(r) = \int_{r-q\alpha}^{r+q\alpha} g_{\alpha,p}(r-r')e(r')\,dr' \tag{9.64}$$

$$g_{\alpha,q} = \begin{cases} \frac{A_q}{\alpha}\exp\left[-\frac{r^2}{\alpha^2}\right], & |r| \leq q\alpha \\ 0, & |r| > q\alpha \end{cases} \tag{9.65}$$

$$A_q = \frac{1}{\int_{-q}^{q} \exp[-r'^2]\,dr'}; \quad \int_{-q\alpha}^{q\alpha} g_{q,\alpha}(r)\,dr = 1 \tag{9.66}$$

with α (called the Mollification radius) defining the width of the Gaussian kernel and $q\alpha$ specifies the range over which it is applied, $\alpha > 0$ and $q > 0$. Recalling that the Fourier transform of a Gaussian distribution is also a Gaussian distribution, the width of the distribution determines the frequency cutoff of such kernel, if it were applied in the frequency domain. One can argue, therefore, that a Gaussian mollifier is no more that a Gaussian low-pass (due to its frequency cut off) filter with optimized width.

Murio (2003) proved that:

$$\|I_\alpha e^{(0)}(r) - e^{(0)}(r)\| \leq K\alpha \tag{9.67}$$

$$\|I_\alpha e^{(0)}(r) - I_\alpha e^{(\delta)}(r)\| \leq \delta \tag{9.68}$$

$$\|I_\alpha e^{(\delta)}(r) - e^{(0)}(r)\| \leq K\alpha + \delta \tag{9.69}$$

with K being a constant. The first condition ensures that the mollified function converges to error-free data as $\alpha \to 0$. The second condition shows that mollified noisy measurements approach mollified error-free measurements as $\delta \to 0$. According to the third condition, the mollified noisy measurements converge uniformly to the exact measurements, as $\alpha \to 0$ and $\delta \to 0$. Without smoothing at all, $\alpha = 0$, the mollifier becomes a Dirac delta function, and the measurements would stay as they were without any filtering. Therefore, mollification converts the optimization of Eq. (9.63) into a well-posed problem that does not propagate error.

The optimization of the cost function of Eq. (9.63) amounts to finding the best α value for a given set of measurements. The generalized cross validation method can be used for this purpose. It is an extension of the ordinary validation method of

Section 9.4.4, and is discussed in Section 10.5.5. It amounts in this situation to the minimization of the functional (Murio, 2003):

$$V(\alpha) = \frac{\boldsymbol{e}_{ext}^{\mathrm{T}}(J_\alpha - \boldsymbol{G}_\alpha)^2 \boldsymbol{e}_{ext}}{\mathrm{trace}[(J_\alpha - \boldsymbol{G}_\alpha)^2]} \tag{9.70}$$

where trace denotes the trace of matrix (the sum of its diagonals) $\boldsymbol{e}_{ext}$ is the measurement vector extended from a length of M to a length of $M+2M_e$, to cover intervals $[-q\alpha R, 0]$ and $[R, R(1+q\alpha)]$ to accommodate the width of the Gaussian distribution at the two ends of the data stream ($\boldsymbol{e}$ sampled originally over the distance interval $[0,R]$), $\boldsymbol{J}$ is an $M \times (M+2M_r)$ matrix whose element $J_{ij} = 1$ for $i, j \leq M$, and otherwise is zero, and $\boldsymbol{G}_\alpha$ is a circulant kernel matrix with an element, $G_{\alpha_{1,j}}$, in its first row such that:

$$G_{\alpha_{1,j}} = \begin{cases} \displaystyle\int_{r_{j-1}}^{r_j} g_{\alpha,p}(-r')\, dr, & j \leq M \\ 0 & j > M \end{cases} \tag{9.71}$$

The value q is taken typically to be equal to three ($q=3$), since larger values do not introduce much difference in results (Murio, 2003). The values added to create the extended measurement set are constant values that minimize $\|I_\alpha e_{ext} - e\|$ in the near-boundary intervals: $[0, q\alpha R, 0]$ and $[1 - q\alpha R, 1]$. Murio (2003) provided a MATLAB program that shows how α is obtained by minimizing Eq. (9.70), and how the mollified values, $\bar{\boldsymbol{r}}$ are arrived at.

References

D. Allen, "The relationship between variable selection and data augmentation and a method for prediction," *Technometrics*, vol. 16, pp. 125–127, 1974.

H. H. Barrett, *Radiological Imaging*. New York: Academic Press, 1981.

J. Ben Rosen, H. Park, J. Glick, and L. Zhang, "Accurate solution to overdetermined linear equations with errors using L_1 norm minimization," *Computational Optimization and Applications*, vol. 17, pp. 329–341, 2000.

A. Buja, T. Hadtie, and R. Tibshirani, "Linear smoothers and additive models (with discussion)," *Annals of Statistics*, vol. 26:1, pp. 1826–1856, 1989.

P. Craven and G. Wahba, "Smoothing noisy data with spline functions," *Numerische Mathematik*, vol. 31, pp. 377–403, 1979.

F. R. de Hoog and M. F. Hutchinson, "An efficient method for calculating smoothing splines using orthogonal transformations," *Numerische Mathematik*, vol. 50, pp. 311–319, 1987.

L. W. Goldman, "Principles of CT and CT technology," *Journal of Nuclear Medicine Technology*, vol. 35, pp. 115–128, 2007.

D. Murio, "Mollification and space marching," in *Inverse Engineering Handbook*, K. A. Woodbury, Ed. Boca Raton: CRC Press, 2003, ch. 4.

D. A. Murio, *The Mollification Method and the Numerical Solution of Ill-Posed Problems*. New York: John Wiley & Sons, Interscience Division, 1993.

J. G. Proakis and D. G. Manolakis, *Digital Signal Processing: Principles, Algorithms, and Applications*, 4th ed. Upper Saddle River: Prentice-Hall, 2007.
S. S. Rao, *Engineering optimization: Theory and practice*. New York: John Wiley & Sons, 1996.
RSICC, *FORIST. Neutron spectrum unfolding code*, Computer Codes for Unfolding the Pulse-Height Distribution of Neutrons Measured with Organic Scintillators, Radiation Safety Information Computational Center, Oak Ridge National Laboratory, 1978, PSR-0092/01.
S. W. Smith, *The Scientist and Engineer's Guide to Digital Signal Processing*. San Diego: California Technical Publishing, 2003, (`http://www.dspguide.com`).
D. M. Trujillo and H. R. Busby, *Practical Inverse Analysis in Engineering*. Boca Raton: CRC Press, 1997.
G. Wahba, *Spline Models for Observational Data*. Philadelphia: Society for Industrial and Applied Mathematics, 1990.
G. Wahba and S. Wold, "A completely automatic french curve." *Communications in Statistics*, vol. 4, pp. 1–17, 1975.

10 Matrix-Based Methods

When the inverse problem is in the matrix form presented in Section 8.1, direct or indirect (iterative) inversion of the system's forward-mapping matrix provides a solution for the inverse problem, as Eq. (8.1) shows. If the forward model generates a linear matrix, direct inversion guarantees the existence and uniqueness of image reconstruction, when the matrix is nonsingular i.e. invertible. However, as indicated in Section 7.6, the image reconstruction problem is always ill-posed. It is, therefore, essential to analyze the susceptibility of the system's matrix to error propagation by determining its rank and condition number. The design of an imaging system can be revisited, if it is found that the resulting matrix is too ill-conditioned. Even with the best system design, the problem will still be ill-posed because of the discretization process associated with creating a matrix from what is physically a continuous forward model. Therefore, a modified (by regularization) matrix needs to be inverted. Direct inversion is, however, computationally demanding (in terms of computer storage and floating-point operations), and an iterative solution is usually implemented, as discussed in Section 10.6. Iteration is also required when the image reconstruction problem is nonlinear, as in scatter imaging. These aspects of solution of a matrix-based system are presented here, starting with linear problems.

10.1 Error Propagation

The norm, $\|\boldsymbol{A}\|$, of a linear (solution-independent) matrix, $\boldsymbol{A}$, defined according to Eq. (7.6), provides a measure of the magnification of $\boldsymbol{c}$, as it is mapped to $\boldsymbol{e}$ by $\boldsymbol{A}\boldsymbol{c} = \boldsymbol{e}$. Therefore, $\|\boldsymbol{A}\|$ also measures the magnification of a change, $\delta\boldsymbol{c}$, in $\boldsymbol{c}$ in the computation: $\boldsymbol{A}\,\delta\boldsymbol{c} = \delta\boldsymbol{e}$. Similarly, the magnification of, hence susceptibility to, error of the matrix-inversion problem of Eq. (8.1): $\boldsymbol{A}^{-1}\boldsymbol{e} = \boldsymbol{c}$, is measured by:

$$\|\boldsymbol{A}^{-1}\| = \sup \frac{\|\boldsymbol{A}^{-1}\delta\boldsymbol{e}\|}{\|\delta\boldsymbol{e}\|}; \quad \delta\boldsymbol{e} \neq 0 \tag{10.1}$$

Accordingly,

$$\mathcal{N} = \|\boldsymbol{A}^{-1}\|\,\|\boldsymbol{A}\| = \sup \frac{\|\boldsymbol{A}^{-1}\delta\boldsymbol{e}\|}{\|\delta\boldsymbol{e}\|} \times \sup \frac{\|\boldsymbol{A}\boldsymbol{c}\|}{\|\boldsymbol{c}\|}; \quad \boldsymbol{c} \neq 0; \quad \delta\boldsymbol{e} \neq 0, \tag{10.2}$$

Computed Radiation Imaging. DOI: 10.1016/B978-0-12-387777-2.00010-0

is a measure of the maximum amount a relative error, $\left|\frac{\delta e}{e}\right|$, in $\boldsymbol{e}$, is magnified into a relative error, $\left|\frac{\delta c}{c}\right|$, in $\boldsymbol{c}$. Note here that vector division implies element-by-element division.

The norm of matrix $\boldsymbol{A}$ is related to its eigenvalues, λ's, obtained from $\boldsymbol{A}\boldsymbol{v} = \lambda \boldsymbol{v}$, where $\boldsymbol{v}$'s are eigenvectors, by the fact that:

$$\|\boldsymbol{A}\| \geq \sup \frac{\|\boldsymbol{A}\boldsymbol{v}\|}{\|\boldsymbol{v}\|} = \sup \frac{|\lambda| \; \|\boldsymbol{v}\|}{\|\boldsymbol{v}\|} = \max |\lambda| = \rho(\boldsymbol{A}); \quad \boldsymbol{v} \neq 0 \tag{10.3}$$

where $\rho(\boldsymbol{A})$, being the largest value of $|\lambda|$, is known as the spectral radius. The maximum of the absolute of the eigenvalues of $\boldsymbol{A}$ is equal to the minimum of the absolute of the eigenvalues of $\boldsymbol{A}^{-1}$, and vice versa. Therefore, $\mathcal{N}$ in Eq. (10.2), as defined by Eq. (7.17), is the condition number of the system matrix, $\boldsymbol{A}$. In other words, error propagation in a matrix-based system is determined by the ratio between the largest and smallest magnitudes of the eigenvalues of the system matrix. A large value of $\mathcal{N}$ indicates an ill-conditioned matrix. If the smallest magnitude of the eigenvalue is zero, the matrix is singular and error will propagate uncontrollably. Singularity can be avoided by having $\boldsymbol{A}$ as a full-rank matrix, which can be accomplished by avoiding duplicate or parallel (providing essentially same information) measurements, and by ensuring that all measurements are independent from each other. However, even a strictly full-rank matrix can appear to be rank-deficient, if noise causes measurements to overlap in magnitude. On the other hand, small changes introduced by noise can make otherwise parallel measurements appear to be independent, leading to singularity. Adding additional measurements, even though they may not be completely independent from previously acquired ones, can overcome the effect of noise, and help bring the matrix to a rank-sufficient status by the additional and affirmative information they bring. Round-off error also has a similar effect on the rank of a matrix. Therefore, in assessing a matrix vulnerability to noise and error, one should rely on the condition number of the matrix, not its rank.

10.2 Singular Value Decomposition

The singular value decomposition (SVD) of a linear matrix is a useful tool, not only in analyzing the basic features of a matrix, but also in inverting a matrix since the calculation of the singular values[1] is highly conditioned. This is accomplished by positioning the singular values, s_i's, of $\boldsymbol{A}\boldsymbol{A}^{\mathrm{T}}$, arranged in a nonincreasing magnitude, into the diagonal of a matrix $\boldsymbol{S}$. Then, $\boldsymbol{A}$ is decomposed to:

$$\boldsymbol{A} = \boldsymbol{U}\boldsymbol{S}\boldsymbol{V}^{\mathrm{T}} \tag{10.4}$$

[1] Singular values are the square roots of the eigenvalues of the square matrix $\boldsymbol{A}\boldsymbol{A}^{\mathrm{T}}$, where $\boldsymbol{A}^{\mathrm{T}}$ is the transpose of matrix $\boldsymbol{A}$. Note that $\boldsymbol{A}^{\mathrm{T}}\boldsymbol{A}$ has also the same singular values as $\boldsymbol{A}\boldsymbol{A}^{\mathrm{T}}$. The number of singular values is determined by the rank of the matrix, which is the same for the two square matrices $\boldsymbol{A}\boldsymbol{A}^{\mathrm{T}}$ and $\boldsymbol{A}^{\mathrm{T}}\boldsymbol{A}$.

where $\boldsymbol{U}$ is an orthogonal matrix whose columns are formed by the eigenvectors, $\boldsymbol{u}_i$'s of $\boldsymbol{A}\boldsymbol{A}^{\mathrm{T}}$, and $\boldsymbol{V}$ is an orthogonal matrix consisting of the eigenvectors, $\boldsymbol{v}_i$'s, of $\boldsymbol{A}^{\mathrm{T}}\boldsymbol{A}$. The matrices $\boldsymbol{U}$ and $\boldsymbol{V}$, being orthogonal (also orthonormal), have the properties:

$$\begin{aligned} &\boldsymbol{u}_i^{\mathrm{T}}\boldsymbol{u}_j = \delta_{ij}, && \boldsymbol{v}_i^{\mathrm{T}}\boldsymbol{v}_j = \delta_{ij}, \\ &\boldsymbol{U}\boldsymbol{U}^{\mathrm{T}} = \boldsymbol{I}, && \boldsymbol{V}\boldsymbol{V}^{\mathrm{T}} = \boldsymbol{I}, \\ &\boldsymbol{A}\boldsymbol{v}_i = s_i\boldsymbol{u}_i, && \boldsymbol{A}^{\mathrm{T}}\boldsymbol{u}_i = s_i\boldsymbol{v}_i \end{aligned} \tag{10.5}$$

where $\boldsymbol{I}$ is the identity matrix, and δ_{ij} is the Kronecker delta (= 1 when $i = j$, otherwise = 0). With $\boldsymbol{A}$ being an $M \times N$ matrix, then when $M \geq N$, $\boldsymbol{U}$ is an $M \times M$ matrix, $\boldsymbol{S}$ is $M \times M$, and $\boldsymbol{V}$ is $N \times M$ ($\boldsymbol{V}^{\mathrm{T}}$ is $M \times N$), while when $M < N$, $\boldsymbol{U}$ becomes an $M \times N$ matrix, $\boldsymbol{S}$: $N \times N$, and $\boldsymbol{V}$: $N \times N$. Note that there are other ways for formulating a SVD, but they all aim at the same goal of decomposing the original matrix in terms of its singular values.

With the aid of the formulation of Eqs. (10.4) and (10.5), the solution of $\boldsymbol{A}\boldsymbol{c} = \boldsymbol{e}$ can be expressed as:

$$\hat{\boldsymbol{c}} = \boldsymbol{A}^{\dagger}\hat{\boldsymbol{e}} = \boldsymbol{V}\boldsymbol{S}^{\dagger}\boldsymbol{U}^{\mathrm{T}}\hat{\boldsymbol{e}} \tag{10.6}$$

where $\boldsymbol{S}^{\dagger}$ is a diagonal matrix whose elements are the reciprocal of the corresponding diagonal elements of $\boldsymbol{S}$; except when the elements of the latter are zero or very close to zero where the elements of $\boldsymbol{S}^{\dagger}$ are equated to those of $\boldsymbol{S}$. When $\boldsymbol{A}$ is not a square matrix, then the inversion, $\boldsymbol{A}^{\dagger}$, given Eq. (10.6), is a pseudoinverse (also called a generalized or Moore-Penrose inverse), but it assures that $\boldsymbol{A}\boldsymbol{A}^{\dagger}\boldsymbol{A} = \boldsymbol{A}$, $\boldsymbol{A}^{\dagger}\boldsymbol{A}\boldsymbol{A}^{\dagger} = \boldsymbol{A}^{\dagger}$, $(\boldsymbol{A}\boldsymbol{A}^{\dagger})^{\mathrm{T}} = \boldsymbol{A}\boldsymbol{A}^{\dagger}$, and $(\boldsymbol{A}^{\dagger}\boldsymbol{A})^{\mathrm{T}} = \boldsymbol{A}^{\dagger}\boldsymbol{A}$. Note that with Eq. (10.6), a solution for the inverse problem is found, even when the problem is underdetermined (i.e. $N < M$), also called an incomplete problem, see Chapter 14. The matrix $\boldsymbol{A}^{\dagger}$ is sometimes referred to as the influence matrix (Trujillo and Busby, 1997); but $\boldsymbol{A}\boldsymbol{A}^{\dagger}$ is also called the influence matrix (Vogel, 2002).

If the relationship between $\hat{\boldsymbol{e}}$ and $\boldsymbol{c}$ remains consistent, in spite of the noise, then the solution of Eq. (10.6) will satisfy the inverse mapping. If there is no consistency, then the solution given by Eq. (10.6) is the solution that provides a minimum residual error, i.e. when substituted back into the forward problem produces the closest matching to the measurements, i.e. $\min(\boldsymbol{A}\hat{\boldsymbol{c}} - \hat{\boldsymbol{e}})$. For a rank-deficient matrix of rank, k, one can obtain an approximate solution by considering only k singular values.

The main benefit of SVD is that it always produces a solution (never fails). It also provides direct calculation of all the eigenvalues, hence the condition number of $\boldsymbol{A}\boldsymbol{A}^{\mathrm{T}}$. This enables direct assessment of system design, and allows examination of the effect of design changes on the conditioning of the system. Also, SVD enables direct control of the error in the measurements, $\boldsymbol{e}$, by setting a cutoff value for the smallest eigenvalues in $\boldsymbol{S}$ that will not be inverted when calculating $\boldsymbol{S}^{\dagger}$. This, therefore, constitutes a form of solution regularization. The cutoff value is determined empirically, depending on the nature of the imaging system and the quality of recorded measurements. By trying different values, one can determine the appropriate cutoff value. Singular value decomposition is also useful in analyzing other regularization methods, as discussed in Section 10.4.

10.3 Least Squares

The solution $\hat{c} = A^{\dagger}e$ that minimizes the residual error when expressed by the Euclidean distance (norm), $\|Ac - \hat{e}\|_2$, is the least-squares solution of $Ac = \hat{e}$, since the minimizing the residual norm becomes equivalent to minimizing χ^2 of Eq. (8.2). The generalized inverse of the matrix A is then given by:

$$A^{\dagger} = [A^{T}A]^{-1}A^{T} \tag{10.7}$$

This solution can also be arrived at using probability theory with the maximum-likelihood principle, see Appendix 8.A. The least-squares solution of Eq. (8.3) is a weighted form of the solution provided by $A^{\dagger}\hat{e}$. The least-squares solution is a special case of that obtained with singular-value decomposition, but it is numerically easier to formulate. The effect of low singular values on increasing error propagation is also observed in the least-squares solution. While, the pseudoinversion by SVD enables direct elimination of low singular values, the same explicit control is not possible in the least-squares solution. This is achieved, however, by regularization as discussed below.

10.4 Regularization Methods

10.4.1 Approach

One approach to control the propagation of error in an ill-posed inverse problem is to replace the original problem, $A^{-1}\hat{e}$, with a problem that constrains error propagation, leading to a stable solution. This process is known as *regularization* and is accomplished by tolerating some residual error in the solution.[2] Regularization assists, therefore, in establishing a continuous relationship between $\hat{e}$ and c. It enables a solution to exist, and hopefully be unique for a given regularization arrangement.

Regularization aims at reaching a minimal (but not necessarily the absolute minimum) error measured by the residual norm (also known as the data fidelity norm):

$$\chi(c) = \|Ac - \hat{e}\|_2 \tag{10.8}$$

by imposing some constraints. The residual error is defined by the Euclidean norm, since the residual error is random and there is no reason to expect one random error to

[2] The process of discretizing a continuous mapping introduces a regularization effect by limiting the sharpness of gradients in the continuous problem. This is referred to as projection regularization, or discretization regularization. It does not, however, ensure control of the magnification of the round-off error introduced by the high condition number of the associated system matrix.

depend on another. Recall that minimizing $\chi^2(\boldsymbol{c})$ without any constraints is the least-squares minimization of Eq. (10.7). By imposing constraints on the minimization of $\chi^2(\boldsymbol{c})$, one in effect is attempting to "smooth" error magnification in the ill-posed problem, i.e. reduce error propagation at the expense of smoothing the image. Regularization is, therefore, often referred to as smoothing. In addition, by avoiding exact matching between the forward model, $\boldsymbol{Ac}$, and the measurements, $\hat{\boldsymbol{e}}$, one is also acknowledging the presence of modeling errors, which will exist even in the absence of measurement errors. In other words, regularization attempts to compensate for the mismatch between forward mapping and actual measurements, the so-called resolving error, caused by statistical variability and/or modeling error. Regularization provides a family of solutions (whose members are determined by the type and degree of regularization) that approach the error-free solution. Nashed (1981) provided a general mathematical framework for regularization for some of the approaches presented here.

In regularization, one of the following two forms of constraints can be employed (Hansen, 1998):

1. $\boldsymbol{c} \in \boldsymbol{S}(\boldsymbol{C})$, where $\boldsymbol{S}(\boldsymbol{C}) \subseteq \boldsymbol{C}$, i.e. restrict the solution to within a subset $\boldsymbol{S}(\boldsymbol{C})$ of the space of solutions, $\boldsymbol{C}$. This can be achieved for example by restricting a solution (density or attenuation coefficient) to be nonnegative and not to exceed some expected maximum value. Such confinement of solution is a form of regularization, that is also based on *a priori* physical knowledge of the range of the expected solution.
2. $\eta(\boldsymbol{c}) \leq \delta$, where $\eta(\boldsymbol{c})$ is some measure to which the size of the residual error, $\chi(\boldsymbol{c})$, is compared, and δ is a designated upper-bound of $\eta(\boldsymbol{c})$. Physically, this amounts to allowing the solution to produce corresponding modeled measurements, $\tilde{\boldsymbol{e}}$, that do not match exactly the actual measurements, $\boldsymbol{e}$. The latter measurements have measurement uncertainties associated with them, while the former ones are subject to modeling errors. Therefore, it is logical not to expect perfect matching between the two.

Alternatively, instead of minimizing $\chi^2(\boldsymbol{c})$ subject to constraints, one can (Hansen, 1998):

1. Minimize $\eta(\boldsymbol{c})$ with the constraint: $\chi^2(\boldsymbol{c}) \leq \gamma^2$, where γ^2 is a pre-assigned positive parameter.
2. Minimize a combination of $\chi(\boldsymbol{c})$ and $\eta(\boldsymbol{c})$, so that the mathematical desirability of minimizing the residual error is combined with the minimization of $\eta(\boldsymbol{c})$ in accordance to some physical constraints. Since both $\chi(\boldsymbol{c})$ and $\eta(\boldsymbol{c})$ are measures of "error", their squared values (representatives of variance) are added. Therefore, regularization is implemented by:

$$\min\left\{[\chi(\boldsymbol{c})]^2 + \alpha^2\left[\|\eta(\boldsymbol{c})\|_2\right]^2\right\} \tag{10.9}$$

where α^2 is a regularization parameter that controls the amount of weighting in the minimization process, given to $\eta(\boldsymbol{c})$ relative to $\chi(\boldsymbol{c})$. The regularization of (10.9) is known as Tikhonov regularization (Tikhonov, 1977)[3]. Notice that when $\alpha = 0$, the minimization of (10.9) resorts to basic least-squares, χ^2, minimization of Eq. (8.2). Therefore, Tikhonov regularization allows the residual error, $\chi(\boldsymbol{c})$, to reach only a "near" minimum value.

[3] Although the method is widely named after A.N. Tikhonov (1963), J.D. Riley (1955), D.L. Phillips (1962), S. Twomey (1963), and K. Miller (1970) introduced variants of this regularization approach.

The result is "smoothness" (at least smoother than the oscillations in the solution due to ill-conditioning) of the solution by adding a "roughness" penalty (a stabilization) term to the minimization of the residual error, $\chi(\mathbf{c})$. One can define $\eta(\mathbf{c})$ in a number of ways, to accommodate various regularization criteria and minimize:

$$\min\left\{[\|\chi(\mathbf{c})\|_2]^2+\alpha_1^2[\|\eta_1(\mathbf{c})\|_2]^2+\alpha_2^2[\|\eta_2(\mathbf{c})\|_2]^2+\cdots\right\} \tag{10.10}$$

where $\eta_1(\mathbf{c})^2$, $\eta_2(\mathbf{c})^2$, etc., designate various regularization functionals and α_1^2, α_2^2, etc., are the associated weights, i.e. regularization parameters.

The problem now is to determine which is the most suitable minimization process, to define $\eta(\mathbf{c})$ and find the optimum values for the parameters, δ, γ^2 or α^2. Let us start first by giving some insight into the mathematical significance of regularization.

10.4.2 Mathematical Significance

In order to arrive at a rationale for developing a regularization functional, let us revisit the SVD solution given by Eq. (10.6). Rewriting this solution by expanding the right-hand-side of Eq. (10.6):

$$\hat{\mathbf{c}}=\mathbf{A}^{\dagger}\hat{\mathbf{e}}=\mathbf{V}\mathbf{S}^{\dagger}\mathbf{U}^{\mathrm{T}}\mathbf{e}=\sum_i\frac{1}{s_i}\left(\mathbf{u}_i{}^{\mathrm{T}}\hat{\mathbf{e}}\right)\mathbf{v}_i \tag{10.11}$$

one has an expression showing that the division by a small singular value, s_i, leads to the magnification of any error in $\mathbf{e}$. Rather than excluding small singular values, as done in arriving at matrix $\mathbf{S}^{\dagger}$ in Eq. (10.6), one can introduce a filter, $g_\alpha(s_i)$, of the eigenvalues $\mathbf{A}\mathbf{A}^{\mathrm{T}}$ so that the solution becomes:

$$\hat{\mathbf{c}}=\sum_i g_\alpha(s_i^2)\frac{1}{s_i}\left(\mathbf{u}_i^{\mathrm{T}}\hat{\mathbf{e}}\right)\mathbf{v}_i \tag{10.12}$$

Arriving at the solution of Eq. (10.6) is equivalent to setting $g_\alpha(s_i)=1$, when $s_i>\alpha$, and otherwise zero. A less abrupt filter is given by:

$$g_\alpha(s_i)=\frac{s_i^2}{s_i^2+\alpha^2} \tag{10.13}$$

This is the Tikhonov filter, which smoothly dampens the effect of small eigenvalues, as it approaches zero when $s_i<<\alpha$, but retains the effect of large eigenvalues by being close to unity when $s_i>>\alpha$. It has the effect of shifting the spectrum of the eigenvalues away from singularity. This method of regularization is, therefore, sometimes referred to as spectral shifting. Substituting the filter of Eq. (10.13) in Eq. (10.12) leads to the solution:

$$\hat{\mathbf{c}}=\sum_i\frac{s_i}{s_i^2+\alpha^2}\left(\mathbf{u}_i^{\mathrm{T}}\hat{\mathbf{e}}\right)\mathbf{v}_i \tag{10.14}$$

Resorting back to the matrix form of the solution, Eq. (10.14) is expressed as:

$$\hat{\boldsymbol{c}} = \boldsymbol{V}\left[\boldsymbol{S}^2 + \alpha^2 \boldsymbol{I}\right]^{-1} \boldsymbol{S}\boldsymbol{U}^{\mathrm{T}}\hat{\boldsymbol{e}} = \left[\boldsymbol{A}^{\mathrm{T}}\boldsymbol{A} + \alpha^2 \boldsymbol{I}\right]^{-1} \boldsymbol{A}^{\mathrm{T}}\hat{\boldsymbol{e}} \tag{10.15}$$

Let us now rewrite the regularization condition of (10.9), making use of Eq. (10.8) and defining $\eta(\boldsymbol{c}) = \boldsymbol{G}\boldsymbol{c}$, so that:

$$\min\left\{[\boldsymbol{A}\boldsymbol{c} - \hat{\boldsymbol{e}}]^2 + \alpha^2[\boldsymbol{G}\boldsymbol{c}]^2\right\} \tag{10.16}$$

Minimization with respect to $\boldsymbol{c}$ yields the solution:

$$\hat{\boldsymbol{c}} = \left[\boldsymbol{A}^{\mathrm{T}}\boldsymbol{A} + \alpha^2 \boldsymbol{G}^2\right]^{-1} \boldsymbol{A}^{\mathrm{T}}\hat{\boldsymbol{e}} \tag{10.17}$$

The solution of Eq. (10.17) requires that the null spaces of $\boldsymbol{A}$ and $\boldsymbol{G}$ not intersect (or interest trivially), i.e. $\mathcal{N}(\boldsymbol{A}) \bigcap \mathcal{N}(\boldsymbol{G}) = 0$, where $\mathcal{N}$ refers to null space,[4] i.e. the space of images that corresponds to zero measurements. This condition is required so that the inverted matrix, $\boldsymbol{A}^{\mathrm{T}}\boldsymbol{A} + \alpha^2\boldsymbol{G}^2$, is never an empty singular matrix. The same condition also implies $\boldsymbol{A}$ has a full rank.[5] This addresses the question of existence of the regularized solution of Eq. (10.17). The linearity of the problem, $\boldsymbol{A}\boldsymbol{C} = \boldsymbol{e}$, assures the solution's uniqueness, for a given regularization functional, since no two solutions will produce exactly the same set of measurements (unless they are so close that they overlap within the range of measurement uncertainties).

Comparing Eq. (10.17) to Eq. (10.15) shows that $\boldsymbol{G} = \boldsymbol{I}$. Therefore, applying a regularization functional $\eta(\boldsymbol{c}) = \boldsymbol{c}$ (i.e. equal to the solution itself) is equivalent to filtering singular values by the Tikhonov filter, Eq. (10.13), without performing SVD of the matrix $\boldsymbol{A}\boldsymbol{A}^{\mathrm{A}}$. Consequently, one can state that in the solution given by Eq. (10.17), the amount of error propagation is controlled via the regularization parameter, α^2. When α^2 is very small, error magnification will be quite high, since small singular values would still be implicitly present in the solution. On the other hand, with a large α^2 value, the solution can be over-smoothed by losing the contribution of higher singular values. An optimum value of α^2 is, therefore, desired. Methods for finding such optimal values are discussed in Section 10.5.

The advantage of Tikhonov regularization with $\boldsymbol{G} = \boldsymbol{I}$, is that it aims at minimizing the magnitude of the solution, $\boldsymbol{c}$, itself. This stabilizes the solution by reducing its variability, preventing any severe oscillations produced by the ill-conditioning of the system matrix, $\boldsymbol{A}$. Also, this approach of regularization assumes no *a priori* knowledge of the nature of the solution. It risks, however, the removal of genuine variations in $\boldsymbol{c}$, in the process of stabilizing the solution.

The regularization functional $\eta(\boldsymbol{c}) = \boldsymbol{c}$, i.e. with $\boldsymbol{G} = \boldsymbol{I}$, measures the size of the residual error, $\chi(\boldsymbol{c})$, in terms of the solution, $\boldsymbol{c}$, with the regularization parameter, α^2,

[4] $\mathcal{N}(\boldsymbol{A})$ is the space of all vectors $\boldsymbol{c}$ such that $\boldsymbol{A}\boldsymbol{c} = 0$.

[5] Recall that the rank of a matrix is the number of its nonzero singular values, and corresponds to the number of linearly independent rows or columns of the matrix.

determining how close $\chi^2(\boldsymbol{c})$ is allowed to reach its minimum value. In other words, the minimization of the residual error, χ, is constrained by $\eta(\boldsymbol{c})$, which also constrains the solution itself. Therefore, generally, the regularization functional, $\eta(\boldsymbol{c})$, is expressed in terms of the solution, $\boldsymbol{c}$, so that:

$$\eta(\boldsymbol{c}) = \boldsymbol{G}\boldsymbol{c} \tag{10.18}$$

where $\boldsymbol{G}$ is an $M \times N$ matrix, called the regularization matrix. Note that $\alpha\boldsymbol{G}$ has same physical units as $\boldsymbol{A}$. The matrix $\boldsymbol{G}$ can take a number of forms, the standard being $\boldsymbol{G} = \boldsymbol{I}$. Other forms for choosing $\boldsymbol{G}$ are discussed later in this section, after introducing a different perspective to regularization, by considering the minimization problems of (10.9) as a constrained optimization problem.

10.4.3 An Optimization Perspective

Let us consider the minimization of (10.9) as an optimization problem, of two functionals: one for the residual error and the second of the regularization functional. Let us first formulate a functional for the residual error that takes into account the fact that in radiation measurements *a priori* estimate of the statistical variance of each measurement is given by the Poisson statistics of radiation counting. Let σ_i^2 be the variance associated with each measurement, e_i. Since one has more confidence in measurements of low variance than those of high variance, one should give more influence to the former measurements because of their good statistics. Let us introduce this preference via a diagonal, $M \times M$, statistical weight matrix, $\boldsymbol{W}$, with elements, W_{ij}, defined as:

$$W_{ij} = \frac{\delta_{ij}}{\sigma_j} \tag{10.19}$$

where δ_{ij} is the Kronecker delta. The function for the residual error can now be defined (and constrained) as:

$$\left[\boldsymbol{W}\left(\boldsymbol{A}\boldsymbol{c} - \hat{\boldsymbol{e}}\right)\right]^{\mathrm{T}}\left[\boldsymbol{W}\left(\boldsymbol{A}\boldsymbol{c} - \hat{\boldsymbol{e}}\right)\right] \leq \delta_\chi^2 \tag{10.20}$$

where δ_χ is an upper limit of the weighted residual error. Similarly, one can restrict the regularization functional of Eq. (10.18) by:

$$(\boldsymbol{G}\boldsymbol{c})^{\mathrm{T}}(\boldsymbol{G}\boldsymbol{c}) \leq \delta_\eta^2 \tag{10.21}$$

The two constraints of (10.20) and (10.21) can be combined into a single cost function (also called the objective function), $\Phi(\boldsymbol{c})$, weighted by the amount of imposed constraints, so that:

$$\Phi(\boldsymbol{c}) = \delta_\chi^2 \times \left[\boldsymbol{W}\left(\boldsymbol{A}\boldsymbol{c} - \hat{\boldsymbol{e}}\right)\right]^{\mathrm{T}}\left[\boldsymbol{W}\left(\boldsymbol{A}\boldsymbol{c} - \hat{\boldsymbol{e}}\right)\right] + \delta_\eta^2 \times (\boldsymbol{G}\boldsymbol{c})^{\mathrm{T}}(\boldsymbol{G}\boldsymbol{c}) \tag{10.22}$$

The minimization of $\Phi(\boldsymbol{c})$ is equivalent to the minimization of (10.9) with $\chi^2(\boldsymbol{c})$ and $\eta^2(\boldsymbol{c})$ defined by the left-hand-sides of Eqs. (10.20) and (10.21), respectively, and $\alpha^2 = \frac{\delta_\eta^2}{\delta_\chi^2}$. Therefore, regularization can be viewed as a relative constraining of the minimization of the residual error, in accordance to some functional defined by the regularization matrix, $\boldsymbol{G}$. The optimization parameter, α^2 is then a measure of the amount of regularization, or smoothing, measured relative to the upper-bound of the residual error.

Minimizing the objective function of Eq. (10.22) yields:

$$\hat{\boldsymbol{c}} = \left[\boldsymbol{A}^{\mathrm{T}}\boldsymbol{W}^2\boldsymbol{A} + \alpha^2\boldsymbol{G}^2\right]^{-1}\boldsymbol{A}^{\mathrm{T}}\boldsymbol{W}^2\hat{\boldsymbol{e}} \tag{10.23}$$

Note that, in the above discussion, a single regularization parameter, α^2, is applied in formulating the regularization functional. This parameter can be made to affect some voxels, or a region-of-interest composed of a cluster of voxels, by replacing α^2 with a diagonal matrix in which an element, α_{ii}^2, represents the regularization effect on the physical parameter, c_i, of voxel i.

In summary, one can state that the purpose of regularization is to replace the originally ill-posed inverse problem by a well-posed problem in its proximity. This is achieved by eliminating low singular values, or equivalently constraining the solution of the inverse problem by adding a regularization functional to the original problem. The effect is a smoothing of the ill-posed inverse problem. There are many possible approximations to the original inverse problem, depending on the choice of the regularization functional and the associated regularization problem. Some *a priori* knowledge of the solution can help in determining the desired nature of regularization, as discussed below.

10.4.4 Minimum Information

The solution given by the standard Tikhonov regularization of (10.16) with $\boldsymbol{G} = \boldsymbol{I}$, is sometimes referred to as the "minimum information solution".[6] This is because with $\boldsymbol{G} = \boldsymbol{I}$, no *a priori* information is involved in arriving at the solution. The lack of correlation in Tikhonov regularization with $\boldsymbol{G} = \boldsymbol{I}$ also implies a lack of interaction between adjacent regions within the imaged medium, or a minimum-energy.[7] Tikhonov regularization is also called quadratic regularization, for the reasons given in Section 10.4.11. Statistically, Tikhonov regularization is equivalent to ridge regression, see Section 10.4.8. Long before Tikhonov regularization, Rutishauser (1968) called the solution obtained by this method, the relaxed solution of the original least-squares (residual minimization) problem of (8.2), since it is a near-minimization of the residual.

[6] It is also called the Occam approach (Constable et al., 1987), after the fourteenth century philosopher who advocated that reasoning should be as simple as possible.

[7] The energy of $\eta(\boldsymbol{c})$ is equal to the square of its Euclidean norm (Bertero and Boccacci, 1998), $\|\eta(\boldsymbol{c})\|_2^2$.

10.4.5 Doubly Relaxed

Rutishauser (1968) proposed a doubly-relaxed solution that further helps overcome the ill-posedness of the problem, accomplished by:

$$\min\left\{\left[(\boldsymbol{A}^{\mathrm{T}}\boldsymbol{W}^2\boldsymbol{A}+\alpha\boldsymbol{G})\boldsymbol{c}-\boldsymbol{A}^{\mathrm{T}}\boldsymbol{W}^2\hat{\boldsymbol{e}}\right]^2+\alpha^2[\boldsymbol{G}\boldsymbol{c}]^2\right\} \tag{10.24}$$

This so-called Rutishauser regularization is a relaxation of the solution obtained by the already relaxed Tikhonov regularized solution of Eq. (10.16), weighted with $\boldsymbol{W}$. Equation (10.24) leads to the solution:

$$\hat{\boldsymbol{c}}=\left[\boldsymbol{A}^{\mathrm{T}}\boldsymbol{W}^2\boldsymbol{A}+\alpha^2\boldsymbol{G}^2+\alpha^2\left(\boldsymbol{A}^{\mathrm{T}}\boldsymbol{W}^2\boldsymbol{A}+\alpha^2\boldsymbol{G}^2\right)^{-1}\right]^{-1}\boldsymbol{A}^{\mathrm{T}}\boldsymbol{W}^2\hat{\boldsymbol{e}} \tag{10.25}$$

The regularization matrix, $\boldsymbol{G}$, hence the regularization functional, $\eta(\boldsymbol{c})$, can be formulated in other ways to take advantage of *a priori* information, or to guide the solution toward desired features. Some of these formulations are presented below. The weight matrix, $\boldsymbol{W}$, of Eq. (10.19) is employed in subsequent formulations, since statistical estimates of the variance of radiation counting measurements can be readily made; if not one should set $\boldsymbol{W}=\boldsymbol{I}$. Note also that the formulations below can be combined according to (10.10).

10.4.6 Estimate of Solution

When some *a priori* estimate of the solution is known, say $\boldsymbol{c}^*$, then the regularization functional can be defined as (Hansen, 1998):

$$\eta(\boldsymbol{c})=\boldsymbol{G}\left(\boldsymbol{c}-\boldsymbol{c}^*\right) \tag{10.26}$$

Leading to the solution:

$$\hat{\boldsymbol{c}}=\left[\boldsymbol{A}^{\mathrm{T}}\boldsymbol{W}^2\boldsymbol{A}+\alpha^2\boldsymbol{G}^2\right]^{-1}\left(\boldsymbol{A}^{\mathrm{T}}\boldsymbol{W}^2\hat{\boldsymbol{e}}+\alpha^2\boldsymbol{G}^2\boldsymbol{c}^*\right) \tag{10.27}$$

This solution is obviously guided toward $\boldsymbol{c}^*$, by minimization of the difference between the solution and its initial estimate. This is useful when the purpose of imaging is to detect the presence of anomalies in an object with a known structure. Note that in Eq. (10.27), when $\alpha=0$, one would obtain a solution that provides fidelity to measurements, while $\alpha=\infty$ would lead to best matching with the *a priori* solution estimate.

10.4.7 Gradient of Solution

Some information about the image to be reconstructed is often known. For example, in many cases the object's physical attributes tend to vary between clusters of voxels, rather from voxel to voxel. Within each cluster, a voxel's attributes tend to be similar to that of its neighbors (cliques); i.e. a voxel tends to resemble its neighbors. Such trends can be incorporated into the solution as *a priori* knowledge, by defining the regularization matrix as a correlation matrix relating one physical attribute of a voxel to those of its neighbors.

The smoothness of a solution, $\boldsymbol{c}$, can be controlled by a regularization functional that incorporates a first or second spatial derivative of $\boldsymbol{c}$. This is equivalent to applying a "low-pass" filter to the solution that minimizes any oscillatory features in the image. The first derivative is employed to maximize flatness, while the second derivative minimizes roughness. Since no spatial relationship is explicitly present in the formulation of $\boldsymbol{c}$, though each element of $\boldsymbol{c}$ corresponds to a point in the spatial space, the spatial derivatives (gradients) are formulated by difference equations, assuming unit spatial intervals (unless image pixels/voxels vary in size within the same image). Also, in formulating these gradients, the spatial location of a voxel relative to its neighbors should be taken into account, since $\boldsymbol{c}$ is considered here as a one-dimensional array while images encompass more than one direction. We assume here that $\boldsymbol{c}$ is structured over a Cartesian mesh, such that its first N_x voxels correspond to the first set of elements in the $\vec{x}$ direction, the second N_x elements to the second set of voxels, and so on to the last set being N_y, so that the second N_xN_y voxels correspond to the second layer of voxels, up to the N_z^{th} layer, where $N = N_xN_yN_z$.

A regularization functional formulated in terms of the first spatial Cartesian derivative of $\boldsymbol{c}$ is suited when a nearly uniform image is anticipated. The regularization functional is then expressed as:

$$\eta(\boldsymbol{c}) = \boldsymbol{Gc}$$
$$G_{ij} = \begin{cases} -1 & \text{for } i=j \\ 1 & \text{for } j = \begin{cases} i+1 & \text{for gradient in } x \\ i+N_x & \text{for gradient in } y \\ i+N_xN_y & \text{for gradient in } z \end{cases} \\ 0 & \text{otherwise} \end{cases} \tag{10.28}$$

The gradients in more than one direction can be incorporated by including more than one regularization functional in the form of Eq. (10.28) in accordance to (10.10).

When one expects a linear variation in the solution, $\boldsymbol{c}$, from a voxel to its neighbor, the regularization functional with the second spatial derivative (Laplacian operator) should be used:

$$\eta(\boldsymbol{c}) = \boldsymbol{Gc}$$
$$G_{ij} = \begin{cases} -4 & \text{for } i=j \\ 1 & \text{for } j = \begin{cases} i \pm 1, i \pm N_x & \text{for Laplacian in } x,y \\ i \pm 1, i = N_xN_y & \text{for Laplacian in } x,z \\ i \pm N_x, i = N_xN_y & \text{for Laplacian in } y,z \end{cases} \\ 0 & \text{otherwise} \end{cases} \tag{10.29}$$

This regularization functional, also known as Phillips regularization, attempts to reach a solution with minimum curvature in the distribution of the image attributes

(Bertero and Boccacci, 1998). It tends, therefore, to enhance the edges of abrupt changes between regions within the image, that would have otherwise been smeared out by a first-gradient smoothing, or would not have been obvious without smoothing.

These regularization functionals can be combined in accordance to Eq. (10.10). The spatial-gradient form of regularization can also be further extended to higher order derivatives. Gradient-based regularization acts as a bounding constraint, by limiting the amount of change from one voxel to its neighbors. As such, while third-order, or higher, gradients will not be effective in smoothing an image, it will prevent the solution from reaching unrealistically high values caused by problem ill-posedness. However, regularization with gradient functionals often tends to produce a "blocky" effect, or spurious resolution, i.e. artificial details. This is because of its tendency to emphasize change, whether such change is genuine or an artifact.

One can apply different gradient schemes to different regions of interest. If a region is known to be nearly homogeneous, a first-order gradient regularization can be applied within this region, while a Laplacian gradient can be applied in its immediate neighboring regions if a gradual change is expected from the region of interest to adjacent regions. Region-of-interest regularization is applied by restricting the regularization functional to the elements of solution, $\boldsymbol{c}$, that correspond to the region of interest, while assigning zero or much smaller regularization functionals in other regions. Care, however, must be taken to ensure that $[\boldsymbol{G}^{\mathrm{T}}\boldsymbol{G}]^{-1}$ is bounded, otherwise the whole exercise of regularization would be futile.

10.4.8 Covariance Matrix

Another interpretation of the regularization functional, $\eta(\boldsymbol{c})$, can be arrived at if one examines the weighted solution of Eq. (10.23) for a forward self-mapping (or mapping of $\boldsymbol{c}$ to its dual), i.e., with $\boldsymbol{A} = \boldsymbol{I}$. Then, given that $(\boldsymbol{W}^2)^{-1}$ is a covariance matrix of the measurements, $\boldsymbol{e}$ (with zero off-diagonal elements due to the assumed independence of measurements), one can consider $(\boldsymbol{G}^2)^{-1}$ to be also a covariance matrix. Given that $\boldsymbol{G}$ operates on the solution $\boldsymbol{c}$ as (10.22) shows, $(\boldsymbol{G}^2)^{-1}$ can be formulated as the covariance matrix of the solution. Statistical variability in the solution can be due to the physical nature of the imaged object itself, e.g. a concrete medium that may appear to be uniform, but its density varies randomly around a mean value. Such *a priori* information about the nature of the imaged object can be accommodated in the regularization process via a covariance (correlation) matrix that can be determined in advance (Tarantola, 1987). This allows the establishment of a regularization functional that incorporates the regularization matrix as a weight matrix of the solution, similar to using $\boldsymbol{W}$ in Eq. (10.23) as a weight matrix for measurements. Assuming a Gaussian (normal) distribution of the image features, one can express such regularization functionals as:

$$\begin{aligned} &\eta(\boldsymbol{c}) = \boldsymbol{G}\boldsymbol{c} \\ &G_{ij}^{-2} = \mathrm{cov}(c_i^*, c_j^*) \end{aligned} \tag{10.30}$$

where the G_{ij}^{-2} is an element of the covariance matrix, $\left(\boldsymbol{G}^2\right)^{-1}$ and "cov" refers to the covariance of the respective argument. The covariance matrix of $\boldsymbol{c}$ is to be determined *a priori* for a reference (datum) object assumed to be typical of the objects for which an image is to be reconstructed. The Gaussian formulation of Eq. (10.30) is reasonable when the error in the reconstructed image is expected to be a random variable that follows a normal distribution (the value of c_i is random and is independent from any other element of $\boldsymbol{c}$).

Obviously if there is no correlation (structural cohesion) between any two points in the image, the off-diagonal ($i \neq j$) elements will be zero, and the covariance matrix will only be constituted of diagonal elements, each corresponding to the variance of a c_i value. If all c_i's values have the same variance, say σ_c^2, one would have $\left(\boldsymbol{G}^2\right)^{-1} = \sigma_x^2 \boldsymbol{I}$. Then, one resorts back to the usual Tikhonov regularization, with $G = I$ (which is known in statistical terms as ridge regression), with the regularization parameter becoming equal to $\frac{\alpha^2}{\sigma_c^2}$. As such, α^2 can be statistically viewed as an adjuster of the spread of the variance of the solution parameters in an uncorrelated image, i.e. it sets an upper-bound on the solution. In fact, when $\sigma_c^2 \longrightarrow \infty$, or equivalently $\alpha^2 \longrightarrow 0$, the image parameter is considered to be contaminated with white noise (infinite variance, null correlation) (Tarantola, 1987).

The formulation of Eq. (10.30) allows the imposition of "soft" bounds on the solution, $\hat{\boldsymbol{c}}$. This can be easily seen by considering an uncorrelated object, where only the diagonal elements of the covariance elements are nonzero. Then the diagonal elements will act as regulators of their corresponding c_i values. More precisely, $\frac{\alpha^2}{\sigma_{c_i}^2}$, where $\sigma_{c_i}^2$ is the variance associated with c_i, becomes an element-by-element (at various values of c_i) regularization parameter. In other words, the statistical smoothing of Eq. (10.30) is an element-by-element form of Tikhonov regularization. This feature can be used in region-of-interest imaging, to deliberately emphasize the importance of a region over others, by assigning it an artificially diminished variance.

10.4.9 Spatial Correlation

Statistical *a priori* information can also be incorporated into the solution through a regularization functional that involves a correlation matrix relating one voxel to its neighbors.

One can formulate the regularization matrix and associated regularization functional, $\boldsymbol{G}$, so that (Jackson, 1979):

$$\begin{aligned} \eta(\boldsymbol{c}) &= \boldsymbol{G}\boldsymbol{c} \\ G_{ij}^{-2} &= \sigma_i \sigma_j \exp\left[-\frac{|\vec{r}_i - \vec{r}_j|^2}{2L^2}\right] \end{aligned} \tag{10.31}$$

where σ_i^2 designates a predetermined variance, $\vec{r}_i$ is the position of voxel i, and L is the correlation length, or smoothing distance (length of neighborhood to which the regularization is to be applied). Then, the regularization parameter, α^2, in the solution of Eq. (10.23), determines the degree of influence this statistical regularization has on the solution. In the formulation of Eq. (10.31), one value, σ^2, can be used for all voxels, instead of $\sigma_i\sigma_j$ (Tarantola, 1987).

The correlation of Eq. (10.31) is designed to capture the homogeneity or near homogeneity of an image, but it can be selectively applied to regions where the image structure is expected to be uniform. Other formulation for $\boldsymbol{G}$, as a covariance matrix, can be employed. For example, by using $|\vec{r}_i - \vec{r}_j|/L$, rather than $|\vec{r}_i - \vec{r}_j|^2/L^2$ in Eq. (10.31), one obtains an exponential covariance matrix with fractals, i.e. a repetition of the inherent features of the image upon zooming into a portion of the image (Tarantola, 1987).

When it is known that each voxel in the image is distinct from its immediate neighbors, the voxel is called "self-similar" (similar only to parts of itself), then the von Karman covariance function should be applied in the neighborhood (Tarantola, 1987):

$$W_{r_{ij}} = K_0\left(\frac{|\vec{r}_i - \vec{r}_j|}{L}\right) \tag{10.32}$$

where K_0 is the modified Bessel function[8] of the second kind of order zero, and L is the correlation length. Taking the zeroth order of the Bessel function ensures the distinctiveness of a voxel from its neighbors. It should be stated that L, in Eqs. (10.31) and (10.32), can vary from voxel to voxel, allowing the definition of some regions-of-interest.

10.4.10 Modeling-Error Compensation

The minimization of the residual-error functional of Eq. (10.8) presumes that the difference between the actual measurements, $\hat{\boldsymbol{e}}$, and the corresponding computed ones, $\boldsymbol{Ac}$, is within the range of the uncertainties in $\hat{\boldsymbol{e}}$. In practice, as indicated in Part I of this book, the forward model is an approximation. As such, the error expressed in Eq. (10.8) by $\|\boldsymbol{Ac} - \hat{\boldsymbol{e}}\|$ is not only a residual error, in the numerical sense, but also includes a modeling error. However, the modeling error is not random, and is in general an additive error, since it is the result of ignoring effects that are complicated to model (such as scattering in transmission and emission imaging and multiple scattering in scatter imaging). The modeling error can be estimated using a reference object as:

$$\boldsymbol{B}^* = \boldsymbol{e}^* - \boldsymbol{Ac}^* \tag{10.33}$$

where it is assumed that the reference measurements, $\boldsymbol{e}^*$, are acquired with low statistical variability, for an object with known physical attributes, $\boldsymbol{c}^*$. Then, as

[8] The modified Bessel function is given by:

$$K_n(z) = \frac{1}{2}\left(\frac{z}{2}\right)^{-n}\sum_{k=0}^{n-1}\frac{(n-k-1)!}{k!}\left(-\frac{z^2}{4}\right)^k + (-1)^{n+1}\ln\left(\frac{z}{2}\right)$$

Eq. (10.33) indicates, the difference between the actual measurements and those modeled by the forward model gives the modeling error, $\boldsymbol{B}^*$ for the reference object. For an object with unknown features, but not expected to be very different from the reference object, the modeling error can be estimated as $\beta\boldsymbol{B}^*$, with $\beta \geq 0$ being an adjustment parameter that can be estimated *a priori*, perhaps by comparing $\|\boldsymbol{e}\|$ for the imaged object to $\|\boldsymbol{e}\|^*$ of the reference object (Arendtsz and Hussein, 1995a). Then, the functional of Eq. (10.8) can be replaced by:

$$\chi(\boldsymbol{c}) = \|\boldsymbol{A}\boldsymbol{c} + \beta\boldsymbol{B}^* - \hat{\boldsymbol{e}}\|_2 \tag{10.34}$$

This becomes the data fidelity norm (actual residual) error, which can be vulnerable to the random variability in $\hat{\boldsymbol{e}}$, as well as to uncertainties in determining β. One can then devise a regularization function to moderate the effect of modeling errors as:

$$\eta(\boldsymbol{c}) = \boldsymbol{B} = \hat{\boldsymbol{e}} - \boldsymbol{A}\boldsymbol{c} \tag{10.35}$$

where $\boldsymbol{B}$ is the unknown modeling error for the problem at hand. Note that the regularization functional now is the residual error of Eq. (10.35), if the modeling error is not taken into account, while the residual error, Eq. (10.34), incorporates some estimate of the modeling error. One now needs to minimize the objective function:

$$\min\left\{[\boldsymbol{W}(\boldsymbol{A}\boldsymbol{c} + \beta\boldsymbol{B}^* - \hat{\boldsymbol{e}})]^{\mathrm{T}}[\boldsymbol{W}(\boldsymbol{A}\boldsymbol{c} + \beta\boldsymbol{B}^* - \hat{\boldsymbol{e}})] + \alpha^2(\hat{\boldsymbol{e}} - \boldsymbol{A}\boldsymbol{c})^2\right\} \tag{10.36}$$

which leads to the solution:

$$\hat{\boldsymbol{c}} = \left[\boldsymbol{A}^{\mathrm{T}}\boldsymbol{W}^2\boldsymbol{A} - \alpha^2\boldsymbol{A}^{\mathrm{T}}\boldsymbol{A}\right]^{-1}\left(\boldsymbol{A}^{\mathrm{T}}\boldsymbol{W}^2(\hat{\boldsymbol{e}} - \beta\boldsymbol{B}^*) - \alpha^2\boldsymbol{A}^{\mathrm{T}}\hat{\boldsymbol{e}}\right) \tag{10.37}$$

10.4.11 Piecewise

While regularization can overcome the ill-posedness of an inverse problem, it comes, as mentioned in Section 10.4.2, at the expense of smoothing the reconstructed image. In many situations, the imaged object consists of clusters of homogeneous regions separated by boundaries or edges. At each edge, there is an abrupt change in the physical attributes. The above methods impose a global regularization function on the entire solution, which does not help in preserving genuine edges and discontinuities in an image. Preferential application of the above mentioned regularization schemes on regions of interest can help bring out the presence of such regions in the reconstructed image. This, however, requires *a priori* knowledge of the location of such regions. Then, regularization methods can be judiciously applied to enhance a reconstructed image, as discussed in Chapter 16. For the detection, during image reconstruction, of unknown edges between otherwise homogeneous regions, one must introduce piecewise (local) regularization functionals, i.e. those that apply to a few voxels at a time.

In order to enable piecewise regularization, the regularization term, $\eta(\boldsymbol{c})$ in (10.9), is replaced by the summation of potential functions defined locally, i.e. piecewise, to emphasize local sharp changes in the image attributes. The cost function of Eq. (10.9) takes the form:

$$\min\left\{[\boldsymbol{Ac}-\hat{\boldsymbol{e}}]^2+\alpha^2\sum_r \phi(G_r\boldsymbol{c})\right\} \tag{10.38}$$

where G_r is a gradient operator acting in the neighborhood of each voxel and applied in various ways (since the summation over r),[9] and ϕ is called the potential function, which characterizes the relationship between a voxel and its neighbors.[10] The potential function is the logarithm of the probability density function, which is a quadratic function for a Gaussian distribution. The potential function is positive-definite symmetric continuous and increases with t, with a unique minimum value at zero.

The minimization of Eq. (10.38) poses a number of problems. The first is the absence of the Euclidean metric in defining the penalty term. The Euclidean metric defines the closest distance between two entities, as such it defines a convex set, for which a minimum exists.[11] However, by choosing a convex function for $\phi(t)$, the cost function of Eq. (10.38) can remain convex. But even then, a second problem remains: the argument t of $\phi(t)$ is a function of the solution, which makes the problem of Eq. (10.38) nonlinear. This necessitates the employment of an iterative solution process, see Section 10.7.

The function $\phi(t)$, where $t=\boldsymbol{G}_r(\boldsymbol{c})$, should be chosen such that at locations where the spatial gradient of c is weak it promotes homogeneity, while when the gradient is

[9] In the $x-y$ plane, the first-derivative operator on $c(x_i,y_j)$ corresponding to a voxel at (x_i,y_j) takes the form (Nikolova and Chan, 2007):

$$G_1(x_i,y_i)\boldsymbol{c}=c(x_i,y_i)-c(x_{i-1},y_i)$$
$$G_2(x_i,y_i)\boldsymbol{c}=c(x_i,y_i)-c(x_i,y_{i-1})$$
$$G_3(x_i,y_i)\boldsymbol{c}=\frac{1}{2}[c(x_i,y_i)-c(x_{i-1},y_{i+1})]$$
$$G_4(x_i,y_i)\boldsymbol{c}=\frac{1}{2}[c(x_i,y_i)-c(x_{i+1},y_{i-1})]$$

Similar formulations can be made in three dimensions. For the second gradient operator (Pan and Reeves, 2006):

$$G_1(x_i,y_i)\boldsymbol{c}=c(x_i,y_{i+1})-2c(x_i,y_i)-c(x_i,y_{i-1})$$
$$G_2(x_i,y_i)\boldsymbol{c}=\frac{1}{2}[c(x_{i-1},y_{i+1})-2c(x_i,y_i)+c(x_{i+1},y_{i-1})]$$
$$G_3(x_i,y_i)\boldsymbol{c}=c(x_{i+1},y_i)-2c(x_i,y_i)+c(x_{i-1},y_i)$$
$$G_4(x_i,y_i)\boldsymbol{c}=\frac{1}{2}[c(x_{i-1},y_{i-1})-2c(x_i,y_i)+c(x_{i+1},y_{i+1})]$$

[10] The penalty function in Eq. (10.38) is used in statistical estimates as a replacement of the least-squares cost function, in the so-called M-estimators technique (Rey, 1983), to overcome the presence of outliers in observations.

[11] A function, $\phi(x)$, is convex within an interval $[a,b]$ if it is such that $\phi([\gamma x+(1-\gamma)y)) \le \gamma\phi(x)+(1-\gamma)\phi(y)$, $0<\gamma<1$, and as such has a valley at the bottom of which a minimum is attained. In a strictly convex function: $\phi([\gamma x+(1-\gamma)y)) < \gamma\phi(x)+(1-\gamma)\phi(y)$. Equivalently, $\phi(x)\ge 0$ for all $a\in[a,b]$.

steep $\phi(t)$ should manifest the presence of the edge. Let us consider G_r to be the spatial gradient operator of c, then $t = G_r c$ $\left(= \frac{\partial c}{\partial x}\text{ in one dimension}\right)$. We will use $\phi'(t) = \frac{\partial c}{\partial t}$, and $\phi''(t) = \frac{\partial^2 c}{\partial t^2}$. In weak-gradient regions, $\phi(t)$ should be such that (Aubert and Kornprobst, 2006):

$$\phi'(0) = 0 \tag{10.39a}$$

$$\lim_{t\to 0^+} \frac{\phi'(t)}{t} = \lim_{t\to 0^+} \phi''(t) = \phi''(0) > 0 \tag{10.39b}$$

The first condition, called influence function, indicates that when there is no spatial gradient in c, there should be no change in the value of ϕ in the considered neighborhood. The second condition, where $\frac{\phi'(t)}{t}$, called the weight function, discourages huge changes in ϕ when the spatial gradient approaches zero in any direction. The equality of the three terms in Eq. (10.39b) assures a nearly flat function in the neighborhood of $t = 0$, where the function is minimum and the gradient is zero.

In the presence of an edge (Aubert and Kornprobst, 2006):

$$\lim_{t\to\infty} \phi''(t) = 0 \tag{10.40a}$$

$$\lim_{t\to\infty} \frac{\phi'(t)}{t} = \text{a finite value } > 0 \tag{10.40b}$$

where γ is some positive real number. The condition of Eq. (10.40a) states that when the spatial gradient, t, is extremely large, it should be accompanied by a fixed change in c, to indicate the presence of an edge. On the other hand, the second condition of Eq. (10.40b) ensures the existence of ϕ'', and limits sharp changes in c in the presence of a small gradient. These two conditions of Eq. (10.40) can be mathematically incompatible: $\phi'' = 0$ requires $\phi' =$ constant or zero, and in turn $\lim_{t\to\infty} \frac{\phi'(t)}{t} = 0$. This led to the development of a number of creative formulations for $\phi(t)$ that attempt to partially satisfy the conditions of Eq. (10.40) by taking advantage of the indefinite nature of infinity and allow the two conditions to approach infinity at different rates, i.e. by ensuring that:

$$\lim_{t\to\infty} \frac{\phi''(t)}{\frac{\phi'(t)}{t}} = 0 \tag{10.41}$$

The requirements of Eqs. (10.39) to (10.41), provided that $\phi(t)$ is convex, do deal with the ill-posedness of the problem for which regularization is devised in the first place, while attempting to preserve image features that can be destroyed by a global regularization of the problem.

To relate a voxel to its neighbors, one can take the view that in general, objects are formed by physical phenomena that are at equilibrium or near equilibrium conditions. As such, images of physical objects tend to consist of nearly homogeneous regions. One can, therefore, say broadly that a voxel in an image likes to look like its neighbors, or in other words, a voxel in an image can be considered as being in a state that tends to evolve toward equilibrium. Continuing with this analogy, a voxel can be considered to be at a slightly perturbed state relative to the states of its closest neighboring voxels. This can be described in thermodynamics by Gibbs function, and in statistics this would form a Gaussian (normally distributed) Markov chain.[12] Transitions from one homogeneous (equilibrium) state to another homogeneous, but different, state takes place gradually. Therefore, rather than smoothing the entire image, one should introduce smoothing in regions known *a priori* to be smooth, while accommodating the sharp transitions between states. This is typically used in image enhancement, as discussed in Section 16.5.

Below are some of the potential functions that can be used in piecewise regularization. Table 10.1 provides a list of various piecewise regularization functions and examines how they meet the conditions of Eq. (10.39), Eq. (10.40) and Eq. (10.41), as well as of convexity, $\phi''(t_{min}) > 0$. The reader can plot these function for better understanding of their influence on their neighbors.

Quadratic

Since the potential function of a Gaussian distribution is a quadratic function, which is also a convex function, the most basic convex formulation for $\phi(t)$ is the form (Demoment, 1989)[13]:

$$\phi(t) = t^2 \tag{10.42}$$

This form satisfies the homogeneity condition of Eq. (10.39), and one of the edge preservation conditions, Eq. (10.40b), but it does not fulfill the other condition Eq. (10.40a), nor its substitute, Eq. (10.41). Therefore, $\phi(t)$ of Eq. (10.42), when applied to the N_i neighbors of a voxel i will tend to dampen edges, in the presence of sharp gradients, since its gradient is always finite (being proportional to the difference between the c value of adjacent neighbors). One can argue that though this quadratic potential when applied to the entire image domain (rather than only to the neighborhood of a voxel) is equivalent to the gradient Tikhonov regularization of Section 10.4.7.

Huber

The influence of the quadratic potential can be limited via a scaling parameter, β, determined in advance such that:

$$\phi(t) = \begin{cases} \frac{t^2}{2} & \text{if } 0 \le t \le \beta \\ \beta|t| - \frac{\beta^2}{2} & \text{if } t > \beta \end{cases} \tag{10.43}$$

[12] The logarithm of a distribution defines its potential. Since the Gibbs function is an exponential function of the potential, it resembles a Gaussian distribution. A Markov chain is a random walk governed by conditional probabilities, i.e. the future of a present state is conditionally independent of the past.

[13] This paper also has an excellent overview of some regularization concepts.

Table 10.1 Various Piecewise Regularization Functions, $\phi(t)$, $\beta > 0$ (Nikolova and Chan, 2007; Nikolova et al., 1998; Green, 1990).

$\boldsymbol{\phi(t)}$	**Convex**	**Homogeneity**		**Edge Preservation**				
		(10.39a)		**(10.39b)**		**(10.40a)**	**(10.40b)**	**(10.41)**
	$\phi''(t) \geq 0$	$\phi'(0) = 0$	$\lim_{t\to 0^+} \frac{\phi'(t)}{t} > 0$	$\lim_{t\to 0^+} \phi''(t) > 0$	$\phi''(0) > 0$	$\lim_{t\to\infty} \phi''(t) = 0$	$\lim_{t\to\infty} \frac{\phi'(t)}{t} = \gamma > 0$	$\lim_{t\to\infty} \frac{\phi''(t)}{\frac{\phi'(t)}{t}} = 0$
t^2	yes	yes	yes	yes	yes	no	yes	no
$t^\beta,\ 1 < \beta < 2$	yes	yes	yes	yes	yes	yes	no	yes
$\begin{cases} \frac{t^2}{2}, & 0 \leq t \leq \beta \\ \frac{\lvert t\rvert}{2} - \frac{\beta^2}{2}, & t > \beta \end{cases}$	yes	yes	yes	yes	yes	no	yes	no
$\min(\beta t^2, 1)$	no	yes	yes	yes	yes	yes	no	yes
$\sqrt{\beta + t^2}$	yes	yes	yes	yes	yes	yes	no	yes
$\frac{\beta t^2}{1+\beta t^2}$	no	yes	yes	no	no	yes	yes	no
$1 - \exp(-\beta t^2)$	no	yes	no	no	no	yes	no	yes
$\ln(1 + \beta t^2)$	no	yes	yes	yes	yes	yes	no	no
$\ln(\cosh(\beta t))$	yes	yes	yes	no	no	yes	no	yes
$\frac{t}{\beta} - \ln\left(1 + \frac{t}{\beta}\right)$	yes	yes	yes	no	no	yes	no	yes
$\begin{cases} \sin(\beta t^2), 0 \leq t \leq \sqrt{\frac{\pi}{2\beta}} \\ 1,\ t > \sqrt{\frac{\pi}{2\beta}} \end{cases}$	no	yes	yes	no	no	yes	no	yes
$t\ln(\beta t)$	yes	no	no	yes	yes	yes	no	yes

Then, while the quadratic potential (Tikhonov-like regularization) remains for small values of t, once the t value exceeds β the strength of the potential function is reduced enabling boundaries to evolve. The value of β controls the penalty imposed on the evolution of edges by the quadratic potential, allowing edge preservation. In other words, by maintaining the quadratic term at low gradients (small t), weak variations in c are retained, while larger ones are encouraged by the linear part of the potential at high gradients. However, the transition in the value of the gradient between the quadratic and linear regions of the function is predetermined by the value of β, which should be related to the expected magnitude of change in c at edges. When the expected magnitude of edge changes is not known in advance, or is expected to vary widely, it would then be difficult to assign a single value of β that will not smooth some edges more than others.

The form of regularization associated with Eq. (10.43) is known as Huber regularization (Huber, 1981, 1964). A disadvantage of this function, and similar discontinuous functions involving a threshold value, is that they require *a priori* assignment of the value of β at which an edge is considered to have been formed.

Generalized Gaussian

Bouman and Sauer (1993) generalized the Gaussian (quadratic function) of Eq. (10.42) into the form:

$$\phi(t) = t^{\beta}, \quad 1 < \beta < 2 \tag{10.44}$$

giving $\lim_{t\to\infty} \frac{\phi''(t)}{\frac{\phi'(t)}{t}} = \beta - 1$, which is closer to satisfying condition (10.41), particularly when β is closer to unity. Note that, at $\beta = 1$, $\lim_{t\to\infty} \phi(t) = \infty$, and the condition of Eq. (10.40b) becomes too severe, while $\beta < 1$ leads to a nonconvex function, since $\phi''(0) < 0$. On the other hand, $\beta > 2$ provokes the two conditions of Eq. (10.40). In this so-called generalized Gaussian function, large (close to 2) values of β promote image homogeneity at the expense of damping abrupt edge discontinuities, while a small (close to unity) β allows edges to evolve. Therefore, with β values somewhere in between one and two, edges can be preserved, while not over-amplifying small fluctuations. At the same time, with such β values, ill-posedness can still be overcome by employing a convex potential function. In addition, unlike in the Huber function, Eq. (10.43), one needs not assign in advance a threshold value, beyond which a change in the image attributes is considered to be caused by an edge.

Approximately Quadratic

Another approach is to find an approximation of the Huber function of Eq. (10.43), that (1) moderates the smoothing effect of the quadratic potential function, Eq. (10.42), (2) encourages edge preservation, and at the same time (3) avoids the thresholding of the Huber function. Green (1990) proposed the log-cosh function:

$$\phi(t) = \ln\left[\cosh(\beta t)\right] \tag{10.45}$$

This is a strictly convex function ($\phi > 0$), and is approximately quadratic at small t, and tends to linearity at large values, emulating the Huber function.

Truncated Quadrature

Rather than the linear term beyond the discontinuity, $t > \beta$, of the Huber function of Eq. (10.43), Nikolova et al. (1998) truncated the quadratic function at some threshold value, $t = q_\gamma < \beta$, with q_γ being a truncation parameter. At another threshold value, $t = r_\gamma > \beta$, the potential function was assigned a constant value, β^2. In the transition from the quadratic region, where the image is expected to be homogeneous, to the discontinuity at the edge, a transition state is introduced. Within that undetermined transition state, the potential function is defined by fitted quadratic splines that bridge the continuous homogeneous state to the discontinuous edge state. Therefore, the potential function is defined as:

$$\phi(t) = \begin{cases} t^2, & t < q_\gamma \\ \beta^2 - \gamma(r_\gamma - t)^2, & q_\gamma \leq |t| < r_\gamma \\ \beta^2, & |t| \geq r_\gamma \end{cases} \qquad \begin{array}{l} q_\gamma = \beta\left(1 + \frac{1}{\gamma}\right)^{-\frac{1}{2}} \\ r_\gamma = \beta\left(1 + \frac{1}{\gamma}\right)^{\frac{1}{2}} \end{array} \tag{10.46}$$

where $\gamma > 0$ is called the relaxation parameter that determines the width of the relaxation state. Note that $\gamma = \infty$, gives $q_\gamma = r_\gamma = \beta$, and the transition state ceases to exist; consequently $\gamma = \infty$ creates an abruptly terminated truncated quadratic potential function. On the other hand, a small $\gamma << 1$ value would produce a wide transition state. The potential function within the transition state is concave, which does not assure minimization of the solution's cost function. However, the degree of concavity can be controlled by the value of γ, since $\phi''(t) = -2\gamma, q_\gamma \leq |t| < r_\gamma$. Therefore, Nikolova et al. (1998) proposed a graduated nonconvexity (concavity) iterative algorithm using an exponential relaxation sequence that evolves slowly at the beginning of iteration and faster later on (they used as an example $\gamma_k = 0.25\exp(0.2k)$, where γ_k is the value of γ at iteration k).

Hypersurface Minimal

This is a potential function based on the physical phenomenon of electron transitions in chemical reactions, where two electronic states are degenerate, allowing a hypersurface (minimal energy). These transitions are analogous to transition across edges in an image. The corresponding potential function is of the form (Aubert and Kornprobst, 2006):

$$\phi(t) = \sqrt{1 + t^2} \tag{10.47}$$

This function satisfies conditions Eq. (10.39) to Eq. (10.41), though the limit for Eq. (10.40b) is a small value close to zero. It is also a convex function.

"Fair"

The so-called "Fair" function, given by Rey (1983):

$$\phi(t) = \frac{t}{\beta} - \ln\left(1 + \frac{t}{\beta}\right) \tag{10.48}$$

satisfies the homogeneity conditions of Eq. (10.39), as well as the edge conditions of Eqs. (10.40a) and (10.41). This function has continuous first, second, and third order derivatives, which facilitates convergence in iterative computations.

Nonconvex Functions

A number of edge-preserving nonconvex potential functions are used for regularization. Although such functions do not help with the problem's ill-posedness, their ability to tolerate abrupt edges makes them attractive for use in preserving image discontinuities. Nevertheless, with the first term in the objective function of (10.9) being convex, and given that the effect of potential functions is local and can be controlled by the regularization parameter, α^2 in (10.38), the nonconvex effect of such regularization on the problem's ill-posedness can be diminished.

One such nonconvex potential function, suggested by Geman and McClure (1987) (cited by Nuyts et al. (2002)) is as follows:

$$\phi(t) = \frac{\beta t^2}{1 + \beta t^2} \tag{10.49}$$

where β is again a control parameter. This function satisfies simultaneously the conditions of Eq. (10.40) by having $\phi' \to 0$ as $t \to \infty$. The function of Eq. (10.49) is, however, nonconvex. As a result, it does not serve as a regularization function for the purpose of dealing with ill-posedness. Note that because of the nonconvex nature of the function of Eq. (10.49), a different solution (image) can be arrived at for the same value of β, with a slight variation in measurement error (noise).

10.4.12 Variational

This form of regularization, like the piecewise regularization of (10.38), allows for bounded discontinuities in the solution, such as those encountered at image edges. This is done by introducing regularization via a non-Euclidean term, so that the distance between image parameters in the obtained solution is not necessarily the shortest distance as dictated by the Euclidean metric. Such regularization can be achieved by simply using the L_1 norm, in place of the Euclidean norm for the regularization term in (10.9), producing the so-called total variation (Rudin et al., 1992). This is equivalent to utilizing globally the generalized Gaussian regularization of Eq. (10.44) with $\beta = 1$. Accordingly, the regularization of (10.38) applied globally (rather than piecewise), as was done by Charbonnier et al. (1997), is a form of variational regularization.

10.4.13 Maximum Entropy

Entropy is used in information theory as a measure of the amount of choice one has in selecting an event: the more choice, the less the constraints, the higher the entropy. In other words, entropy is a measure of disorder or randomness, and a measure of loss (or lack) of information; hence a measure of uncertainty. In imaging, lack of information is associated with an insufficient number of measurements (an undetermined problem), or a high measurement uncertainty. In such problems, the same set of measurements could correspond to a number of solutions, i.e. possible image configurations. With no other information available to favor one particular configuration over others, one is compelled to choose the arrangement with maximal entropy (maximum uncertainty). In other words, a maximum-entropy solution is a solution with no unfounded constraints, i.e. a solution that "agrees" with what is known, but does not assume anything that is not known (Jaynes, 1957). Therefore, information entropy provides a means for regularization.

In information theory, the entropy, u, associated with a solution, $\mathbf{c}$, is defined by:

$$u = -\sum_{i=1}^{n^+}\left(\frac{c_i}{c_{\text{sum}}}\right)\left[\ln\left(\frac{c_i}{c_{\text{sum}}}\right) - \frac{\ln\beta_i}{n^+}\right] \tag{10.50}$$

where $c_{\text{sum}} = \sum_{i=1}^{n^+} c_i$ is a normalization constant, the summation is over the n^+ positive terms of $\mathbf{c}$, and β_i is the distribution with respect to which entropy is defined. When $\beta_i = 1$ for all i, entropy is defined with respect to a uniform distribution, which is the default distribution in the absence of any other *a priori* knowledge of $\mathbf{c}$. The normalization of $\mathbf{c}$ provides a probability distribution of its elements. The entropy of Eq. (10.50) is known as the Shannon-Jaynes (Jaynes, 1957; Shannon and Weaver, 1949) (or information) entropy. It is a negative measure of the information content of the image formed by $\mathbf{c}$. The maximum entropy regularization minimizes the negative of the entropy, along with the minimization of the fidelity (residual) norm, that is:

$$\min\left\{[A\mathbf{c} - \hat{\mathbf{e}}]^2 + \alpha^2\sum_{i=1}^{n^+}\left(\frac{c_i}{c_{\text{sum}}}\right)\left[\ln\left(\frac{c_i}{c_{\text{sum}}}\right) - \frac{\ln\beta_i}{n^+}\right]\right\} \tag{10.51}$$

Since the Euclidean norm does not appear in the regularization term, (10.51) can be considered as a form of variational regularization. The nonlinear nature of the entropy term dictates an iterative solution of the problem, using one of the methods described in Section 10.6.

A solution obtained with maximum-entropy regularization produces an image that best matches available measurements, but with minimal unnecessary correlations between voxels. It is equivalent to a basic Tikhonov regularization: both are minimum-information methods. However, the maximum-entropy solution has two unique features: the ability to provide an unbiased solution from an incomplete set of measurements, i.e. an underdetermined problem (see Chapter 14), and the nonnegativity of solution (i.e. it is a Tikhonov regularization with positivity). The first feature arises from its favoring of a solution with maximum entropy (i.e. uncertainty), among many possible solutions. The nonnegativity comes from the fact that entropy is only defined for positive and additive parameters, since order cannot be restored by subtracting or adding negative attributes. This feature allows the use of maximum-entropy regularization as a stand-alone form of regularization (Landl and Anderssen, 1996), or as a follow-up solution to refine the results obtained using other means of regularization, as suggested by Chiang et al. (2005). In summary, a solution based on (10.51) would be most consistent with available measurements, nonnegative (because of the logarithm), and most noncommittal about missing information, i.e. the obtained solution is most "objective" or "maximally uncommitted" to missing information (Hansen, 1998).

Each term in the summation of Eq. (10.50) can be considered as a potential function: $\phi(t) = t\ln(\beta t) + \text{constant}$, analogous to those defined in Section 10.4.11. This is also a convex function, but while it does not fully satisfy the homogeneity conditions of Eq. (10.39), it satisfies the two edge requirements of Eq. (10.40a) and Eq. (10.41). Therefore, maximum-entropy regularization is edge preserving. A major difference between this entropy-based potential function and those of Section 10.4.11 is that the latter functions reach their minimum at zero, but $\phi(t) = t\ln(\beta t)$ has a minimum at $\frac{\exp(-\beta)}{\beta} > 0$. Its minimization, therefore, promotes positivity of solution by pushing it away from zero.

An *a priori* estimate of the solution, $\boldsymbol{c}^*$, can be used to define the distribution of β_i in Eq. (10.50) so that:

$$u = -\sum_{i=1}^{n^+} \left(\frac{c_i}{c_{\text{sum}}}\right)\left[\ln\left(\frac{c_i}{c_{\text{sum}}}\right) - \ln\left(\frac{c_i^*}{c_{\text{sum}}^*}\right)\right] \tag{10.52}$$

Regularization with this relative entropy will seek a solution closest to $\boldsymbol{c}^*$. Equation (10.52) is known as the Kullback-Leibler metric (Kullback and Leibler, 1951) (called simply the Kullback distance) or the cross-entropy (Shore and Johnson, 1980). This distance, being obviously different from the Euclidean distance, makes the maximum-entropy regularization of (10.51) a variational regularization. Note that the Kullback distance in noncommutive, i.e. the distance between c_i and c^* is not equal to the distance between c_i^* and c_i, since $c_i \ln(c_i/c_i^*) \neq c_i^* \ln(c_i^*/c_i)$. It is, therefore, pathological (Demoment, 1989), in the sense that it is a measure of deviation from "normal" state.

Another modified form of the entropy is given by (Dudík et al., 2007; Engl et al., 2000; Hofmann and Krämer, 2005):

$$u = \sum_{i=1}^{n^+} \left(\frac{c_i}{c_{\text{sum}}}\right)\left[\ln\left(\frac{c_i}{c_{\text{sum}}}\right) - \ln\left(\frac{c_i^*}{c_{\text{sum}}^*}\right) - \left(\frac{c_i}{c_{\text{sum}}} - \frac{c_i^*}{c_{\text{sum}}^*}\right)\right] \tag{10.53}$$

The inclusion of $\boldsymbol{c}^*$ helps in stabilizing the solution process (Chiang et al., 2005).

10.4.14 Solution Bounding

The regularization methods discussed in Sections 10.4.7 to 10.4.13 aimed at smoothing a reconstructed image, based on some *a priori* knowledge of its nature. This was done without any regard to the nature of the acquired measurements, except by incorporating the weight matrix, $\boldsymbol{W}$, in the solution. However, more information can be extracted from the measurements themselves, or from supplementary measurements.

Given the forward model: $\boldsymbol{Ac} = \boldsymbol{e}$, one can estimate an upper bound, q_j, for the value of c_j at a given voxel by:

$$q_j = \min_i \left\{ \frac{e_i + \sigma_i}{A_{ij}} \right\}; \quad \text{for } A_{ij} > 0 \tag{10.54}$$

where σ_j^2 is the variance associated with the measurement e_j. Equation (10.54) gives an upper bound, when $A_i j > 0$, because it does incorporate the contribution of all corresponding elements of $\boldsymbol{A}$, and because it takes into account the elevating effect of any statistical fluctuations by adding σ_i to each e_i. However, this is a weak upper-bound[14] because the minimization in Eq. (10.54) restricts the value of q_j. By compiling all the estimates of q_j's into an $N \times N$ diagonal matrix, $\boldsymbol{Q}$, one can introduce the following regularization functional as a solution constraint:

$$\eta(\boldsymbol{c}) = \boldsymbol{Q}^{-1}\boldsymbol{c} \tag{10.55}$$

$$Q_{ij} = q_j \delta_{ij} \tag{10.56}$$

with δ_{ij} being the Kronecker delta. With the regularization functional of Eq. (10.55), the solution of Eq. (10.23) takes the form:

$$\hat{\boldsymbol{c}} = \left[\boldsymbol{A}^{\mathrm{T}}\boldsymbol{W}^2\boldsymbol{A} + \alpha^2\boldsymbol{Q}^{-2}\right]^{-1}\boldsymbol{A}^{\mathrm{T}}\boldsymbol{W}^2\hat{\boldsymbol{e}} \tag{10.57}$$

Solution bounding can also be accomplished by obtaining information from supplementary measurements not employed in the image reconstruction process. For instance, in scatter imaging, transmission measurements can be employed to provide an estimate of the sum of the attenuation coefficients along certain directions. This information, for M' supplementary measurements, can be put in the matrix form:

$$\boldsymbol{Hc} \leq \boldsymbol{T} \tag{10.58}$$

where $\boldsymbol{H}$ is an $M' \times N$ constraint matrix (e.g. the linearized forward model for transmission measurements) and $\boldsymbol{T}$ is a diagonal $M' \times M'$ matrix in which each diagonal element corresponds to a supplementary measurement plus its statistical standard deviation, to provide an upper-bound for the constraint and to take into account the

[14] This approach was adopted by the FERDOR and FORIST codes (Oak Ridge National Laboratory, code collection PSR-17 and PSR 92, respectively) for neutron spectrum unfolding, and used by the author of this book in scatter-imaging problems (Arendtsz and Hussein, 1995b; Hussein et al., 1986a,b).

statistical quality of the measurements. The regularization functional with Eq. (10.58) takes then the form:

$$\eta(\boldsymbol{c}) = \boldsymbol{T}^{-1}\boldsymbol{H}\boldsymbol{c} \tag{10.59}$$

The resulting solution is:

$$\hat{\boldsymbol{c}} = \left[\boldsymbol{A}^{\mathrm{T}}\boldsymbol{W}^2\boldsymbol{A} + \alpha^2\boldsymbol{H}^{\mathrm{T}}\boldsymbol{T}^{-2}\boldsymbol{H}\right]^{-1}\boldsymbol{A}^{\mathrm{T}}\boldsymbol{W}^2\hat{\boldsymbol{e}} \tag{10.60}$$

Note that in Eq. (10.58), when $\boldsymbol{H} = \boldsymbol{I}$, i.e., no mapping of the supplementary information, $\boldsymbol{T}$ becomes an upper-bound of the solution. With $\boldsymbol{T} = c_{\mathrm{up}}\boldsymbol{I}$, where c_{up} is a some upper-bound that any element in $\boldsymbol{c}$ cannot exceed, Eq. (10.60) becomes identical to the basic Tikhonov regularization of Eq. (10.17) with $\boldsymbol{G} = \boldsymbol{I}$, but with $\frac{\alpha^2}{c_{\mathrm{up}}^2}$ being now the regularization parameter, instead of α^2 in Eq. (10.17). Therefore, an explicit bounding of the solution, determined from *a priori* information, can be formulated as:

$$\boldsymbol{c} \leq c_{\mathrm{up}} \tag{10.61}$$

leading to the regularization functional:

$$\eta(\boldsymbol{c}) = \frac{1}{c_{\mathrm{up}}}\boldsymbol{c} \tag{10.62}$$

and the solution:

$$\hat{\boldsymbol{c}} = \left[\boldsymbol{A}^{\mathrm{T}}\boldsymbol{W}^2\boldsymbol{A} + \frac{\alpha^2}{c_{\mathrm{up}}}\boldsymbol{I}\right]^{-1}\boldsymbol{A}^{\mathrm{T}}\boldsymbol{W}^2\hat{\boldsymbol{e}} \tag{10.63}$$

Regularization to bound the solution to nonnegative values can also be achieved via the maximum-entropy regularization method, described in Section 10.4.13. In addition, applying regularization using the Kullback distance of Eq. (10.52) gives some rationale for the establishment the nonnegative solution, by relating it to an *a priori* estimate of the solution. Note that constraints can be applied to bound the solution within explicit lower and upper limits by nonlinear programming, see Lawson and Hanson (1995), Stark and Parker (1995), Rao (1996), and Varah (1979).

10.5 Regularization-Parameter Determination

Regularization is introduced to control error propagation due to measurement uncertainties. However, regularization also introduces its own error, by virtue of preventing absolute minimization of the residual error, $\|\boldsymbol{A}\boldsymbol{c} - \hat{\boldsymbol{e}}\|$. Let $\hat{\boldsymbol{c}}_\alpha$ be the solution obtained by a regularized solution, with a certain regularization function with a regularization parameter, α^2. Assuming that the true solution, $\boldsymbol{c}$, is given by the error introduced by regularization plus that caused by the propagation of measurement uncertainties, $\boldsymbol{\delta}_e$, the error, $\boldsymbol{\delta}_{\hat{\boldsymbol{c}}_\alpha}$, in a solution subjected to regularization is then given by:

$$\boldsymbol{\delta}_{\hat{\boldsymbol{c}}_\alpha} = (\boldsymbol{A}_\alpha^\dagger \boldsymbol{A}\boldsymbol{c} - \boldsymbol{c}) + \boldsymbol{A}_\alpha^\dagger \boldsymbol{\delta}_e \tag{10.64}$$

where $\boldsymbol{A}_\alpha^\dagger$ is the inversion operator that led to the solution $\hat{\boldsymbol{c}}_\alpha$.

The choice of α^2 is a balance between the desire to temper the ill-posedness of the inverse problem and the need to provide a solution that best matches the measurements. As $\alpha^2 \to \infty$, the effect of measurements on the solution diminishes, which is either meaningless or produces a default solution. On the other hand, as $\alpha^2 \to 0$, the role of regularization in overcoming ill-posedness is weakened.

An optimal value of α^2 is, therefore, sought. A number of methods can be used to select a proper value for α^2. These methods are presented below. While the ensuing discussion focuses on α^2, it is equally applicable to the selection of other regularization parameters, such as β in piecewise regularization.

10.5.1 Convergence Regularization

Obviously, if there were no measurement uncertainties, there would be no need for regularization. Since measurement uncertainties are inevitable, one should ensure that as these uncertainties approach zero, the error in the solution also vanishes, even in the presence of regularization. This becomes the basis for selecting the regularization parameter, α^2. Using the solution expressed in terms of the singular values, in Section 10.4.2, for the basic Tikhonov regularization, and following the analysis of Vogel (2002), the solution error of Eq. (10.64) can be expressed as:

$$\boldsymbol{\delta}_{\hat{\boldsymbol{c}}_\alpha} = \sum_i \left[[g_\alpha(s_i) - 1] \left(\boldsymbol{v}_i^{\mathrm{T}} \boldsymbol{c} \right) \boldsymbol{v}_i \right] + \left[\frac{g_\alpha(s_i)}{s_i} \left(\boldsymbol{u}_i^{\mathrm{T}} \boldsymbol{\delta}_e \right) \boldsymbol{v}_i \right] \tag{10.65}$$

where use is made of Eqs. (10.11) and (10.13). The error introduced by regularization vanishes when $g_\alpha(s_i) \to 1$, since then in accordance to Eq. (10.13) $\alpha^2 \to 0$. The propagation of $\boldsymbol{\delta}_e$ is controlled by the term: $\frac{g_\alpha(s_i)}{s_i} = \frac{s_i}{s_i^2+\alpha^2}$, which for $s > \alpha$ approaches $\frac{1}{s} < \frac{1}{\alpha}$; for $s < \alpha$ is approximately equal to $\frac{s}{\alpha^2} < \frac{1}{\alpha}$; and for $s = \alpha$ is equal to $\frac{1}{2\alpha}\frac{1}{\alpha}$. Therefore, the inequality $\frac{g_\alpha(s_i)}{s_i} < \frac{1}{\alpha}$ is always valid. As a result $\left\| \frac{g_\alpha(s_i)}{s_i} \left(\boldsymbol{u}_i^{\mathrm{T}} \boldsymbol{\delta}_e \right) \boldsymbol{v}_i \right\| \leq \frac{\|\boldsymbol{\delta}_e\|}{\alpha}$. When there is no uncertainty in the measurements, i.e. when $\|\boldsymbol{\delta}_e\| \to 0$, the associated error propagation term, which is always $\leq \frac{\|\boldsymbol{\delta}_e\|}{\alpha}$, should also vanish. Therefore, one must choose $\alpha = \|\boldsymbol{\delta}_e\|^r$, with $r < 1$, so that $\frac{\|\boldsymbol{\delta}_e\|}{\alpha} = \|\boldsymbol{\delta}_e\|^{1-r} \neq 1$ and is bounded. With $r > 0$, one can also assure that the error introduced by regularization also vanishes as $\alpha^2 \to 0$. Therefore,

$$\boldsymbol{\delta}_{\hat{\boldsymbol{c}}_\alpha} \to 0, \quad \text{as} \quad \|\boldsymbol{\delta}_e\| \to 0, \quad \text{if} \quad \alpha^2 = \|\boldsymbol{\delta}_e\|^{2r}, \qquad 0 < r < 1 \tag{10.66}$$

This condition leads to a *convergent* regularization, since its effect vanishes with the disappearance of measurement uncertainties. While this selection for α^2 is suited for the basic Tikhonov regularization, with $\boldsymbol{G} = I$, and by extension the SVD solution of Eq. (10.12), it is more difficult to apply to other regularization methods. It illustrates, however, the need to assure that the regularization parameter is chosen with the convergence criteria in mind.

10.5.2 Discrepancy Principle

Rather than assigning up front a specific value for α^2, it can be assigned *a posteriori*, i.e. after a solution estimate is arrived at. Use is then made of the discrepancy principle (Morozov, 1984), which essentially indicates that there is no point in obtaining a solution that produces corresponding estimated measurements, $\boldsymbol{A}\hat{\boldsymbol{c}}$, with an error greater than $\boldsymbol{\delta}_e$ (the error in available measurements). Therefore, one should seek a value for α^2 that produces a solution, $\hat{\boldsymbol{c}}_\alpha$, such that:

$$\|\boldsymbol{A}\hat{\boldsymbol{c}}_\alpha - \hat{\boldsymbol{e}}\| = \gamma \|\boldsymbol{\delta}_e\| \tag{10.67}$$

where γ is slightly greater than one, e.g. $\gamma = 1.01$, with $\|\boldsymbol{\delta}_e\| \ll \|\hat{\boldsymbol{e}}\|$ (otherwise the uncertainty in the measurements is too excessive to allow a meaningful solution). This requires an iterative solution with gradually increasing (from zero) value of α^2, with Eq. (10.67) used as the stopping criteria for the iterative process. In radiation measurements, an estimate of $\|\boldsymbol{\delta}_e\|$ can be obtained using the Poisson counting statistics, see Section 15.2. However, when the forward model does not accurately simulate the measurements, as almost always the case (see Part I of this book), and in the presence of other systematic errors, a solution produced by Eq. (10.67) can be under-regularized. This, however, can be overcome by generalizing the discrepancy principle to include the modeling error, as defined by $\boldsymbol{B}^*$ in Eq. (10.33). Then the regularization parameter is chosen so that:

$$\|\boldsymbol{A}\hat{\boldsymbol{c}}_\alpha - \hat{\boldsymbol{e}}\| = \gamma(\|\boldsymbol{\delta}_e\| + \gamma_m\|\boldsymbol{B}^*\|) \tag{10.68}$$

10.5.3 L-Curve

After obtaining a regularized solution, $\hat{\boldsymbol{c}}_\alpha$, with a certain parameter, α^2, one can calculate the regularization norm, $\|\boldsymbol{G}\hat{\boldsymbol{c}}_\alpha\|$, and the residual norm, $\|\boldsymbol{A}\hat{\boldsymbol{c}}_\alpha - \hat{\boldsymbol{e}}\|$. If for various values of α^2, $\ln\|\boldsymbol{G}\hat{\boldsymbol{c}}_\alpha\|$ is plotted verses $\ln(\|\boldsymbol{A}\hat{\boldsymbol{c}}_\alpha - \hat{\boldsymbol{e}}\|)$, one typically obtains the L-shaped of Figure 10.1 (Hansen, 1992; Hansen and O'Leary, 1993). At small values of α^2, the solution is quite susceptible to the measurement error, $\|\boldsymbol{\delta}_e\|$, and consequently quite sensitive to changes in α^2, hence the vertical portion of the curve. As α^2 becomes large, the problem tends to be over-regularized to the extent that the solution is more dependent on regularization than on measurements; hence the horizontal portion of the curve. This can be substantiated by Eq. (10.65), which shows that for small values of α^2, $g_\alpha(s_i) \approx 1$, the solution error, $\boldsymbol{\delta}_{\hat{\boldsymbol{c}}_\alpha}$ becomes dominated by propagated measurement error, $\|\boldsymbol{\delta}_e\|$, leading to large $\|\boldsymbol{G}\hat{\boldsymbol{c}}_\alpha\|$ values. At high α, $g_\alpha(s_i) \ll 1$, the regularization error dominates, taking $\boldsymbol{A}\hat{\boldsymbol{c}}$ further away from $\hat{\boldsymbol{e}}$, leading to the increase in $\|\boldsymbol{A}\hat{\boldsymbol{c}}_\alpha - \hat{\boldsymbol{e}}\|$. The use of ln-ln scale emphasizes the changes in the values of these

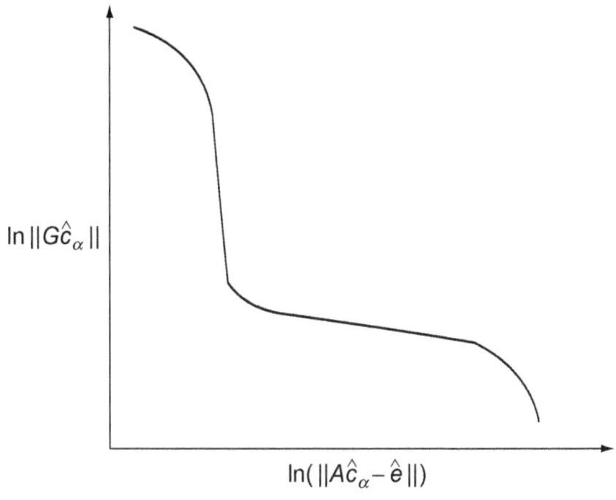

Figure 10.1 Typical L-curve showing the general relationship between the logarithms of regularization norm, $\|G\hat{c}_\alpha\|$, and that of the residual norm, $\ln(\|A\hat{c}_\alpha - \hat{e}\|)$.

norms, and makes it easier to locate the optimal value of α^2, which obviously is at the corner of the L-curve, where there is no further gain by introducing more regularization. This curve, therefore, can be used to select the proper value of α^2. Note that selecting α^2 according to the discrepancy principle, Eq. (10.67), corresponds to choosing a value of $\|A\hat{c}_\alpha - \hat{e}\|$ just to the right of the corner of the L-curve, i.e. before regularization starts to affect the solution significantly, making it insensitive to $\|\delta_e\|$.

10.5.4 Minimum Predictive Error

The best regularization parameter, α^2, can also be considered to be the value that minimizes the predictive error defined by (Vogel, 2002):

$$p_\alpha = A\hat{c}_\alpha - Ac \tag{10.69}$$

since this will provide the closest solution, $\hat{c}$, to the true solution, c. This error cannot be explicitly computed, but it can be estimated as:

$$\begin{aligned} p_\alpha &= A\hat{c}_\alpha - Ac = (A\hat{c}_\alpha - \hat{e}) + (\hat{e} - Ac) \\ &= (A\hat{c}_\alpha - \hat{e}) + (2AA^\dagger_\alpha - I)\delta_e \end{aligned} \tag{10.70}$$

with $\hat{e} = Ac + (2AA^\dagger_\alpha - I)\delta_e$ to incorporate the effect of regularization (note that with no regularization: $A^\dagger_\alpha = A^{-1}$, $2AA^\dagger_\alpha - I = I$). One can recognize that the first term in the estimate of Eq. (10.70), $A\hat{c}_\alpha - \hat{e}$, is the residual error, which is a deterministic term that is not affected by randomness, once a certain value of $\hat{e}$ is utilized. On the other hand, the second term, $(2AA^\dagger_\alpha - I)\delta_e$, is a random variable, since it contains the statistical variability of the measurements via δ_e. The expected value, $E(\|p_\alpha\|_2^2)$, is then given by (Vogel, 2002):

$$E(\|\boldsymbol{p}_\alpha\|) = \|\boldsymbol{A}\hat{\boldsymbol{c}}_\alpha - \hat{\boldsymbol{e}}\|_2^2 + 2\|\hat{\boldsymbol{e}}\|_2 \operatorname{trace}(\boldsymbol{A}\boldsymbol{A}_\alpha^\dagger) - \|\hat{\boldsymbol{e}}\|_2 \tag{10.71}$$

where the trace[15] of $\boldsymbol{A}\boldsymbol{A}_\alpha^\dagger$ is used to estimate its expected value. In Eq. (10.71), the variance of a radiation counting measurement is assumed to be equal to its mean (measured) value, in accordance to the Poisson statistics described in Section 15.2. Vogel (2002) provided an analogous expression to Eq. (10.71), albeit for a normally distributed white noise with a certain variance. The expected value of the predictive error is given by $E(\|\boldsymbol{p}_\alpha\|)$, which is an estimate of the risk in predicting a wrong solution. Therefore, $E(\|\boldsymbol{p}_\alpha\|)$ is called the *unbiased predictive risk estimator*. Obviously, the regularization parameter, α^2, should be chosen to minimize Eq. (10.71), in order to obtain the lowest expected predictive error.

Thompson et al. (1991) proposed finding the regularization parameter by minimizing the total predicted mean squared error defined as:

$$\begin{aligned} E(\|\boldsymbol{A}\boldsymbol{c} - \boldsymbol{A}(\boldsymbol{A}_\alpha^\dagger\hat{\boldsymbol{e}})\|_2^2) &= E(\|\boldsymbol{A}\boldsymbol{c} - \boldsymbol{A}\boldsymbol{A}_\alpha^\dagger(\boldsymbol{A}\boldsymbol{c} + \boldsymbol{\delta}_e)\|_2^2) \\ &= \|(\boldsymbol{I} - \boldsymbol{A}\boldsymbol{A}_\alpha^\dagger)\boldsymbol{A}\boldsymbol{c}\|_2^2 + \|\hat{\boldsymbol{e}}\|_2 \operatorname{trace}(\boldsymbol{A}_\alpha^\dagger)^2 \end{aligned} \tag{10.72}$$

where again the variance in $\hat{\boldsymbol{e}}$ is taken to be equal to $\|\hat{\boldsymbol{e}}\|$ according to Poisson statistics. However, the minimization of Eq. (10.72) requires knowledge of the true value of $\boldsymbol{c}$, but an estimate of $\boldsymbol{c}$ can be used to facilitate finding a value for α^2.

A heuristic quasi-optimization approach for finding α^2 was proposed by Hanke and Raus (1996). It is based on minimizing the difference between two regularized solutions, one ($\hat{\boldsymbol{c}}_\alpha$) with the usual regularization, and the second solution ($\hat{\boldsymbol{c}}_{\alpha\alpha}$), using the first solution, $\hat{\boldsymbol{c}}_\alpha$, as a solution estimate, as in Eq. (10.27). The difference between the two solutions is then given by:

$$\|\hat{\boldsymbol{c}}_\alpha - \hat{\boldsymbol{c}}_{\alpha\alpha}\|_2^2 = \|(\boldsymbol{A}\hat{\boldsymbol{c}}_\alpha - \hat{\boldsymbol{e}}) - (\boldsymbol{A}\hat{\boldsymbol{c}}_{\alpha\alpha} - \hat{\boldsymbol{e}})\|_2^2 \approx \left\|\alpha^2\left(\boldsymbol{A}\boldsymbol{A}^{\mathrm{T}} + \alpha^2\boldsymbol{I}\right)\boldsymbol{A}^{\mathrm{T}}\hat{\boldsymbol{e}}\right\|_2^2 \tag{10.73}$$

10.5.5 Generalized Cross Validation

The predictive error of Eq. (10.70) can also be expressed as:

$$\boldsymbol{p}_\alpha = (\boldsymbol{A}\hat{\boldsymbol{c}}_\alpha - \hat{\boldsymbol{e}}) - (\boldsymbol{I} - \boldsymbol{A}\boldsymbol{A}_\alpha^\dagger)\boldsymbol{\delta}_e + \boldsymbol{A}\boldsymbol{A}_\alpha^\dagger\boldsymbol{\delta}_e \tag{10.74}$$

With the view that regularization aims at controlling error propagation, irrespective of the random variability of measurements, the predictive error can be minimized by minimizing:

$$V(\alpha) = \frac{\|\boldsymbol{A}\hat{\boldsymbol{c}}_\alpha - \hat{\boldsymbol{e}}\|_2^2}{[\operatorname{trace}(\boldsymbol{I} - \boldsymbol{A}\boldsymbol{A}_\alpha^\dagger)]^2} \tag{10.75}$$

[15] The trace of a square matrix is the sum of its diagonal elements.

The function $V(\alpha)$ is also an estimator of the mean-squared norm of the predictive error. Choosing an α^2 value that minimizes Eq. (10.74) reduces then the risk of obtaining an erroneous solution. The function $V(\alpha)$ of Eq. (10.75) is called the generalized cross validation functional (Vogel, 2002), first introduced by Craven and Wahba (1979).

10.5.6 Minimum Bound

In this method, the regularization parameter, α^2, is chosen to minimize an approximate upper-bound of the solution error, $\hat{\boldsymbol{c}} - \boldsymbol{c}$. This upper-bound is determined by a functional, $B(\alpha)$, that combines the residual error and the error that would have been obtained if true (error-free measurements) were available and were used to obtain a regularized solution. Since the latter error is not known, its squared value is estimated to be at least equal to the measurement variance divided by the regularization parameter. With the variance of the measurements taken to be equal to the measurement itself (in accordance to the Poisson statistics of counting), one has (Lukas, 1998; Vogel, 2002):

$$B(\alpha) = \left\| (\boldsymbol{A}\hat{\boldsymbol{c}}_\alpha - \hat{\boldsymbol{e}})^{\mathrm{T}} (\boldsymbol{A}\boldsymbol{A}^{\mathrm{T}})^{\dagger} (\boldsymbol{A}\hat{\boldsymbol{c}}_\alpha - \hat{\boldsymbol{e}}) \right\|_2^2 + \frac{\gamma}{\alpha^2} \|\hat{\boldsymbol{e}}\|_2 \tag{10.76}$$

for any $\gamma \geq 1$. The parameter γ allows for adjusting the upper-bound in terms of the measurement variance. The division of the variance term by α^2 in Eq. (10.76) reduces the upper-bound with increased regularization.

10.5.7 Statistical Hypothesis Testing

Rather than focusing on finding a regularization parameter, α^2, that minimizes the residual error while overcoming ill-posedness, or minimizing the predictive error or some upper-bound of solution error, one can aim at find an α^2 value that preserves the profile (distribution) of the measurements. This is a powerful rationale given that, as indicated in Part I of this book, the forward problem is itself approximate, and as such best matching with measurements, or minimizing some measure of solution error, may not be the most appropriate means of reaching an acceptable solution. Vogel (2002) proposed using this profile-preservation approach, originally introduced by Veklerov and Llacer (1987), as a stopping rule in an iterative solution, to chose a regularization parameter which produces modeled measurements, $\boldsymbol{A}\hat{\boldsymbol{c}}_\alpha$, that are Poisson distributed (given Poisson-distributed measurements). To test whether a calculated set of measurements, $\boldsymbol{A}\hat{\boldsymbol{c}}_\alpha$, follows Poisson statistics, the Pearson's chi-square (χ^2) test can be applied, as was done by Veklerov and Llacer (1987). Care should be taken, however, to restore into the computed measurements any normalization[16] or transformation that

[16] As indicated in Section 15.2, the variance of a Poisson distribution is equal to its mean. With the value of measurement taken as an estimate of the mean, the variance depends on the magnitude of a measurement. Scaling or normalizing the latter changes the absolute value of the variance.

may have been introduced to the original measurements, in order to correctly apply the χ^2 test.

The χ^2 test involves calculating the relative difference between the observed frequency, h_{o_k}, and the expected (theoretical) frequency, h_{t_k}, of a distribution of an experimental outcome, k:

$$\chi^2 = \sum_{k=1}^{K} \frac{h_{o_k} - h_{t_k}}{h_{t_k}} \tag{10.77}$$

where K is the total number of outcomes of the experiment. The hypothesis of Poisson distribution is rejected if the obtained χ^2 value exceeds a certain critical value determined by $K-1$ (the number of degrees of freedom) and a desired χ^2 probability (significance level, typically taken to be 0.05 or less).

The procedure suggested by Veklerov and Llacer (1987) involves distributing, in increasing order, the values of actual, $\hat{\boldsymbol{e}}$, and modeled, $\tilde{\boldsymbol{e}} = \boldsymbol{A}\hat{\boldsymbol{c}}_\alpha$, measurements amongst K bins of equal intervals, so that all have the same theoretical probability, $p_k = \frac{1}{K}$, for $k = 1, \ldots K$. Then any observed measurement, $\hat{e}_i$, is assigned to the kth bin according to the Poisson distribution, with a computed measurement, $\tilde{e}_i$, taken as the mean value. Using Eq. (15.1), we have the two probabilities:

$$P_a = \sum_{n=0}^{\hat{e}_i - 1} \frac{\tilde{e}_i^n}{n!} \exp(-\tilde{e}_i)$$

$$P_b = \sum_{n=0}^{\hat{e}_i} \frac{\tilde{e}_i^n}{n!} \exp(-\tilde{e}_i)$$

representing two cumulative probabilities from zero to $\hat{e}_i - 1$ and zero to $\hat{e}_i$, respectively. A random number, ρ, uniformly distributed between P_a and P_b is generated using:

$$\rho = P_a + \xi(P_b - P_a) \tag{10.78}$$

where ξ is a random number uniformly distributed in $[0, 1]$. Then the measurement, $\hat{e}_i$ is assigned to the kth bin, with k being the smallest integer such that: $k \geq \rho K$.

Applying the above procedure to all values in the vector $\hat{\boldsymbol{e}}$, one obtains a histogram distribution, of h_k, the number of measurements belonging to bin i. Taking $h_k = h_{o_k}$ in Eq. (10.77), then the theoretical outcome becomes equal to $h_{t_k} = \frac{M}{K}$, where M is the number of available measurements, which are assumed to be equally distributed among the K bins. The χ^2 test can now be applied, provided that M is sufficiently a large number to make this test meaningful.

If the above χ^2 test fails for a given value of the regularization parameter, α^2, then another value for α^2 should be used to obtain another regularized solution, and the above test is repeated, until an acceptable value for α^2 is obtained. This exercise may yield more than one acceptable α^2 value, in which case one should select the value that best satisfies one of the regularization parameter selection methods described above. On the other hand, if no acceptable value is produced by the χ^2 test, then this should be taken as an indication of the inappropriateness of the *a priori* information, or the regularization method used in solving the inverse problem, but it could also be due to inadequacy of the forward model itself.

10.5.8 Regularization by Iteration

Iterative solution methods discussed in Section 10.6 also introduce a regularization effect, i.e. they help overcome the effect of uncertainties in the measurements (Bertero and Boccacci, 1998). A converging iterative solution will approach the true solution, within the range of error caused by measurement uncertainties. If an iterative process is stopped within that error, one in effect obtains a regularized solution. In essence, the number of iterations, k, becomes inversely proportional to the regularization parameter, α^2. That is, as $k \to \infty$, the effect of regularization vanishes, while at the beginning of iteration a large amount of regularization is introduced. Then, the number of iterations at which a satisfactory solution is arrived determines the effective amount of regularization introduced into the solution.

10.6 Iterative Methods

In a matrix-based linear formulation, direct matrix inversion is only possible when dealing with a matrix of a manageable size, within the available computing capacity. In large-scale problems, matrix inversion is avoided altogether. An iterative scheme is employed, even when the problem is linear, to avoid the increasing storage and manipulation demands of computing a direct inversion.

An iterative solution starts with an initial estimate (guess) of the solution, $\hat{c}_0$. Obviously, if *a priori* estimate of the solution is available, it should be used as the initial guess. On the other hand, if a reasonable *a priori* knowledge of the nature of the expected image is not available, a uniform image of some guessed attribute, or even an empty imaging domain, can serve as the initial guess. Of course, the initial guess must be physically sound, and within the lower and upper bounds of the expected solution; for instance an initial guess that has negative densities or attenuation coefficients is not acceptable.

Progression from one solution estimate, $\hat{\boldsymbol{c}}^{(k)}$, at iteration k, to the next, with $k = 0$ being the initial guess, is achieved via a number of schemes, summarized in the ensuing sections. However, one must ensure that each obtained solution estimate, $\hat{\boldsymbol{c}}^{(k)}$, like the initial guess, corresponds to a physically acceptable value within the expected solution bounds; otherwise one may not arrive at the end of the iteration process to an acceptable solution.

The problem's solution is arrived at when the iterative process is stopped in accordance to some predesignated criterion. Let $\hat{\boldsymbol{c}}^{(K)}$ be the solution at the last iteration, K. Intuitively, one would want $\hat{\boldsymbol{c}}^{(K)}$ to be equal to the true solution, $\boldsymbol{c}$, which is unknown.

Therefore, some metrics are employed for the stopping criteria, and are discussed in Section 10.6.5, after introducing various iterative schemes.

10.6.1 Geometric Approach

For an approximation $\boldsymbol{Ac}^{(k)}$ of $\boldsymbol{e}$, all vectors $\boldsymbol{c}_i^{(k)}$, satisfying:

$$\sum_j A_{ij} c_j^{(k)} = \text{constant} \tag{10.79}$$

define a hyperplane,[17] since the values of the elements of $\boldsymbol{c}^{(k)}$ become restricted by the value of the constant. As such, the vector loses one degree of freedom, and the dimensionality of its vector space is reduced by one. Note that when the constant in Eq. (10.79) becomes equal to e_i, one reaches the hyperplane of the sought solution. An iterative scheme that "depends only on the orientation of hyperplanes and not on a specific algebraic representation of the hyperplanes" is a geometric algorithm, in accordance to Gordon and Mansour (2007). If both sides of Eq. (10.79) are normalized, say with the squared value of the Euclidean norm of its A_{ij} coefficients, then one would still arrive at the same hyperplane, since the right-hand-side of the equation remains constant. However, this normalization process removes any dependence on specific values of the elements of the matrix, $\boldsymbol{A}$, and if applied before the implementation of any non-geometric iterative scheme, renders it into a geometric one. Now, one can devise an iterative scheme as follows:

$$\boldsymbol{Ac}^{(k+1)} = \boldsymbol{Ac}^{(k)} + \boldsymbol{A}\Delta\boldsymbol{c}^{(k)} \tag{10.80}$$

where $\Delta\boldsymbol{c}^k$ is some corrective vector for $\boldsymbol{c}^{(k)}$. Equation (10.80) indicates that the iterative process moves from a set of hyperplanes to another. When $\boldsymbol{A}\Delta\boldsymbol{c}^{(k)} = \boldsymbol{0}$, the iterations stabilize at the same set of hyperplanes, and the solution is reached.

In linear systems, $\boldsymbol{A}$ does not change from one iteration to another and Eq. (10.80) is reduced to:

$$\boldsymbol{c}^{(k+1)} = \boldsymbol{c}^{(k)} + \Delta\boldsymbol{c}^{(k)} \tag{10.81}$$

This can be applied to one voxel at a time so that:

$$c_j^{(k+1)} = c_j^{(k)} + \sum_{i=1}^{M_j} \Delta c_{ji}^{(k)} \tag{10.82}$$

[17] A hyperplane is a generalization of a two-dimensional plane, in a three-dimensional space, into many dimensions in a multidimensional space. A k-dimensional hyperplane is defined in an n-dimensional space ($k < n$) with $k+1$ points (a line is a one-dimensional hyperplane in a three-dimensional space, and is defined by two points). A hyperplane described with a single linear equation: $\sum_1^n a_i x_i = b$, where a_i's and b are constants and x_i's are variables, is a linear plane with $n-k=1$, as the x_i values have lost one degree of freedom. A set of linear planes given by $\boldsymbol{Ax} = \boldsymbol{b}$, where $\boldsymbol{A}$ is $(n-k)\times n$ matrix, $\boldsymbol{x}$ is an $n\times 1$ vector and $\boldsymbol{b}$ is an $(n-k)\times 1$ vector, represents a set of linear k-dimensional hyperplanes in an n-dimensional space. For more information, consult a book on algebraic geometry.

where $\Delta c_{ij}^{(k)}$ is a correction applied to pixel j from its contribution to measurement e_i at iteration k, and M_j is the total number of measurements to which c_j contributes. The corrections can be (1) conducted simultaneously to all voxels, without any updates until the end of the iteration, (2) applied one voxel at a time, with all corrections effected before proceeding to other voxels, or (3) performed for one set of measurements (typically a projection) at a time, and applied to all voxels. The latter method is known as the Algebraic Reconstruction Technique (ART), and is one of the efficient methods used in early tomographic reconstruction systems (Herman, 1980).

The correction factor $\Delta c_{ji}^{(k)}$ in Eq. (10.82) can be estimated from the difference between actual measurements and the corresponding estimated ones based on a current estimate of $\boldsymbol{c}$. This difference is converted to a corresponding Δc by normalization with the corresponding $\boldsymbol{A}$-matrix elements, so that:

$$\Delta c_{ij}^{(k+1)} = \frac{\left(e_i - \sum_{j=1}^{N_i} A_{ij} c_j^{(k)}\right) A_{ij}}{\sum_{j=1}^{N_i} A_{ij}^2} \tag{10.83}$$

where N_i is the number of voxels contributing to a measurement e_i. Recall that in case of transmission tomography the measurements applied in image reconstruction are the ray-sums, as indicated in Chapter 4. The correction of Eq. (10.83) is known as the additive correction. This correction procedure can also be weighted by the inverse of measurement variance to give more influence to more accurate measurements, see for example Barrett (1981).

A multiplicative correction process can also be applied:

$$\Delta c_{ij}^{(k+1)} = \frac{\left(e_i - \sum_{k=1}^{N_i} A_{ik} c_k^{(k)}\right)}{\sum_{k=1}^{N_i} A_{ik} c_k^{(k)}} \times c_j^{(r)} \tag{10.84}$$

where the superscript r refers to the most recently available value of c_j. It can be easily shown that Eq. (10.84) distributes the residual error for a measurement in proportion to the contribution of a pixel to the estimated measurements. One can show that $c_j^{(r)} + \Delta c_{ij}^{(k)} = c_j^{(r)} \frac{e_i}{\sum_{j=1}^{N_i} A_{ij} c_j^{(k)}}$, which indicates that the ratio between the measured and modeled measurement is used in the multiplicative correction process of Eq. (10.84).

10.6.2 Successive Approximation

A natural iterative course is to proceed from one approximate solution, $\hat{\boldsymbol{c}}^{(k)}$, to another one, $\hat{\boldsymbol{c}}^{(k+1)}$, that is closer to the final solution, $\hat{\boldsymbol{c}}^{(K)}$. This requires a mapping of $\boldsymbol{c}$ into itself, say $\boldsymbol{c} = T(\boldsymbol{c})$, so that one can set the iterative scheme as:

$$\hat{\boldsymbol{c}}^{(k+1)} = T\left(\hat{\boldsymbol{c}}^{(k)}\right) \tag{10.85}$$

with the mapping, T, such that the successive approximation process converges to a *fixed point* in the solution space, $\boldsymbol{C}$, when $\hat{\boldsymbol{c}}^{(K)} = T\left(\hat{\boldsymbol{c}}^{(K)}\right)$ as $K \to \infty$.

The mapping, T, can be constructed for the generalized least-squares solution of Eq. (10.23) as (Bertero and Boccacci, 1998):

$$T(\boldsymbol{c}) = \boldsymbol{c} + \tau\left\{\boldsymbol{A}^{\mathrm{T}}\boldsymbol{W}^2\hat{\boldsymbol{e}} - \left[\boldsymbol{A}^{\mathrm{T}}\boldsymbol{W}^2\boldsymbol{A} + \alpha^2\boldsymbol{G}^2\right]\boldsymbol{c}\right\} \tag{10.86}$$

where τ is a *relaxation* (damping) parameter selected to ensure that the iterative process of Eq. (10.85) converges. Since the solution of Eq. (10.23) is based on minimizing the objective function (10.22), upon reaching this minimum, the difference between $T(\boldsymbol{c})$ and $\boldsymbol{c}$ is also minimized. Then the iterative scheme of Eq. (10.85) reaches its fixed-point.

In order for this successive approximation process of Eq. (10.86) to reach a fixed point, $T(\boldsymbol{c})$, it must be a contraction mapping. Let $\bar{\boldsymbol{A}} = \left[\boldsymbol{A}^{\mathrm{T}}\boldsymbol{W}^2\boldsymbol{A} + \alpha^2\boldsymbol{G}^2\right]\boldsymbol{c}$, and $\bar{\boldsymbol{e}} = \boldsymbol{A}^{\mathrm{T}}\boldsymbol{W}^2\hat{\boldsymbol{e}}$, then the incorporation of Eq. (10.86) into Eq. (10.85) becomes:

$$\begin{aligned}
\hat{\boldsymbol{c}}^{(k+1)} &= T\left(\bar{\boldsymbol{c}}^{(k+1)}\right) = \tau\bar{\boldsymbol{e}} + (\boldsymbol{I} - \tau\bar{\boldsymbol{A}})\hat{\boldsymbol{c}}^{(k)} \qquad (10.87)\\
\hat{\boldsymbol{c}}^{(1)} &= = \tau\bar{\boldsymbol{e}} + (\boldsymbol{I} - \tau\bar{\boldsymbol{A}})^2\hat{\boldsymbol{c}}^{(0)}\\
\hat{\boldsymbol{c}}^{(2)} &= \tau\bar{\boldsymbol{e}} + (\boldsymbol{I} - \tau\bar{\boldsymbol{A}})\hat{\boldsymbol{c}}^{(1)} = \tau[\boldsymbol{I} + (\boldsymbol{I} - \tau\bar{\boldsymbol{A}})]\bar{\boldsymbol{e}} + (\boldsymbol{I} - \tau\bar{\boldsymbol{A}})\hat{\boldsymbol{c}}^{(0)}\\
\cdots &\quad \cdots\cdots\cdots\cdots\\
\hat{\boldsymbol{c}}^{(k)} &= \tau[\boldsymbol{I} + (\boldsymbol{I} - \tau\bar{\boldsymbol{A}}) + (\boldsymbol{I} - \tau\bar{\boldsymbol{A}})^2 + \cdots + (\boldsymbol{I} - \tau\bar{\boldsymbol{A}})^{k-1}]\bar{\boldsymbol{e}} + (\boldsymbol{I} - \tau\bar{\boldsymbol{A}})^k\hat{\boldsymbol{c}}^{(0)}
\end{aligned} \tag{10.88}$$

For this series to converge, one must have:

$$\varrho = \|\boldsymbol{I} - \tau\bar{\boldsymbol{A}}\| < 1 \tag{10.89}$$

which ensures that:

$$\left\|T(\boldsymbol{c}^{(k+1)}) - T(\boldsymbol{c}^{(k)})\right\| \leq \varrho\left\|\boldsymbol{c}^{(k+1)} - \boldsymbol{c}^{(k)}\right\| \tag{10.90}$$

The relaxation parameter τ, must then be chosen to ensure that the condition of Eq. (10.89) is satisfied. Such a contraction mapping will lead to a single fixed point $\boldsymbol{c} \in \boldsymbol{C}$ provided that $\hat{\boldsymbol{c}}^{(k+1)} \in C$ for all k. If there were two fixed points, say $\boldsymbol{c}_1$ and $\boldsymbol{c}_2$, then in accordance to Eqs. (10.85) and (10.90): $\|\boldsymbol{c}_1 - \boldsymbol{c}_2\| = \|T(\boldsymbol{c}_1) - T(\boldsymbol{c}_2)\| \leq \varrho\|\boldsymbol{c}_1 - \boldsymbol{c}_2\|$, which necessitates that $\boldsymbol{c}_1 = \boldsymbol{c}_2$, since $\varrho < 1$. Therefore, contraction mapping leads to a fixed point: the solution of Eq. (10.23).

The mapping of Eq. (10.85) is only a contraction mapping if $\boldsymbol{A}^{\mathrm{T}}\boldsymbol{A}$ is a full-rank matrix (Bertero and Boccacci, 1998), i.e. all its columns and rows are linearly independent, or equivalently the eigenvalues of $\boldsymbol{A}^{\mathrm{T}}\boldsymbol{A}$ have a positive lower-bound. If this condition is not satisfied, then the mapping becomes non-expansive, i.e. $\varrho = 1$, and it is then difficult to guarantee that the sequence of Eq. (10.88) converges. Another difficulty is to ensure that $\boldsymbol{c}^{(k)} \in \boldsymbol{C}$, for all values of k. However, with *a priori* knowledge of the lower and upper bounds of c, out-of-bound values can be set equal to their closest bounds.

The successive approximation method is also known as the Landweber method. It is a gradient method, in the sense that the iterative process proceeds in the direction of the gradient of the cost function (10.22). It can, therefore, be expressed as:

$$\boldsymbol{c}^{(k+1)} = \boldsymbol{c}^{(k)} + \tau \boldsymbol{r}^{(k)} \tag{10.91}$$

where,

$$\boldsymbol{r}^{(k)} = \boldsymbol{A}^{\mathrm{T}} \boldsymbol{W}^2 \hat{\boldsymbol{e}} - \left[\boldsymbol{A}^{\mathrm{T}} \boldsymbol{W}^2 \boldsymbol{A} + \alpha^2 \boldsymbol{G}^2\right] \boldsymbol{c}^{(k)} \tag{10.92}$$

with $\alpha^2 = \frac{\delta_\eta^2}{\delta_\chi^2}$. The gradient of the cost function with respect to $\boldsymbol{c}$ is proportional to $\boldsymbol{A}^{\mathrm{T}} \boldsymbol{W}^2 \left(\boldsymbol{A}\boldsymbol{c}^{(k)} - \hat{\boldsymbol{e}}\right) + \alpha^2 \boldsymbol{G}^2 \boldsymbol{c}^{(k)} = -\boldsymbol{r}^{(k)}$. Therefore, the successive approximation process is driven in the direction opposite to the gradient of the convex cost function, i.e. toward its minimum. Other gradient-based iterative processes are given below.

10.6.3 Steepest Descent

In this method, as in the successive approximation method, iterations are driven in the direction of the gradient of the solution's cost function, except that the relaxation parameter changes from one iteration to another, in an attempt to match the actual, i.e. the steepest, gradient of the cost function. The iterative process of Eq. (10.91) takes now the form:

$$\boldsymbol{c}^{(k+1)} = \boldsymbol{c}^{(k)} + \tau^{(k)} \boldsymbol{r}^{(k)} \tag{10.93}$$

with $\boldsymbol{r}^{(k)}$ defined by Eq. (10.92). This method is also known as the Cauchy method.

The value of τ_k is selected to maximize the change in the cost function from one iteration to another. Returning to the notation $\bar{\boldsymbol{A}} = \left[\boldsymbol{A}^{\mathrm{T}} \boldsymbol{W}^2 \boldsymbol{A} + \alpha^2 \boldsymbol{G}^2\right] \boldsymbol{c}$, and $\bar{\boldsymbol{e}} = \boldsymbol{A}^{\mathrm{T}} \boldsymbol{W}^2 \hat{\boldsymbol{e}}$, minimizing the cost function of Eq. (10.22) is equivalent to minimizing $(\bar{\boldsymbol{A}}\boldsymbol{c} - \bar{\boldsymbol{e}})^2$. With a convex cost function, one would desire $\left(\bar{\boldsymbol{A}}\boldsymbol{c}^{(k+1)} - \hat{\boldsymbol{e}}\right)^2 < \left(\bar{\boldsymbol{A}}\boldsymbol{c}^{(k)} - \hat{\boldsymbol{e}}\right)^2$. Therefore, proceeding from iteration k to iteration $k+1$, τ_k is determined by maximizing:

$$\begin{aligned}
\left(\bar{\boldsymbol{A}}\boldsymbol{c}^{(k)} - \hat{\boldsymbol{e}}\right)^2 - \left(\bar{\boldsymbol{A}}\boldsymbol{c}^{(k+1)} - \hat{\boldsymbol{e}}\right)^2 &= -2\tau^{(k)} \left(\bar{\boldsymbol{A}}\boldsymbol{c}^{(k)} - \bar{\boldsymbol{e}}\right)^{\mathrm{T}} \bar{\boldsymbol{A}}\boldsymbol{r}^{(k)} - \left(\tau^{(k)}\right)^2 \left(\bar{\boldsymbol{A}}\boldsymbol{r}^{(k)}\right)^2 \\
&= 2\tau^{(k)} \boldsymbol{r}^{(k)\mathrm{T}} \bar{\boldsymbol{A}}\boldsymbol{r}^{(k)} - \left(\tau^{(k)}\right)^2 \left(\bar{\boldsymbol{A}}\boldsymbol{r}^{(k)}\right)^2 \\
&= 2\tau^{(k)} \left(\sqrt{\bar{\boldsymbol{A}}}\boldsymbol{r}^{(k)}\right)^2 - \left(\tau^{(k)}\right)^2 \left(\bar{\boldsymbol{A}}\boldsymbol{r}^{(k)}\right)^2
\end{aligned} \tag{10.94}$$

where use is made of the fact that $\boldsymbol{r}^{(k)} = \bar{\boldsymbol{e}} - \bar{\boldsymbol{A}}\boldsymbol{c}^{(k)}$. Maximizing (10.94) gives:

$$\tau^{(k)} = \frac{\|\boldsymbol{r}^{(k)}\|^2}{\|\sqrt{\bar{\boldsymbol{A}}}\boldsymbol{r}^{(k)}\|^2} \tag{10.95}$$

Note that with $\sqrt{\bar{A}} = AW$, i.e. with no regularization ($\alpha = 0$), and with $W = I$, Eq. (10.95) becomes equal to the usual expression reported in the literature, such as in Bertero and Boccacci (1998).

The dynamic change of the relaxation parameter, τ, from one iteration to another, is needed to provide the steepest descent toward the minimum of the solution's cost function. This should accelerate the iterative process toward the solution. However, as the iterative process approaches the solution, it may tend to oscillate back and forth around the minimum of the cost function. Moreover, the iterative step size, $\tau^{(k)}\boldsymbol{r}^{(k)}$, tends to shrink near the solution, which also tends to increase the number of iterations required for solution convergence. Therefore, a method that drives the iterative process directly toward the minimum of the cost function, such as the conjugate gradient method, is favored.

10.6.4 Conjugate Gradient

The conjugate gradient generalizes the definition of orthogonality of one vector to another, in terms of an operator: a positive definite matrix. In minimizing the cost function of Eq. (10.22), one can define "conjugacy" using $\bar{A} = \left[A^{\mathrm{T}} W^2 A + \alpha^2 G^2\right] c$, to relate two iterations gradients, $\boldsymbol{d}^{(k+1)}$ and $\boldsymbol{d}^{(k)}$, to each other, so that:

$$\left(\boldsymbol{d}^{(k+1)}\right)^{\mathrm{T}} \bar{A} \boldsymbol{d}^{(k)} = 0 \tag{10.96}$$

It is then said that the gradient $\boldsymbol{d}^{(k+1)}$ is $\bar{A}$-conjugate to the gradient $\boldsymbol{d}^{(k)}$. If $\boldsymbol{d}^{(k)}$ is simply orthogonal to the normal gradient, $\boldsymbol{r}^{(k)}$, i.e. $(\boldsymbol{d}^{(k)})^{\mathrm{T}}\boldsymbol{r}^{(k)} = 0$, then $\boldsymbol{d}^{(k+1)}$ will be directed toward the minimum of the solution's cost function (Bertero and Boccacci, 1998). This is an effective way of guiding the iterative process toward a solution.

The iterative process starts in the direction of the steepest descent, i.e. with $\boldsymbol{d}^{(k)} = \boldsymbol{r}^{(0)}$, with the latter determined using Eq. (10.92) with $k = 0$. All subsequent, $\bar{A}$-conjugate gradients are then determined by the recursive relationship:

$$\boldsymbol{d}^{(k+1)} = \boldsymbol{r}^{(k)} + \tau_r^{(k)} \boldsymbol{d}^{(k)} \tag{10.97}$$

with $\tau_r^{(k)}$ determined, for example as:

$$\tau_r^{(k)} = \frac{\|\boldsymbol{r}^{(k+1)}\|^2}{\|\boldsymbol{r}^{(k)}\|^2} \tag{10.98}$$

The iterative process then proceeds as:

$$\boldsymbol{c}^{(k+1)} = \boldsymbol{c}^{(k)} + \tau^{(k)} \boldsymbol{d}^{(k)} \tag{10.99}$$

with

$$\tau_{(k)} = \frac{\left\| \left(\boldsymbol{d}^{(k)}\right)^{\mathrm{T}} \boldsymbol{r}^{(k)} \right\|}{\left\| \left(\boldsymbol{d}^{(k)}\right)^{\mathrm{T}} \bar{A} \boldsymbol{r}^{(k)} \right\|} \tag{10.100}$$

The relaxation parameters, $\tau_r^{(k)}$ and $\tau_{(k)}$, are chosen to ensure the orthogonality of $\boldsymbol{r}^{(k+1)}$ to $\boldsymbol{r}^{(k)}$ and the $\bar{\boldsymbol{A}}$-conjugate orthogonality of $\boldsymbol{d}^{(k+1)}$ and $\boldsymbol{d}^{(k)}$. The conjugate gradient method typically converges much faster than other iterative methods, since the conjugate gradient drives the iterative process towards the desired solution. For more details, see a book on numerical analysis or optimization, such as (Golub, 1996; Walsh, 1975).

10.6.5 Convergence Metrics and Stopping Criteria

In order to ensure that an iterative process does not go astray, a maximum number of allowed iterations should always be designated. The value of this number can be based on some prior experience with the problem, or when dealing with a new problem, on a value that does not permit excessive computational time. It is also advisable to monitor the execution time, and to stop iterations if that time exceeds a predesignated computing (cpu) processing time. An iterative process is terminated by either the maximum number of iterations, or maximum cpu time, when none of the other stopping criteria are met. An inspection of the value and trend of the metrics for these criteria should then be conducted, to determine whether the iterative process is converging to an acceptable solution or not. If the iterative process was proceeding toward a solution, then the problem should be resumed with the last obtained solution, $\hat{\boldsymbol{c}}^{(K)}$, used as an initial guess. Subsequently, the predesignated values for the maximum number of iterations and allowed cpu time should be reevaluated. If none of the metrics for the other stopping criteria are still satisfied, one should reexamine the problem and the iterative scheme used. Even if a convergent solution is reached, one should also restart the problem with a different initial guess, in the hope of reaching a solution within a smaller number of iterations and execution time. However, a robust solution should be independent of the initial guess.

An iterative process should stop when some metric reaches a sufficiently small predesignated value. In addition, if the stopping metric stabilizes, i.e. ceases to decrease, or decreases very slowly, the iterative process should also stop, since there is no benefit in continuing iteration. Then, the value of the obtained metric is likely the smallest possible value one would arrive at, with the given set of measurements. It is advisable to repeat the solution with a number of distinct initial guesses, in order to ensure that the solution is independent of the initial guess; if not the problem would not have a unique solution and the problem formulation should be reexamined. A number of stopping metrics are given below. The order of applying these stopping criteria should be as follows: (1) allocated cpu reached or exceeded, (2) maximum number of iterations is attained, (3) residual-metric stopping criterion is satisfied, and (4) minimum of convergence metric is reached. Another criterion, discussed at the end of this section, is only applicable when testing an iterative scheme with an image of known attributes, to see whether the true solution is reached. This stopping criterion should be applied after the iterative process survives the above four criteria. In all cases, the attained value of the stopping parameter should be reported at the end of the iterative process, and preferably the value of the metrics should be retained for plotting as a function of the iteration number, to examine the convergence behavior of the iterative scheme.

Residual Metric

Given a measurement uncertainty, $\boldsymbol{\delta}_e$, the best one can hope for is a solution, $\hat{\boldsymbol{c}}^{(k)}$, that produces modeled measurements, $\boldsymbol{A}\hat{\boldsymbol{c}}^{(k)}$, that deviate from the given measurements, $\hat{\boldsymbol{e}}$, by no more than $\boldsymbol{\delta}_e$. Therefore, a stopping criteria can be devised based on the residual norm so that:

$$\|\boldsymbol{A}\hat{\boldsymbol{c}}^{(k)} - \hat{\boldsymbol{e}}\| \leq \|\boldsymbol{\delta}_e\| \tag{10.101}$$

with the norm, $\|\cdot\|$, taken as the Euclidean norm, a measure of the most probable error, see Appendix 7.A.

A number of variants to this stopping criteria can be applied (Barrett et al., 1994). For instance, when $\|\boldsymbol{\delta}_e\|$ cannot be estimated, or when its estimate is unreliable, the following stopping criterion can be applied:

$$\|\boldsymbol{A}\hat{\boldsymbol{c}}^{(k)} - \hat{\boldsymbol{e}}\| \leq \varepsilon_e \|\hat{\boldsymbol{e}}\| \tag{10.102}$$

where ε_e is an estimate of the overall relative uncertainty in $\hat{\boldsymbol{e}}$, also called a stopping tolerance.

When $\|A\|$ is available, a less restrictive stopping criterion is:

$$\|\boldsymbol{A}\hat{\boldsymbol{c}}^{(k)} - \hat{\boldsymbol{e}}\| \leq \varepsilon_e (\|A\| \, \|\hat{\boldsymbol{c}}^{(k)}\| + \|\hat{\boldsymbol{e}}\|) \tag{10.103}$$

In all the above cases, the error in $\boldsymbol{c}$ is $\|\boldsymbol{\delta}_c\| \leq \|\boldsymbol{A}^{\dagger}\| \, \|\boldsymbol{A}\hat{\boldsymbol{c}}^{(k)} - \hat{\boldsymbol{e}}\|$; recall that $\boldsymbol{A}^{\dagger}$ is the inverted matrix. Therefore, if an estimate of $\|\boldsymbol{A}^{\dagger}\|$ is known, then the following stopping criterion may also be applied:

$$\|\boldsymbol{A}\hat{\boldsymbol{c}}^{(k)} - \hat{\boldsymbol{e}}\| \leq \varepsilon_e \frac{\|\hat{\boldsymbol{c}}^{(k)}\|}{|\boldsymbol{A}^{\dagger}\|} \tag{10.104}$$

which assures that the relative error in $\hat{\boldsymbol{c}}^{(k)}$ is less than ε_e.

Convergence Metric

Giving a machine's precision, ε_{mp}, one should stop an iterative process when the distance between two successive estimates of a solution is less than or equal to some value ε_v, i.e.:

$$\|\hat{\boldsymbol{c}}^{(k)} - \hat{\boldsymbol{c}}^{(k-1)}\| \leq \varepsilon_v \|\hat{\boldsymbol{c}}^{(k)}\| \tag{10.105}$$

with $\varepsilon_v \geq \varepsilon_{mp}$. The relationship of Eq. (10.105) can be used as a stopping criterion, even when a residual-based stopping criterion is not reached, since there is no point of obtaining images with precision greater than what is considered to be practically acceptable. However, such stopping criterion should be applied only if the convergence of the solution algorithm is assured in previous testing.

An iterative process that converges to a solution should have a contracting distance between iterations, that is:

$$\|\hat{\boldsymbol{c}}^{(k)} - \hat{\boldsymbol{c}}^{(k-1)}\| < \|\hat{\boldsymbol{c}}^{(k-1)} - \hat{\boldsymbol{c}}^{(k-2)}\| \tag{10.106}$$

In a converging solution, $\|\hat{\boldsymbol{c}}^{(k)} - \hat{\boldsymbol{c}}^{(k-1)}\|$ should continue to decrease as the solution is approached. Significant irregular, oscillatory, or stalled behavior of this metric toward the end of the solution may be caused by attempting to converge to a precision beyond that of the computing machine; otherwise it may due to instabilities in the iterative process. The latter can be addressed with improved regularization. One should expect, however, some irregular behavior in the convergence norm at the beginning of iteration, particularly if the initial guess is far off from the true solution.

Testing Metric

When testing a convergence scheme, one should start with a problem of known solution, $\boldsymbol{c}$, and monitor the distance $\|\hat{\boldsymbol{c}}^{(k)} - \boldsymbol{c}\|$. Obviously, if the initial guess is equal to $\boldsymbol{c}$, the solution should be attained at the first iteration, if not the computer algorithm and its coding should be checked, and the number and quality of measurements should be assessed. Otherwise, the iterative scheme should stop when the following criterion is satisfied:

$$\|\hat{\boldsymbol{c}}^{(k)} - \boldsymbol{c}\| \leq \varepsilon_v \|\boldsymbol{c} \tag{10.107}$$

There is no point in proceeding further once the true solution is reached. The other stopping metrics would then give information on the residual error and the convergence error of the problem. On the other hand, if a solution is arrived at without fulfilling the stopping criterion of Eq. (10.107), while converging in accordance to the other metrics, one should reassess the problem and determine whether the measurements used have a high level of uncertainty, the forward model is too simplistic, or whether more effective regularization is needed. Note that using a forward-model based on analytical calculations, say over a continuous test object such as the Shepp-Logan phantom (see Section 15.6), and utilizing the estimated measurements to reconstruct an image, enables one to determine the discretization error caused by dividing the image into a certain number of pixels/voxels. The magnitude of this discretization error increases with the coarseness of the image. Then, the testing metric will not converge to a zero value, but to a value corresponding to the degree of discretization.

10.7 Nonlinear Problems

When the forward mapping matrix, $\boldsymbol{A}$, is a function of $\boldsymbol{c}$, as in scatter imaging (see Chapter 6), the cost function of Eq. (10.22) becomes nonlinear. The problem must then be solved iteratively, by successive approximation or by gradient-based iterative methods. The latter approach requires the determination of the gradient of the cost function, which is not as straightforward as in the case of the linear problem discussed in Section 10.6, where the gradient is defined by Eq. (10.92). Nevertheless, the gradient can be determined, if not analytically, at least numerically using the finite-difference method to calculate the Jacobian matrix of $\boldsymbol{A}$. The book by Fletcher (1987) is an excellent source for treating nonlinear least-squares problems. Some of the common methods are presented here.

For nonlinear problems more than one solution that minimizes the cost function, Eq. (10.22) can exist. There is a tendency for iterative schemes to converge to a local minimum, typically one that is close to the initial solution guess, $\boldsymbol{c}^{(0)}$. It is, therefore, vital to examine the obtained solution to ensure that it is physically acceptable. An iterative scheme that produces a local minimum would be usually sensitive to $\boldsymbol{c}^{(0)}$, producing different answers for different starting points. Bounding the solution, to be between the physically anticipated low and upper bounds, can guide the iterative process toward conversion. The choice of the regularization method and the value of its associated parameter can also steer the iterative process toward the desired solution. Regularization also helps in addressing ill-posedness that may be caused by the nonlinear nature of the problem, even in the absence of noise. For a discussion on the above-mentioned issues see Hussein et al. (1986b) and Snieder (1998).

10.7.1 Quasi-Linearization

In weakly nonlinear problems, a solution can be found by linearizing the forward problem around a reference solution, $\boldsymbol{c}^*$, so that (Tarantola, 1987):

$$\begin{aligned}\hat{\boldsymbol{e}} &= \boldsymbol{A}\boldsymbol{c} \simeq \boldsymbol{A}\boldsymbol{c}^* + \nabla(\boldsymbol{A}\boldsymbol{c})|_{c=c^*}(\boldsymbol{c}-\boldsymbol{c}^*)\\ &= \boldsymbol{A}\boldsymbol{c}^* + \left[\boldsymbol{A} + \nabla\boldsymbol{A}|_{c=c^*}\boldsymbol{c}^*\right](\boldsymbol{c}-\boldsymbol{c}^*)\\ &= \left[\boldsymbol{A} + \nabla\boldsymbol{A}|_{c=c^*}\boldsymbol{c}^*\right]\boldsymbol{c} - [\nabla\boldsymbol{A}|_{c=c^*}\boldsymbol{c}^*]\boldsymbol{c}^* \end{aligned} \tag{10.108}$$

where the gradient, $\nabla\boldsymbol{A}$, is with respect to $\boldsymbol{c}$. The semi-equal sign is used to indicate that the second-order derivatives are neglected. Then direct inversion, or one of the iterative methods of Section 10.6, can be used to solve Eq. (10.108) for $\boldsymbol{c}$.

10.7.2 Successive Approximation

The successive approximation method, discussed in Section 10.6.2 for linear problems, can also be applied to nonlinear problems. The self-mapping, T, of Eq. (10.86) is structured such that the nonlinear terms in the matrix $\boldsymbol{A}(\boldsymbol{c})$ are evaluated using the current approximation, $\hat{\boldsymbol{c}}^{(k)}$, so that:

$$T(\boldsymbol{c}^{(k)}) = \boldsymbol{c}^{(k)} + \tau\left\{\boldsymbol{A}^{\mathrm{T}}(\boldsymbol{c}^{(k)})\boldsymbol{W}^2\hat{\boldsymbol{e}} - \left[\boldsymbol{A}^{\mathrm{T}}(\boldsymbol{c}^{(k)})\boldsymbol{W}^2\boldsymbol{A}(\boldsymbol{c}^{(k)}) + \alpha^2\boldsymbol{G}^2\right]\boldsymbol{c}^{(k)}\right\} \tag{10.109}$$

where $\boldsymbol{A}(\boldsymbol{c})$ indicates that $\boldsymbol{A}$ is a function of $\boldsymbol{c}$. Then the iterative process resumes as in Eq. (10.85):

$$\hat{\boldsymbol{c}}^{(k+1)} = T(\hat{\boldsymbol{c}}^{(k)}) \tag{10.110}$$

with the mapping T such that the successive approximation process converges to a *fixed point* in the solution space, $\boldsymbol{C}$, when $\hat{\boldsymbol{c}}^{(K)} = T(\hat{\boldsymbol{c}}^{(K)})$ as $K \to \infty$. Again, the relaxation parameter, τ, must be selected to ensure the contraction mapping of Eq. (10.110).

The obvious advantage of the successive approximation method is that it does not require the calculation of the derivative of the cost function, hence the Jacobian of the matrix $\boldsymbol{A}(\boldsymbol{c})$. However, care must be taken to ensure that the conditions of contraction mapping are met at all stages of the solution, since the condition of Eq. (10.89) ($\varrho < 1$) becomes difficult to satisfy due to the change in $\boldsymbol{A}$ from one iteration to another. In addition to satisfying the conditions of contraction mapping, it must also be assured that $\boldsymbol{c}^{(k)} \in \boldsymbol{C}$, by bringing any estimate $\boldsymbol{c}^{(k)}$ outside $\boldsymbol{C}$ back into $\boldsymbol{C}$. The latter may divert from the contraction of the mapping. This successive approximation process was applied for solving the inverse problem of scatter imaging (Arendtsz and Hussein, 1995b; Hussein et al., 1986a,b).

10.7.3 Newton-Raphson

A cost function, $\Phi(\boldsymbol{c})$, can be approximated by a Taylor's series expansion around some estimated solution, $\boldsymbol{c}^{(k)}$ as:

$$\Phi(\boldsymbol{c}) = \Phi(\boldsymbol{c}^{(k)}) + \nabla\Phi_k^{\mathrm{T}}(\boldsymbol{c} - \boldsymbol{c}^{(k)}) + \frac{1}{2}(\boldsymbol{c} - \boldsymbol{c}^{(k)})\nabla^2\Phi_k(\boldsymbol{c} - \boldsymbol{c}^{(k)}) + \mathrm{H.O.T.} \tag{10.111}$$

Ignoring the higher order terms (H.O.T.), and setting $\nabla\Phi = 0$ to maximize the cost function, then:

$$\nabla\Phi = \nabla\Phi_k + \nabla^2\Phi_k(\boldsymbol{c} - \boldsymbol{c}^{(k)}) = 0 \tag{10.112}$$

which leads to the iterative scheme:

$$\boldsymbol{c}^{(k+1)} = \boldsymbol{c}^{(k)} - \tau_k\left[\nabla^2\Phi_k\right]^{-1}\nabla\Phi_k \tag{10.113}$$

where the relaxation parameter, τ_k, is introduced to avoid convergence to a saddle point (where $\nabla^2\Phi_k = 0$). This relaxation parameter is evaluated by minimizing: $\Phi\left(\boldsymbol{c}^{(k)} - \tau_k\left[\nabla^2\Phi_k\right]^{-1}\right)$ (Rao, 1996). Note that when Φ is a quadratic function, as is the case in Eq. (10.22) when $\boldsymbol{A}$ is linear, the iterative scheme of the Newton method becomes identical to that of the steepest descent method, discussed in Section 10.6.3.

The evaluation of $\nabla^2\Phi_k$ involves the computation of the Jacobian matrix of the function derivatives (the Hessian). This and the calculation of $\nabla\Phi_k$ are computationally demanding. The scheme is quite effective when $\boldsymbol{c}^{(k)}$ is close to the solution. If the initial guess is far from the solution, Newton's method would be quite slow, and it is advisable to employ the successive approximation process. The Levenberg-Marquard method, discussed in (Section 10.7.5) combines the two schemes.

10.7.4 Gauss-Newton

This method, also known as the quasi-Newton method, approximates the Hessian matrix by using only the first partial derivatives. The derivative of the residual term, $\left[\boldsymbol{W}\left(\boldsymbol{A}\boldsymbol{c}-\hat{\boldsymbol{e}}\right)\right]^{\mathrm{T}}\left[\boldsymbol{W}\left(\boldsymbol{A}\boldsymbol{c}-\hat{\boldsymbol{e}}\right)\right]$ in the cost function of Eq. (10.22) is:

$$\begin{aligned}\nabla\left\{\left[\boldsymbol{W}\left(\boldsymbol{A}\boldsymbol{c}-\hat{\boldsymbol{e}}\right)\right]^{\mathrm{T}}\left[\boldsymbol{W}\left(\boldsymbol{A}\boldsymbol{c}-\hat{\boldsymbol{e}}\right)\right]\right\} &= \nabla\left\{[\boldsymbol{R}(\boldsymbol{c})]^{\mathrm{T}}\boldsymbol{R}(\boldsymbol{c})\right\} \\ &= 2[\nabla\boldsymbol{R}(\boldsymbol{c})]^{\mathrm{T}}\boldsymbol{R}(\boldsymbol{c}) \qquad (10.114)\end{aligned}$$

where $\boldsymbol{R}_i$ is the ith vector of the matrix $\boldsymbol{R}=\boldsymbol{W}\left(\boldsymbol{A}\boldsymbol{c}-\hat{\boldsymbol{e}}\right)$. The Hessian matrix, i.e. the second derivatives is:

$$\begin{aligned}\nabla^2\left\{[\boldsymbol{R}(\boldsymbol{c})]^{\mathrm{T}}\boldsymbol{R}(\boldsymbol{c})\right\} &= \nabla^2\left\{\left[\boldsymbol{W}^2\left(\boldsymbol{A}\boldsymbol{c}-\hat{\boldsymbol{e}}\right)\right]^{\mathrm{T}}\left[\boldsymbol{W}^2\left(\boldsymbol{A}\boldsymbol{c}-\hat{\boldsymbol{e}}\right)\right]\right\} \\ &= 2[\nabla\boldsymbol{R}(\boldsymbol{c})]^{\mathrm{T}}\nabla\boldsymbol{R}(\boldsymbol{c})+2\sum_{i=1}^{M}\boldsymbol{R}_i\nabla^2\boldsymbol{R}_i(\boldsymbol{c}) \qquad (10.115) \\ &\approx 2[\nabla\boldsymbol{R}(\boldsymbol{c})]^{\mathrm{T}}\nabla\boldsymbol{R}(\boldsymbol{c}) \qquad (10.116)\end{aligned}$$

When the residual is small, the last term in Eq. (10.115), which is computationally demanding, can be ignored. This, in turn, simplifies the application of the Newton's iterative process, Eq. (10.113), leading to the Quasi-Newton method (also known as the Gauss-Newton method, in the absence of regularization) (Tarantola, 1987). Note that if the Hessian is approximated by the identity matrix, the iterative scheme of Eq. (10.113) becomes equivalent to that the steepest descent method, Eq. (10.93).

10.7.5 Levenberg-Marquardt

This method combines the steepest descent method with the Newton-Raphson's method by modifying than latter scheme, Eq. (10.113), to (Rao, 1996):

$$\boldsymbol{c}^{(k+1)}=\boldsymbol{c}^{(k)}-\tau_k\left[\nabla^2\Phi_k+\alpha_k\boldsymbol{I}\right]^{-1}\nabla\Phi_k \qquad (10.117)$$

where the Lagrange multiplier $\alpha_k>0$ ensures that $\nabla^2\Phi_k+\alpha_k\boldsymbol{I}$ is positive definite, even if $\nabla^2\Phi_k$ is not. When $\alpha_k\gg 1$, the effect of $\nabla^2\Phi_k$ in Eq. (10.117) diminishes, and the iterative process is driven toward the steepest descent, while for small α_k the $\nabla^2\Phi_k$ terms influence more the iterative scheme and it becomes more like the Newton-Raphson method. Since the latter is more suited when $\boldsymbol{c}^{(k)}\to\boldsymbol{c}$, iterations should start with a large α_k value, gradually decreasing as the iterative process progresses, eventually reaching zero. This in practice is determined by defining the so-called *trust-region radius*. The value of Δ_k is compared versus the metric $\|\Phi_k\|$. As long as $\|\Phi_k\|\geq\Delta_k$, a positive value of α_k is applied, but when the condition $\|\Phi_k\|<\Delta_k$ is reached, $\alpha_k=0$ is imposed (Moré, 1977). Notice the resemblance between $\alpha_k\boldsymbol{I}$ in Eq. (10.117) and the corresponding term in the standard Tikhonov regularization, Eq. (10.17) with $\boldsymbol{G}=\boldsymbol{I}$, indicating the regularizing effect of the Levenberg-Marquard scheme, which adds to its effectiveness.

10.7.6 Conjugate Gradient

This method, discussed in Section 10.6.4, is also applicable to nonlinear problems (Rao, 1996). The method is known as the Fletcher-Reeves method, after the two workers who generalized the conjugate gradient method to nonlinear problems (Fletcher and Reeves, 1964). Its attraction is that it does not require the Hessian, as in the other Newton-based methods discussed above. Then the residual, $\boldsymbol{r}^{(k)}$ in Eq. (10.97), is replaced by the negative of the gradient, and $\tau_{(k)}$ of Eq. (10.100) is chosen so that the gradient is orthogonal to the search direction.

10.8 Software

There is a number of computer routines that can be utilized for solving the matrix-based problems discussed in this chapter. The Netlib repository `http://www.netlib.org/` contains freely available software. Of interest here are the LAPACK (`http://www.netlib.org/lapack/`) and the MINPAK (`http://www.netlib.org/minpack/`) collections; the former for linear problems and the latter for nonlinear ones. The Association for Computing Machinery (ACM) posts software associated with papers in its journal on `http://calgo.acm.org/`, some of which deals with the linear and nonlinear least-squares problems addressed in this chapter. *The Numerical Recipes* book by Press (2007) (`http://www.nr.com/`) also offers a number of useful routines. The Hansen's package (`http://www2.imm.dtu.dk/~pch/`) has a set of MATLAB regularization tools. Commercial vendors, such as Visual Numerics, Inc. (IMSL: `http://www.vni.com/products/imsl/`), Wolfram Research, Inc. (`http://www.wolfram.com/`), and the Numerical Algorithms Group (`http://nag.com/`) have also similar useful routines.

References

N. V. Arendtsz and E. M. A. Hussein, "Energy-spectral scatter imaging. Part I: Theory and mathematics," *IEEE Transactions on Nuclear Science*, vol. 42, pp. 2155–2165, 1995.

——, "Energy-spectral scatter imaging. Part II: Experiments," *IEEE Transactions on Nuclear Science*, vol. 42, pp. 2166–2172, 1995.

G. Aubert and P. Kornprobst, *Mathematical Problems in Image Processing*. New York: Springer Science, 2006.

H. H. Barrett, *Radiological Imaging*. New York: Academic Press, 1981.

R. Barrett, M. Berry, T. F. Chan, J. Demmel, J. Donato, J. Dongarra, V. Eijkhout, R. Pozo, C. Romine, and H. V. der Vorst, *Templates for the Solution of Linear Systems: Building Blocks for Iterative Methods*, 2nd ed. Philadelphia: Society for Industrial and Applied Mathematics, 1994, `http://www.netlib.org/linalg/html_templates/report.html`.

M. Bertero and P. Boccacci, *Introduction to Inverse Problems in Imaging*. Bristol: Institute of Physics, 1998.

C. Bouman and K. Sauer, "A generalized Gaussian image model for edge-preserving map estimation," *IEEE Trans. Image Processing*, vol. 2, pp. 296–310, 1993.

P. Charbonnier, L. Blanc-Féraud, G. Aubert, and M. Barlaud, "Deterministic edge-preserving regularization in computed imaging," *IEEE Trans. Image Process.*, vol. 6, pp. 298–311, 1997.

Y.-W. Chiang, P. P. Borbat, and J. H. Freed, "Maximum entropy: A complement to Tikhonov regularization for determination of pair distance distributions by pulsed ESR," *Journal of Magnetic Resonance*, vol. 177, pp. 184–196, 2005.

S. Constable, R. Parker, and C. Constable, "Occam's inversion: A practical algorithm for generating smooth models from electromagnetic sounding data," *Geophysics*, vol. 52, pp. 289–300, 1987.

P. Craven and G. Wahba, "Smoothing noisy data with spline functions," *Numerische Mathematik*, vol. 31, pp. 377–403, 1979.

G. Demoment, "Image reconstruction and restoration: Overview of common estimation structure and problems," *IEEE Trans. Acoust. Speech Signal Process*, vol. ASSP-37, pp. 2024–2036, 1989.

M. Dudík, S. J. Phillips, and S. J. Phillips, "Maximum entropy density estimation with generalized regularization and an application to species distribution modeling," *Journal of Machine Learning Research*, vol. 8, pp. 1217–1260, 2007.

H. W. Engl, M. Hanke, and A. Neubauer, *Regularization of Inverse Problems*. Dordrecht: Kluwer Academic Publishing, 2000.

R. Fletcher and C. M. Reeves, "Function minimization by conjugate gradients," *Computer Journal*, vol. 7, pp. 149–154, 1964.

R. Fletcher, *Practical Methods of Optimization*, 2nd ed. New York: John Wiley & Sons, 1987.

S. Geman and D. E. McClure, "Statistical methods for tomographic image reconstruction," *Bull. Int. Stat. Inst.*, vol. 52, pp. 5–21, 1987.

G. H. Golub, *Matrix Computations*. Baltimore: Johns Hopkins University Press, 1996.

D. Gordon and R. Mansour, "A geometric approach to quadratic optimization: An improved method for solving strongly underdetermined systems in CT," *Inverse Problems in Science and Engineering*, vol. 15, pp. 811–826, 2007.

P. J. Green, "Bayesian reconstructions from emission tomogrpahy data using a modified EM algorithm," *IEEE Trans. Med. Imag*, vol. 9, pp. 84–93, 1990.

M. Hanke and T. Raus, "A general heuristic for choosing the regularization parameter in ill-posed problems," *SIAM Journal on Scientific Computing*, vol. 17, pp. 956–972, 1996.

P. C. Hansen, "Analysis of discrete ill-posed problems by means of the L-curve," *SIAM Review*, vol. 34, pp. 561–580, 1992.

P. C. Hansen and D. P. O'Leary, "The use of the L-curve in the regularization of discrete ill-posed problems," *SIAM J. Sci. Comput.*, vol. 14, pp. 1487–1503, 1993.

P. C. Hansen, *Rank-Deficient and Discrete Ill-Posed Problems: Numerical Aspects of Linear Inversion*. Philadelphia: Society for Industrial and Applied Mathematics, 1998.

G. Herman, *Image Reconstruction From Projections: The Fundamentals of Computerized Tomography*. New York: Academic Press, 1980.

B. Hofmann and R. Krämer, "On maximum entropy regularization for a specific inverse problem of option pricing," *J. Inv. Ill-Posed Problems*, vol. 13, pp. 41–63, 2005.

P. J. Huber, *Robust Statistics*. New York: John Wiley & Sons, 1981.

——, "Robust estimation of a location parameter," *Ann. Math. Stats.*, vol. 35, pp. 73–101, 1964.

E. M. A. Hussein, D. A. Meneley, and S. Banerjee, "On the solution of the inverse problem of radiation scattering imaging," *Nuclear Science and Engineering*, vol. 92, pp. 341–349, 1986.

——, "Single exposure neutron tomography of two-phase flow," *International Journal of Multiphase Flow*, vol. 12, pp. 1–36, 1986.

D. D. Jackson, "The use of *a priori* data to resolve non-uniqueness in linear inversion," *Geophys. J. R. Astr. Soc.*, vol. 57, pp. 137–157, 1979.

E. T. Jaynes, "Information theory and statistical mechanics," *Phy. Rev.*, vol. 106, pp. 620–630, 1957.

S. Kullback and R. Leibler, "On information and sufficiency," *The Annals of Mathematics and Statistics*, vol. 22, pp. 79–86, 1951.

G. Landl and R. S. Anderssen, "Non-negative differentially constrained entropy-like regularization," *Inverse Problems*, vol. 12, pp. 35–53, 1996.

C. L. Lawson and R. J. Hanson, *Solving Least Squares Problems*. Philadelphia: Society for Industrial and Applied Mathematics, 1995.

M. A. Lukas, "Asymptotic behaviour of the minimum bound method for choosing the regularization parameter," *Inverse Problems*, vol. 14, pp. 149–159, 1998.

J. J. Moré, "The Levenberg-Marquardt algorithm: Implementation and theory," in *Numerical Analysis*, ser. Lecture Notes in Mathematics, G. Watson, Ed. Berlin: Springer Verlag, 1977, no. 630.

V. A. Morozov, *Methods for Solving Incorrectly Posed Problems*. New York: Springer-Verlag, 1984.

M. Z. Nashed, "Operator-theoretic and computational approaches to ill-posed problems with applications to antenna theory," *IEEE Transactions on Antennas and Propagation*, vol. AP-29, 1981.

M. Nikolova and R. H. Chan, "The equivalence of half-quadratic minimization and the gradient linearization iteration," *IEEE Trans. Image Proces.*, vol. 16, pp. 1623–1627, 2007.

M. Nikolova, J. Idier, and A. Mohammad-Djafari, "Inversion of large-support ill-posed linear operators using a piecewise Gaussian MRF," *IEEE Transactions on Image Processing*, vol. 7, pp. 571–585, 1998.

J. Nuyts, D. Bequé, P. Dupont, and L. Mortelmans, "A concave prior penalizing relative differences for maximum-a-posteriori reconstruction in emission tomography," *IEEE Trans. Nucl. Sci.*, vol. 49, pp. 56–60, 2002.

R. Pan and S. J. Reeves, "Efficient Huber-Markov edge-preserving image restoration," *IEEE Trans. Image Proces*, vol. 15, pp. 3728–3735, 2006.

W. H. Press, *Numerical Recipes: The Art of Scientific Computing*. Cambridge: Cambridge University Press, 2007.

S. S. Rao, *Engineering optimization: Theory and practice*. New York: John Wiley & Sons, 1996.

W. J. Rey, *Introduction to Robust and Quasi-Robust Statistical Methods*. Berlin: Springer-Verlag, 1983.

L. I. Rudin, S. J. Osher, and E. Fatemi, "Nonlinear total variation based noise removal algorithms," *Physica D*, vol. 60, pp. 259–268, 1992.

H. Rutishauser, "Once again: The least squares problem," *Journal of Linear Algebra and its Applications*, vol. 1, pp. 479–488, 1968.

C. F. Shannon and W. Weaver, *Mathematical Theory of Communication*. Urbana: University of Illinois Press, 1949.

J. E. Shore and R. W. Johnson, "Axiomatic derivation of the principle of maximum entropy and the principle of minimum cross-entropy," *Transactions on Information Theory*, vol. 26, pp. 26–37, 1980.

R. Snieder, "The role of nonlinearity in inverse problems," *Inverse Problems*, vol. 14, pp. 387–404, 1998.

P. B. Stark and R. L. Parker, "Bounded-variable least-squares: An algorithm and applications," *Computational Statistics*, vol. 10, pp. 129–141, 1995.

A. Tarantola, *Inverse Problem Theory: Methods for Data Fitting and Model Parameter Estimation*. Amsterdam: Elsevier Science, 1987.

A. Thompson, J. Brown, J. Kay, and D. Titterington, "A study of methods of choosing the smoothing parameter in image restoration by regularization," *IEEE Trans. Pattern Anal. Machine Intell.*, vol. 13, pp. 326–339, 1991.

A. N. Tikhonov, *Solutions of Ill-Posed Problems*. Washington: Halsted Press, 1977.

D. M. Trujillo and H. R. Busby, *Practical Inverse Analysis in Engineering*. Boca Raton: CRC Press, 1997.

J. M. Varah, "A practical examination of some numerical methods for linear discrete ill-posed problems," *SIAM Review*, vol. 21, pp. 100–111, 1979.

E. Veklerov and J. Llacer, "Stopping rule for the MLE algorithm based on statistical hypothesis testing," *IEEE Transactions on Medical Imaging*, vol. MI-6, pp. 313–319, 1987.

C. R. Vogel, *Computational Methods for Inverse Problems*. Philadelphia: Society for Industrial and Applied Mathematics, 2002.

G. R. Walsh, *Methods of Optimization*. London: John Wiley & Sons, 1975.

11 Functional Optimization

11.1 Formulation

With the forward model expressed as a function of the solution, $\boldsymbol{e} = \boldsymbol{f}(\boldsymbol{c})$, the image reconstruction problem can be addressed as an optimization problem of a cost function, $\Phi(\boldsymbol{c})$, with $\boldsymbol{f}(\boldsymbol{c})$ as an operator. The purpose then is to find the vector $\hat{\boldsymbol{c}}$ that minimizes $\Phi(\boldsymbol{c})$, in the presence of some physical constraints, such as the nonnegativity of $\boldsymbol{c}$ and perhaps some designated upper-bound on its values.

Defining a cost function similar to that in Chapter 10, Eq. (10.22), one has the following optimization problem for reconstructing an image, $\boldsymbol{c}$, from measurements $\hat{\boldsymbol{e}}$:

Find $\boldsymbol{c}$ which minimizes

$$\Phi(\boldsymbol{c}) = \left[\boldsymbol{f}(\boldsymbol{c}) - \hat{\boldsymbol{e}}\right]^2 + \alpha^2 (\boldsymbol{G}\boldsymbol{c})^2 \tag{11.1}$$

subject to

$$\boldsymbol{a} \le \boldsymbol{c} \le \boldsymbol{b} \tag{11.2}$$

where $\boldsymbol{a}$ (usually $\boldsymbol{a} = \boldsymbol{0}$) and $\boldsymbol{b}$ are, respectively, lower and upper bounds for $\boldsymbol{c}$. The regularization term, $\alpha^2 (\boldsymbol{G}\boldsymbol{c})^2$, is introduced to overcome the ill-posedness of the inverse problem, using one of the methods discussed in Section 10.4.

The cost function, Eq. (11.1), and the associated constraints of Eq. (11.2), can be combined into a single function, if the inequality constraints are presented as equivalent equality functional constraints. This can be done by adding the nonnegative surplus and slack variables (Rao, 1996), x_j^2 and y_j^2, and constructing the functionals:

$$\Phi_{l_j} = c_j - a_j - x_j^2 = 0, \qquad j = 1, 2, \ldots, N \tag{11.3}$$

$$\Phi_{u_j} = c_j - b_j + y_j^2 = 0, \qquad j = 1, 2, \ldots, N \tag{11.4}$$

with x_j's and y_j's forming the to-be-determined $N \times 1$ vectors, $\boldsymbol{x}$ and $\boldsymbol{y}$, known as the surplus and slack vectors, respectively. The above two functionals, along with the cost

Computed Radiation Imaging. DOI: 10.1016/B978-0-12-387777-2.00011-2

function of Eq. (11.1), formulate the Lagrangian functional:

$$L(\boldsymbol{c},\boldsymbol{x},\boldsymbol{y},\boldsymbol{\lambda}_l,\boldsymbol{\lambda}_u) = \Phi(\boldsymbol{c}) + \sum_{j=1}^{N} \lambda_{l_j}\Phi_{l_j} + \lambda_{u_j}\Phi_{u_j}$$
$$= \Phi(\boldsymbol{c}) + \sum_{j=1}^{N} \lambda_{l_j}(c_j - a_j - x_j^2) + \lambda_{u_j}(c_j - b_j + y_j^2) \tag{11.5}$$

where λ_{l_j}'s and λ_{u_j}'s are the elements of $\boldsymbol{\lambda}_l$ and $\boldsymbol{\lambda}_u$ ($N \times 1$ unknown vectors, called the Lagrange multipliers), which reflect the cost of complying with the constraints. These multipliers have to have positive values to ensure minimization of the Legendre function (Rao, 1996), hence the objective function of Eq. (11.1).

The optimization problem requires the minimization of the Lagrange functional, Eq. (11.5), which in turn necessitates satisfaction of the conditions:

$$\frac{\partial}{\partial c_j} L(\boldsymbol{c},\boldsymbol{x},\boldsymbol{y},\boldsymbol{\lambda}_l,\boldsymbol{\lambda}_u) = \frac{\partial \Phi(\boldsymbol{c})}{\partial c_j} + N(\lambda_{l_j} + \lambda_{u_j}) = 0 \tag{11.6}$$

$$\frac{\partial}{\partial \lambda_{l_j}} L(\boldsymbol{c},\boldsymbol{x},\boldsymbol{y},\boldsymbol{\lambda}_l,\boldsymbol{\lambda}_u) = c_j - a_j - x_j^2 = 0 \tag{11.7}$$

$$\frac{\partial}{\partial \lambda_{u_j}} L(\boldsymbol{c},\boldsymbol{x},\boldsymbol{y},\boldsymbol{\lambda}_l,\boldsymbol{\lambda}_u) = c_j - b_j + y_j^2 = 0 \tag{11.8}$$

$$\frac{\partial}{\partial x_j} L(\boldsymbol{c},\boldsymbol{x},\boldsymbol{y},\boldsymbol{\lambda}_l,\boldsymbol{\lambda}_u) = -2\lambda_{l_j} x_j = 0 \tag{11.9}$$

$$\frac{\partial}{\partial y_j} L(\boldsymbol{c},\boldsymbol{x},\boldsymbol{y},\boldsymbol{\lambda}_l,\boldsymbol{\lambda}_u) = 2\lambda_{u_j} y_j = 0 \tag{11.10}$$

The above five sets of relationships provide in total $5N$ equations for $5N$ unknowns, the components of the $N \times 1$ vectors: $\boldsymbol{c}$, $\boldsymbol{x}$, $\boldsymbol{y}$, $\boldsymbol{\lambda}_l$, and $\boldsymbol{\lambda}_u$. The derivatives: $\frac{\partial \Phi(\boldsymbol{c})}{\partial c_j}, j = 1, 2 \ldots N$, can be approximated by the finite difference method, if their direct analytical evaluation is cumbersome. Note, however, that the condition of Eq. (11.9) is satisfied when either $x_j = 0$ or $\lambda_{l_j} = 0$. With a nil surplus variable, $x_j = 0$, the corresponding solution is zero at the optimal point, i.e. $c_j = 0$, when $a_j = 0$. The constraint is then an *active* one, while $\lambda_{l_j} = 0$ removes the corresponding term from Eq. (11.5), and results in an *inactive* constraint. Similarly, the condition of Eq. (11.10) produces active constraints when $y_j = 0$, i.e. $c_j = b_j$, and inactive ones when $\lambda_{u_j} = 0$ at the optimum point. In other words, when one anticipates at the optimum solution that $c_j = 0$ or $c_j = b_j$, (i.e. c_j equal to exactly the lower or upper bounds of the solution vector), one must have corresponding non-zero Lagrange multipliers. This can be useful for instance when imaging a dense material inserted within a lighter object of known density, or vice versa, where the density of the surroundings can be set to designate a lower bound, or an upper bound, as appropriate. In general, however, one should assume that all constraints are active. Then, zero Lagrange multiplier values at the optimum solution will produce the lower and upper bound values. The optimization problem represented by Eqs. (11.6) to (11.10) can be solved by nonlinear programming; see Lawson and Hanson (1995), Stark and Parker (1995), Rao (1996), and Varah (1979).

The constrained optimization problem of Eqs. (11.1) and (11.2) can be transformed to an unconstrained optimization problem by using a new variable, ζ_j, so that an element c_j in $\boldsymbol{c}$ becomes: (Rao, 1996):

$$\begin{gathered} c_j = a_j + (b_j - a_j)\sin^2\zeta_j, \quad \text{or} \quad c_j = a_j + (b_j - a_j)\cos^2\zeta_j, \text{ or} \\ c_j = a_j + \frac{(b_j - a_j)\exp(\zeta_j)}{\exp(\zeta_j) + \exp(-\zeta_j)}, \quad \text{or} \quad c_j = a_j + \frac{(b_j - a_j)\zeta_j^2}{1 + \zeta_j^2} \end{gathered} \tag{11.11}$$

with ζ_j's forming an $N \times 1$ vector, $\boldsymbol{\zeta}$. If the only constraint is to ensure nonnegative values of $\boldsymbol{c}$, one of the following transformation can be used:

$$c_j = |\zeta_j|, \quad \text{or} \quad c_j = \zeta_j^2, \quad \text{or} \quad c_j = \exp(\zeta_j) \tag{11.12}$$

With one of these transformations, the optimization problems of (11.1) and (11.2) are transformed to the unconstrained optimization problem:

Find $\boldsymbol{\zeta}$ which minimizes

$$\Phi(\boldsymbol{\zeta}) = \left[\boldsymbol{f}(\boldsymbol{\zeta}) - \hat{\boldsymbol{e}}\right]^2 + \alpha^2(\boldsymbol{G\zeta})^2 \tag{11.13}$$

Once $\boldsymbol{\zeta}$ is found, $\boldsymbol{c}$ is accordingly determined. The discussion in this chapter focuses, therefore, on the transformed unconstrained optimization problem of Eq. (11.13). We will refer, however, to $\boldsymbol{c}$ instead of $\boldsymbol{\zeta}$, with the understanding that when the constraints are imposed, the transformed variables are used in the cost function. Readers interested in methods that deal with constrained optimization problems can consult a book on optimization, such as that of (Rao, 1996).

11.2 Effect of Number of Measurements

The optimization formulations discussed in Section 11.1 do not explicitly show the effect of available data points (measurements, $\hat{\boldsymbol{e}}$) on the solution. These measurements appear in the cost function via the residual term (as in Eq. (8.4)), which can be written as:

$$\chi^2 = \sum_{i=1}^{M}\left[\hat{e}_i - \sum_{j=1}^{N} f_j(c_j)\right]^2 \tag{11.14}$$

where $f_i(c_j)$ is the function (part of the forward model) that relates a measurement, e_i, to the a physical parameter, c_j. It is obvious that the larger the value of M, the higher the degree of complexity of χ^2, and the corresponding cost function $\Phi(\boldsymbol{c})$ of Eq. (11.1) (which also incorporates the regularization term). This in turn increases the complexity of the optimization process, which will include more terms as the value of M increases. At the same time, the complexity of the optimization function imposes further consistency on the solution to meet the increasing demands of optimizing the cost function. If M is a small number ($M \leq N$), the optimization process will likely lead to artificial values for the elements of $\boldsymbol{c}$ that do not correspond to reality, unless supplemented by some *a priori* information, see Chapter 14. On the other hand, if M is sufficiently larger than N, i.e. for an overdetermined problem, the optimization process can accommodate fluctuations in $\hat{\boldsymbol{e}}$ due to statistical fluctuations, by the increased demand of consistency imposed on the optimization process. Simply said, the more available is the data, the better the ability of an optimization scheme to produce a solution that matches the data; just like in curve fitting, the more data points, the better the curve fit.

11.3 Sensitivity to Measurement Uncertainty

Analyzing the susceptibility of the system to error propagation in terms of the functional $\boldsymbol{f}(\boldsymbol{c})$, requires the determination of the sensitivity of the obtained optimum solution to changes in $\hat{\boldsymbol{e}}$. Let us consider the effect of the change in one measurement, e_i, on the optimized solution. Minimization of the cost function Eq. (11.1) requires:

$$\frac{\partial \Phi(\boldsymbol{c})}{\partial c_j} = 0 \tag{11.15}$$

Let us also assume that the change in e_i is so small that this derivative remains equal to zero at the optimum solution. The *total* derivatives of Eq. (11.15) with respect to e_i enables examining the effect of a small change in its value on the optimized parameters (Rao, 1996):

$$\frac{\partial^2 \Phi(\boldsymbol{c})}{\partial c_j \partial e_i} + \sum_{k=1}^{N} \frac{\partial^2 \Phi(\boldsymbol{c})}{\partial c_j \partial c_k} \frac{\partial c_k}{\partial e_i} = 0 \tag{11.16}$$

There are N equations for this form, for the N sensitivity coefficients: $\frac{\partial c_j}{\partial e_i}, j = 1, 2, \ldots, N$. This enables the determination of the value of these sensitivity coefficients at the optimum solution. Then the error in an estimated optimum solution, $\boldsymbol{c}$, due to an uncertainty, Δe_i in a measurement, $\hat{e}_j$ can be estimated as:

$$\Delta \boldsymbol{c} = \frac{\partial \boldsymbol{c}}{\partial e_i} \Delta e_i \tag{11.17}$$

The above process can be repeated for all measurements, $e_i, i = 1, 2, \ldots, M$. This can, however, be a tedious process, but it is worth performing at least once as it can assist in adjusting the regularization process to reduce the ill-posedness of the problem.

11.4 Minimization

The minimization of a cost function, $\Phi(\boldsymbol{c})$, requires that its derivatives be equal to zero, as indicated by Eq. (11.15). In addition, a solution, $\hat{\boldsymbol{c}}$, is assured to be a *local* minimum if the Hessian matrix of $\Phi(\boldsymbol{c})$ is positive-definite, i.e.,

$$\left.\frac{\partial^2\Phi(\boldsymbol{c})}{\partial c_i \partial c_j}\right|_{\boldsymbol{c}=\hat{\boldsymbol{c}}} > 0 \tag{11.18}$$

These derivatives can be approximated by the finite-difference method, if their direct analytical evaluation is cumbersome. Satisfying the conditions of Eqs. (11.15) and (11.18) does not ensure a *global* minimum for the objective function. Ensuring that a global minimum is attained requires that $\Phi(\hat{\boldsymbol{c}}) \leq \Phi(\boldsymbol{c})$ for all values of $\boldsymbol{c}$, while a local minimum requires $\Phi(\hat{\boldsymbol{c}}) \leq \Phi(\hat{\boldsymbol{c}} \pm \delta\boldsymbol{c})$ for all values of $\delta\boldsymbol{c} << \boldsymbol{c}$, i.e. the satisfaction of Eqs. (11.15) and (11.18) only near $\hat{c}$. If the cost function, $\Phi(\boldsymbol{c})$, is a convex function, then it has zero derivatives only at the bottom of its valley, and the problem then has only one minimum, the global minimum. In a nonlinear image reconstruction problem, even if the problem has more than one local minimum, the bounding of the solution and the coherence of the image (consistency of measurements) typically (but not necessarily) lead to one physically acceptable solution.

The unrestricted minimization of the cost function of (11.1), or its transformed form (11.13) which incorporated the constraints, can be arrived at using the iterative methods described in Section 10.7 for matrix-based systems. The iterative schemes are formulated here in terms of the cost function, $\Phi(\boldsymbol{c})$, but the reader should refer to Section 10.7 for more information on these methods.

11.5 Search Methods

Searching for the minimum of a cost (objective) function, Φ, can be performed iteratively, by proceeding from one estimate of the solution, $\boldsymbol{c}^{(k)}$, to another. Some of these iterative methods are summarized here.

11.5.1 Sequential Linearization

Similar to the quasi-linearization scheme of Section 10.7.1, the optimization problem can be solved by the solution of a sequence of linear problems. Applying Eq. (11.15) to the cost function of Eq. (11.1) yields:

$$\left[\frac{\partial}{\partial c_j} f(\boldsymbol{c})\right]^{\mathrm{T}} [f(\boldsymbol{c}) - \hat{e}] + \alpha^2 [\boldsymbol{G}\boldsymbol{c}]^{\mathrm{T}} \frac{\partial}{\partial c_j} \boldsymbol{G}\boldsymbol{c} = 0; \quad j = 1, 2, \ldots N \tag{11.19}$$

Approximating the functional $f(\boldsymbol{c})$ as:

$$f(\boldsymbol{c}) \simeq f(\boldsymbol{c}^{(k)}) + \left[\nabla f(\boldsymbol{c}^{(k)})\right]^{\mathrm{T}} \left(\boldsymbol{c}^{(k+1)} - \boldsymbol{c}^{(k)}\right) \tag{11.20}$$

where $\nabla f^{\mathrm{T}} = \left[\frac{\partial f}{\partial c_1}, \frac{\partial f}{\partial c_2}, \ldots \frac{\partial f}{\partial c_N}\right]^{\mathrm{T}}$, the set of equations represented in Eq. (11.19) becomes:

$$\left[\frac{\partial}{\partial c_j} f(\boldsymbol{c}^{(k)})\right]^{\mathrm{T}} \left[f(\boldsymbol{c}^{(k)}) + \left[\nabla f(\boldsymbol{c}^{(k)})\right]^{\mathrm{T}} \left(\boldsymbol{c}^{(k+1)} - \boldsymbol{c}^{(k)}\right) - \hat{\boldsymbol{e}}\right]$$
$$+ \alpha^2 \left[\boldsymbol{G}\boldsymbol{c}^{(k)}\right]^{\mathrm{T}} \frac{\partial}{\partial c_j} \boldsymbol{G}\boldsymbol{c}^{(k)} = 0; \quad j = 1, 2, \ldots N \tag{11.21}$$

The above linear set of equations can be solved, given an estimate, $\boldsymbol{c}^{(k)}$, to obtain a new estimate $\boldsymbol{c}^{(k+1)}$, and the process is repeated until it converges to a solution. If the constraints of (11.2) are imposed along with the equations represented by Eq. (11.21), the problem is a linear programming problem, see for example Rao (1996).

11.5.2 Gradient Driven

The gradient, $\nabla\Phi$, of the cost function, Φ, is simply a vector of the partial derivatives, $\frac{\partial\Phi}{\partial c_j}, j = 1, 2, \ldots, N$. It can be evaluated either analytically or numerically using the finite-difference method. Setting $\boldsymbol{r} = -\nabla\Phi$ in Eq. (10.93), the iterative scheme of the steepest descent (Cauchy) method of Section 10.6.3 can be utilized. The Hessian matrix, $\nabla^2\Phi$, being the matrix of second partial derivatives, can also be found by differentiating $\nabla\Phi$ analytically or numerically. Then the Newton-Raphson iterative scheme of Eq. (10.113) can be directly applied. If $\nabla^2\Phi_k$ is nonsingular, the iterative process would converge to the solution that minimizes Φ, provided that the initial solution guess is sufficiently close to the solution (Rao, 1996).

The Hessian matrix can also be approximated using the residual, $R(\boldsymbol{c}) = f(\boldsymbol{c}) - \hat{\boldsymbol{e}}$, as in Eq. (10.116). The iterative scheme is then known as the Gauss-Newton method. The availability of ∇_k^{Φ} and $\nabla^2\Phi_k$ enables the use of the Levenberg-Marquardt method of Section 10.7.5, with its ability to undertake large iterative steps initially and smaller steps as the final solution is approached. Other search methods can be found in optimization textbooks, such as that of Rao (1996).

11.6 Genetic Evolution

A genetic algorithm can also be used to arrive in an evolutionary manner at the optimum solution for minimizing the cost function of (11.1) and its constraints of (11.2) (Liu and Han, 2003). In its most basic form, a genetic algorithm encodes each image parameter, an element c_j of $\boldsymbol{c}$, by a finite-length string of numbers ("genes"), typically binary numbers. The combination of the strings of all elements of $\boldsymbol{c}$ form a "chromosome" array. An evolutionary process analogous to that of the evolution of biological systems is then applied to find the "fittest chromosome," i.e. the one that minimizes the cost function (or maximizes a fitness function proportional to the reciprocal of the cost function), under the imposed constraints.

The evolution process commences by the random selection of a population of parent chromosomes. The fitness of each chromosome is then determined by the value of the corresponding fitness function. Three evolutionary reproduction processes are then employed to produce the fittest "offspring" in the next generation. First, a selection process is applied to give higher-fitness parents a larger probability to produce the next generation. Secondly, a "cross-over" operation is performed on the selected parent chromosomes in which parts of the genes are exchanged between two chromosomes to produce an offspring for the next generation. Thirdly, a "mutation" process is applied to some chromosomes with a small probability to avoid gene stagnation between generations, i.e. prevent all chromosomes in a generation from having the same gene values at the same location of their strings. The performance of these three basic operations produces a new generation. The gene string of each chromosome is then decoded, and its fitness measured by the fitness function. The evolution processes is repeated until the fittest chromosome is found. The gene string of this chromosome is decoded to produce the optimal solution.

More details on the evolution process of genetic algorithms can be found in a number of textbooks, which the reader can find via a library search. We focus here on a simple method to illustrate the concept. Let us start by a population of randomly L generated images. In order to have an evolution process that emulates a natural biological process, L has to be a large number. This by itself indicates that genetic algorithms are computationally demanding, in terms of computer storage and the associated manipulations. The method, is therefore, not well suited for image reconstruction on routine basis. However, genetic algorithms do not require the evaluation of gradients or matrix inversion, making them appealing for use in nonlinear problems.

Let us define a fitness function of a member, $\boldsymbol{c}$, of the population in terms of the cost function (11.1) so that:

$$\Psi(\boldsymbol{c}) = \frac{1}{1 + \left[f(\boldsymbol{c}) - \hat{\boldsymbol{e}}\right]^2 + \alpha^2 (\boldsymbol{Gc})^2} \tag{11.22}$$

The fitness of each of the L randomly generated images is then calculated using Eq. (11.22), and ranked in a descending order according to their fitness. A certain fraction, γL, of the most fit members is then selected for further processing (with $0 < \gamma < 1$). The remaining $(1-\gamma)L$ members are discarded. The surviving γL members form the "mating" pool that will generate new $(1-\gamma)L$ possible images.

As part of the encoding process, let us accommodate the constraints of (11.2) by making use of the lower limit, a_j, and upper limit, b_j, of each element, c_j, in the member, $\boldsymbol{c}$, of the population, so that:

$$\zeta_j = \frac{c_j - a_j}{b_j - a_j} \tag{11.23}$$

The only concern now, in dealing with the optimization problem of Eqs. (11.1) and (11.2), is to maximize the fitness function of Eq. (11.22). If in any subsequent generation, the constraints of (11.2) are provoked, the offending value is set equal to its closest limit.

Converting the decimal numbers associated with the fractions of Eq. (11.23) accomplishes the encoding process. Let us designate the coded binary value for a variable, c_j, by η_j. The sequential combination of these binary strings, i.e., $\eta_1, \eta_2, \ldots \eta_N$, produces a chromosome corresponding to a possible image, $\boldsymbol{c}$.

The chromosomes corresponding to the selected γL members are then randomly paired by sampling from a distribution representing their relative fitness with respect to the overall fitness of all members. A total of $(\gamma - 1)L$ pairs are to be selected to replace the discarded chromosomes in the selection process. Each paired chromosome is then subjected to a crossover process. A simple crossover operation is the uniform scheme, which generates an offspring from two parents selected randomly.

The available generation of chromosomes, the *fL* from the previous generation and the newly created $(\gamma - 1)L$ chromosomes, are subjected to mutation. The mutation process is applied randomly to very few genes in each chromosome, and converts a gene to its complement value, i.e. a zero value becomes one and vice versa.

Following the mutation process, a new generation of chromosomes is born. These chromosomes are decoded to obtain the new population of $\boldsymbol{c}$ values. Each element in each new member of the population is checked against the constraints of Eq. (11.2), and those that violate these limits are adjusted to the closest constraint value. The fitness of each member is evaluated, and the member with the highest fitness value is considered as a solution candidate. The entire set of operations is repeated again, and so on. With the evolution from one generation to another, the fitness of solution candidates should increase, while the overall fitness of the population should also improve. Like an iterative process, a converging evolution process should lead to a converging value of the fitness of solution candidates. Upon convergence, the solution candidate will have the highest fitness value and is considered the final solution.

Although genetic algorithms are computationally demanding, Ali et al. (1999) proposed their use in computing tomography when the number of projections is limited, i.e. for underdetermined problems; see Chapter 14. The method is also proposed for use in image enhancement, as discussed in Chapter 16.

11.7 Simulated Annealing

This is another optimization process that minimizes the cost function, $\Phi(\boldsymbol{c})$, in a successive manner by emulating a physical phenomenon; namely the annealing of metals. Metal annealing is a hardening process which allows dislocated atoms in a metal to settle in their most stable location by heating the metal and then allowing it to slowly cool down. Heating gives the atoms kinetic energy, and the cooling process allows the moving atoms to settle in a stable thermal-equilibrium arrangement. Simulated annealing is governed by the Maxwell-Boltzmann distribution:

$$p(E) = \frac{1}{Z} \exp\left[-\frac{E}{\mathrm{k}T}\right] \tag{11.24}$$

where $p(E)$ is the probability density function of having atoms with an energy E, T is the medium's absolute temperature, k is Boltzmann's constant ($\mathrm{k} = 1.381 \times$

10^{-23} J/K), and Z is a normalization constant. At the lowest possible temperature, $T = 0$, all atoms will be motionless and the medium will be at its most stable (lowest) potential energy. While, reaching an absolute zero is not possible in practice, the lower the temperature, the more stable the atomic structure. Moreover, the thermal motion of atoms during the annealing process does not allow "spotty" minimum energy regions within the medium to remain for a long period, as atoms will tend to move out. These spotty regions would correspond to local minima in a nonlinear optimization process. As such by simulating this annealing process in optimization, a global minimum would tend to be attained. Measurement uncertainties would prevent a solution from reaching a zero temperature, but a reasonably optimal solution is likely to be arrived at.

Like any iterative scheme, the solution starts with some solution guess, $\boldsymbol{c}^{(0)}$. The iterative process proceeds from one iteration, k, to the next, $k+1$, randomly, e.g. (Parks, 1990):

$$\boldsymbol{c}^{(k+1)} = \boldsymbol{c}^{(k)} + \boldsymbol{D}^{(k)}\boldsymbol{\xi} \tag{11.25}$$

where $\boldsymbol{\xi}$ is a vector of random numbers in the interval $[-1, 1]$ and $\boldsymbol{D}$ is a diagonal matrix that defines the maximum allowed change in $\boldsymbol{c}$. For the first iteration one can take: $D_{ii}^{(0)} = 0.5(b_i - a_i)$ where b_i and a_i are the elements of the upper and lower bounds of $\boldsymbol{c}$. If an element in $\boldsymbol{c}^{(k+1)}$ does not satisfy the constraints of (11.2), a new random number is selected until the constraints are satisfied. This random sampling process emulates the effect of heat on metals, which gives kinetic energy to the atoms and enables them to move around. This random (Monte Carlo) Markov chain simulation process of atoms movement does not reflect physical reality, as it may lead to two atoms occupying the same point in space, or become too close to each other (in defiance of the interaction forces between atoms that keep spatial separation between neighboring atoms). These unrealistic possibilities are overcome by imposing a state of thermal equilibrium, in which the kinetic energy is governed by the Maxwell-Boltzmann distribution of Eq. (11.24), which distributes the energy around a mean value (corresponding to T). In simulated annealing, the change in the cost function, $\Delta\Phi(\boldsymbol{c})$, from one solution estimate to another, replaces E in the Maxwell-Boltzmann distribution of Eq. (11.24), while k becomes a scaling factor and T acts as a control parameter. With this distribution, some states are allowed to have an energy higher than the mean energy, preventing these states from overlapping. Simulated annealing incorporates this physical effect via the algorithm of Metropolis et al. (1953), which checks the energy (cost function) of each obtained solution element, $c_j^{(k+1)}$. If this solution element results in a decrease in the value of the cost function, $c_j^{(k+1)}$ is accepted as a successful indication of system "cooling" and its value is retained. On the other hand, if $c_j^{(k+1)}$ causes an increase in the value of the cost function, the solution is accepted if it causes a small increase in energy, to allow for the Maxwell-Boltzmann distribution of energy; otherwise the solution is rejected and a new value of $c_j^{(k+1)}$ is randomly sampled, and the acceptability criteria is repeated over the newly sampled value. The probability of accepting an increase in the cost function, $\Phi(\boldsymbol{c})$, is expressed, in analogy with Eq. (11.24), as:

$$p(\Delta\Phi(\boldsymbol{c})) = \exp\left[-\frac{\Delta\Phi(\boldsymbol{c})}{\mathrm{k}T}\right] \tag{11.26}$$

The scale factor, k, controls the amount of permitted change in energy. This condition is implemented by sampling a random number, ξ, in the interval [0,1]. If $\xi < \exp\left[-\frac{\Delta\Phi(c)}{\mathrm{k}T}\right]$, then the increase in the cost function is accepted, otherwise it is rejected.

The random sampling and the associated processes are repeated at each temperature, T, until the average value of the cost function reaches a stable value, at which point equilibrium is considered to have been attained at this T value. The iterative process starts with a high temperature, and after reaching equilibrium, the temperature is lowered, and the random sampling process is repeated starting from the previous equilibrium solution. The temperature lowering process continues from one equilibrium state to another until a sufficiently low temperature is reached, below which no significant change in the obtained solution occurs. This solution is considered to be the optimum solution for the problem.

The choice of the initial temperature, the permitted degree of increase in energy via the scaling parameter, k, in Eq. (11.26), the manner of "cooling" the system, and the final temperature, define the so-called "cooling schedule." A number of schemes can be used, see van Laarhoven and Aarts (1987). However, the initial temperature should be sufficiently high to permit complete mobility, as in allowing the melting of a metal, with a gradual temperature decrease toward a completely immobile (frozen) system. Random sampling should be devised to permit a minimum number of accepted energy increases, with the number decreasing as the system approaches the freezing point. The length of the Markov chain, i.e. the number of sampling operations, should be limited to a pre-determined maximum number, to avoid excessive unnecessary computations. The final temperate is typically reached when no improvement in the cost function is realized in an entire random sampling process at a certain temperature.

Simulated annealing was considered for use in image reconstruction with parallel processors to accommodate its computational demands (Girodias et al., 1991; Shoemaker et al., 1991). The cost function for use in CT image reconstruction was analyzed in Haneishi et al. (1990). The method was also introduced for image reconstruction in single photon emission computed tomography (SPECT) (Webb, 1989).

11.8 Neural Networks

Another optimization method that relies on emulating a natural process is the method of neural networks, which mimics the nervous system. The main advantage of this method is that it does not require an explicit formulation of the forward and inverse mappings. It rather establishes the relationship between measurements and sought parameters via a learning process similar to that accomplished by the brain. Sensed inputs (measured signals) are fed into a network of neurons, with each neuron processing the signal and passing it to other neurons, and so on until a final output (result) is arrived at. A network is described by the manner in which its neurons are connected, the weight it gives to the interconnections between neurons, and the transformation function (from input to output) it associates with each neuron. A number of network architectures are available through a variety of software packages, see for example the MATLAB Neural Network Toolbox (`http://www.cpc.cs.qub.ac.uk/summaries/ACGV_v2_0.html`), among many others.

The learning (training) process aims at finding the optimal weights of the interconnections that relate a given set of measurements to known outputs. The learning schemes are based on iteratively minimizing the residual error between a known outcome and that produced by the weight assigned to the network. The usual optimization methods of the steepest descent, Newton, quasi-Newton and conjugate gradients methods, discussed in Chapter 10, can be used as strategies to update the guessed weights. Once the optimal weights are found, the relationship between the measurements and the sought parameters, i.e. the inverse mapping, is considered to have been established, and the neutral network can be used efficiently and quickly to find the parameters for a new set of measurements. For more information, readers can consult a textbook on neural networks, such as Fausett (1994). An example of using neural networks for imaging reconstruction is reported in Paschalis et al. (2004), for a SPECT system.

References

F. E. A. F. Ali, Z. Nakao, and Y.-W. Chen, "An evolutionary approach to CT image reconstruction," *Bulletin of the Faculty of Engineering, University of the Ryukyus*, vol. 57, pp. 81–84, 1999.

L. Fausett, *Fundamentals of Neural Networks: Architecture, Algorithms, and Applications*. Upper Saddle River, NJ: Prentice Hall, 1994.

K. A. Girodias, H. H. Barrett, and R. L. Shoemaker, "Parallel simulated annealing for emission tomography," *Journal of Physics in Medicine and Biology*, vol. 36, pp. 921–938, 1991.

H. Haneishi, T. Masuda, N. Ohyama, T. Honda, and J. Tsujiuchi, "Analysis of the cost function used in simulated annealing for CT image reconstruction," *Applied Optics*, vol. 29, pp. 259–265, 1990.

C. L. Lawson and R. J. Hanson, *Solving Least Squares Problems*. Philadelphia: Society for Industrial and Applied Mathematics, 1995.

G. R. Liu and X. Han, *Computational Inverse Techniques in Nondestructive Evaluation*. Boca Raton: CRC Press, 2003.

N. Metropolis, A. Rosenbluth, M. R. A. Teller, and E. Teller, "Equation of state calculations by fast computing machines," *Journal of Chemical Physics*, vol. 21, pp. 1087–1092, 1953.

G. T. Parks, "An intelligent stochastic optimization routine for nuclear fuel cycle," *Nuclear Technology*, vol. 89, pp. 233–246, 1990.

P. Paschalis, N. D. Giokaris, A. Karabarbounis, G. K. Loudos, D. Maintas, C. N. Papanicolas, V. Spanoudaki, C. Tsoumpas, and E. Stiliaris, "Tomographic image reconstruction using artificial neural networks," *Nuclear Instruments and Methods in Physics Research A*, vol. 527, no. 211–215, 2004.

S. S. Rao, *Engineering optimization: Theory and practice*. New York: John Wiley & Sons, 1996.

R. L. Shoemaker, H. H. Barrett, and R. H. Seacat, "Trimm: A parallel processor for image reconstruction by simulated annealing," in *Parallel Problem Solving from Nature*, ser. Lecture Notes in Computer Science, H. R. Schwefel and R. Männer, Eds. Berlin: Springer-Verlag, 1991, vol. 496, pp. 242–251.

P. B. Stark and R. L. Parker, "Bounded-variable least-squares: An algorithm and applications," *Computational Statistics*, vol. 10, pp. 129–141, 1995.

P. J. M. van Laarhoven and E. H. L. Aarts, *Simulated Annealing: Theory and Applications*. Dordrecht: Reidel, 1987.

J. M. Varah, "A practical examination of some numerical methods for linear discrete ill-posed problems," *SIAM Review*, vol. 21, pp. 100–111, 1979.

S. Webb, "SPECT reconstruction by simulated annealing," *Physics in Medicine and Biology*, vol. 34, no. 259–281, 1989.

12 Analytic Methods

12.1 Radon Transform

In transmission and emission imaging, measurements are cumulative indications, represented by integrals of image parameters along lines of detector response. In transmission imaging, see Chapter 4, the ray-sum is an integration of the attenuation coefficients, which constitute the image parameters. For emission imaging, discussed in Chapter 5, a measurement represents an integral of radiation emission intensities along a detector's line of response. In scatter imaging, Chapter 6, the problem is not as straightforward, because of its nonlinear nature. An integral measurement can be viewed as constituting an integral transform of image parameters. Such transform, when performed over an imaging plane, is known as the Radon transform, which integrates a function over lines in a plane.

Let us consider first the case of transmission tomography. A projection, p, is given by Eq. (4.2), and is rewritten here in terms of the image parameter, c, as:

$$p(r,\vartheta) = \int c(s)\,ds = \mathcal{R}(r,\vartheta)[c(x,y)] \tag{12.1}$$

$$= \int_{-\infty}^{\infty} c(r\cos\vartheta - s\sin\vartheta, r\sin\vartheta + s\cos\vartheta)\,ds \tag{12.2}$$

$$= \int_{-\infty}^{\infty}\int_{-\infty}^{\infty} c(x,y)\delta(r - x\cos\vartheta - y\sin\vartheta)\,dxdy \tag{12.3}$$

where (r,ϑ) define the polar coordinates, see Figure 12.1, with s being the distance in the polar direction and $\mathcal{R}$ refers to the Radon transform.

The limits of integration in Eq. (12.3) are written to cover the entire spatial domain, but in practice the integration needs to cover only the image domain. Consider a domain consisting of a circle of radius R, as shown in Figure 12.1. A fixed point (x,y) within that domain will produce projections at different angles through lines in the plane of the image. The end points corresponding to these projection lines through (x,y) produce a sinusoidal curve: $r = x\cos\vartheta + y\sin\vartheta$. The Radon transform of a Dirac

Computed Radiation Imaging. DOI: 10.1016/B978-0-12-387777-2.00012-4

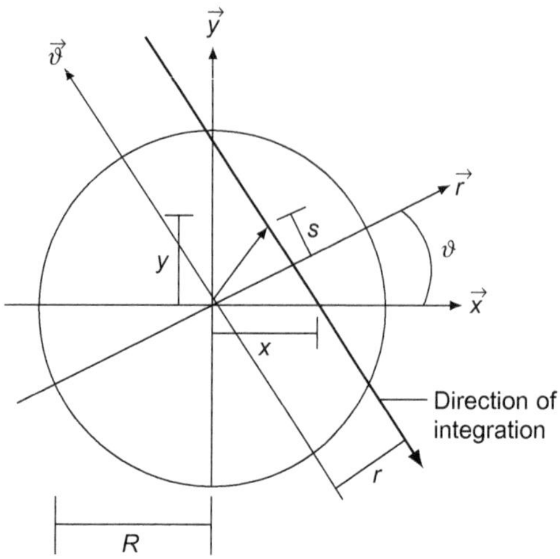

Figure 12.1 Coordinates for Radon transform.

delta function is a distribution with support[1] on the graph of a sine wave. Therefore, a measurement projection is often called a *sinogram*.

The inverse Radon transform of $p(r, \vartheta)$ of Eq. (12.3) gives the sought solution:

$$\mathcal{R}^{-1}(x, y)[p(r, \vartheta)] = c(x, y) = \frac{1}{2\pi} \int_{-\infty}^{\infty} \int_{-\infty}^{\infty} \frac{\partial \mathcal{H}[p(r, \vartheta)]}{\partial r} \, d\vartheta \tag{12.4}$$

where $\mathcal{H}$ defines a Hilbert transform.[2] Equation (12.4) shows that a two-dimensional parameter, $c(x, y)$, can be recovered from one-dimensional projections, $p(r, \vartheta)$, using an analytical formulation.

Two conditions are required to readily apply the Radon transform and its inverse. First, differentiation to obtain $\frac{\partial \mathcal{H}[p(r,\vartheta)]}{\partial r}$ in Eq. (12.4) requires the availability of projections along the $\vec{r}$ direction. With parallel projections, at the same angle ϑ, this derivative can be readily found numerically by the finite difference method or analytically if an analytical formulation of the projections as a function of r can be

[1] A support is the smallest closed set containing the arguments of a function (set closure).

[2] The Hilbert transform of a function $f(x)$ is an integral transform of the form:

$$\mathcal{H}[f(x)] = g(y) = \frac{P_V}{\pi} \int_{-\infty}^{\infty} \frac{f(x)\, dx}{x - y}$$

$$P_V \int_{-\infty}^{\infty} f(x)\, dx = \lim_{R \to \infty} \int_{-R}^{R} f(x)\, dx$$

where P_V is the integral's Cauchy principal value.

established. Secondly, the integration over the polar angle, ϑ in Eq. (12.4), dictates that the inverse Radon transform be conducted over the r-ϑ plan. These two conditions concur with the requirements of parallel-ray transmission tomography. In dealing with fan-beam or three-dimensional transmission tomography, these two conditions are not readily met, and some form of reformulation must be introduced. In transmission imaging, the Radon transform is readily obtained by the projections, see Eq. (4.2). In both emission and scatter imaging, the integral of the Radon transform, Eq. (12.3), is modulated by the attenuation and divergence terms, see Chapters 5 and 6. Rather than dealing with these complications here, we will address them when considering the Fourier transform discussed in the ensuing section; because the Radon transform is closely related to that of Fourier, while the latter is more suited for numerical implementation. We start first with the parallel-beam transmission tomography arrangement, expand the concepts to accommodate other forms of transmission tomography, then consider emission and scatter imaging.

12.2 Two-Dimensional Fourier Transforms

Section 8.3.1 presented a formulation for the linear inverse problem of transmission tomography by expressing the solution in terms of a Fourier series, i.e. the superposition of sinusoidal waves expressed in complex exponentials. The coefficients of this expansion are determined by the inverse Fourier transform, which is readily provided by the efficient, but approximate, Fast Fourier Transform (FFT) computing algorithms (for a free software see: `http://www.fftw.org/`). The basic formulation of Section 8.3.1 is applied to the two-dimensional tomographic imaging problem with the aid of some basic concepts introduced below.

12.2.1 Fundamental Equations

Let us consider the two-dimensional transmission-tomography imaging problem of determining the image parameter, c, (attenuation coefficient), at every image point (x, y) on a cross section in the x-y plane. The Fourier integral of Eq. (8.5) can be expressed as:

$$c(x,y) = \int_{-\infty}^{\infty}\int_{-\infty}^{\infty} \mathcal{C}(k_x,k_y)\exp[2\pi i(k_x x + k_y y)]dk_x\, dk_y \tag{12.5}$$

where k_x and k_y are frequencies in the x and y directions, respectively, and $\mathcal{C}(k_x, k_y)$ is the corresponding Fourier coefficient. The expansion coefficients are determined by the inverse transform of the Fourier integral:

$$\mathcal{C}(k_x,k_y) = \int_{-\infty}^{\infty}\int_{-\infty}^{\infty} c(x,y)\exp[-2\pi i(k_x x + k_y y)]dx\, dy \tag{12.6}$$

where $\mathcal{C}(k_x, k_y)$ is the two-dimensional Fourier transform of $c(x, y)$. Note that in Eq. (12.6), c vanishes outside the image domain. However, the infinite limits of the integrals are preserved here for the sake of mathematical rigor.

Since tomographic imaging requires multidirection exposure, of the imaged section to radiation from various directions, it is more convenient to consider the problem in the polar system of coordinates, (r-ϑ), with:

$$\tan\vartheta = \frac{k_y}{k_x}; \quad r = x\cos\vartheta + y\sin\vartheta \tag{12.7}$$

Then Eq. (12.6) becomes:

$$\mathcal{C}(k_x, k_y) = \mathcal{C}(k, \vartheta) = \int_{-\infty}^{\infty}\int_{-\infty}^{\infty} c(r, \vartheta)\exp(-2\pi ikr)rd\vartheta\, dr \tag{12.8}$$

$$= \int_{-\infty}^{\infty}\left\{\int_{-\infty}^{\infty} c(r, \vartheta)rd\vartheta\right\}\exp(-2\pi ikr)\, dr$$

$$= \int_{-\infty}^{\infty} p(r, \vartheta)\exp(-2\pi ikr)\, dr = \mathcal{P}(k, \vartheta) \tag{12.9}$$

with $k^2 = k_x^2 + k_y^2$, $p(r, \vartheta)$ being the projection (ray sum) at r in the direction $\vec{\vartheta} = \vec{s}$ (normal to $\vec{r}$). Notice that k is not a function of x or y, which made it possible to exchange the order of integrals in the above formulation.

Figure 12.2 shows the projection, p, for a uniform object at some angle, ϑ, in parallel-beam projections, along with the associated modulus of the Fourier transform, $|\mathcal{P}|$. Notice here that the projection, p, is formed by a set of Radon-transformed measurements i.e. ray-sums, (see Section 4.1), and $\mathcal{P}$ is a transform of the entire profile of p.

Combining Eqs. (12.8) and (12.9) gives:

$$\mathcal{C}(k, \vartheta) = \mathcal{P}(k, \vartheta) \tag{12.10}$$

which indicates that the Fourier coefficient, $\mathcal{C}(k, \vartheta)$, of the sought parameter, $c(k, \vartheta)$, is equal to the Fourier coefficient, $\mathcal{P}(k, \vartheta)$, of the measurement projection, $p(r, \vartheta)$, with respect to r. This is the so-called *central slice theorem*, or the Fourier slice theorem (also called the projection theorem), as its states that the Fourier transform of parallel projections of transmission measurements gives a "slice" of the two-dimensional transform of the sought solution. It relates the one-dimensional Fourier transform of a parallel projection to the two-dimensional Fourier transform of the image parameters. In other words, the Fourier transform of a projection is a cross section of the Fourier transform of the image at its center. The central slice theorem is fundamental to the

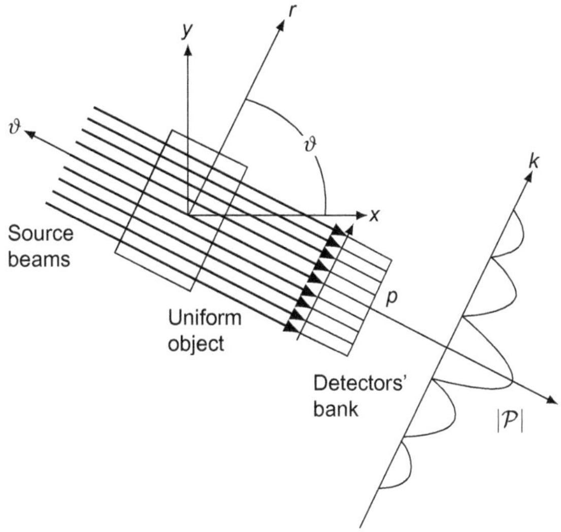

Figure 12.2 A schematic showing the projection, p, of a uniform rectangular object at a single direction, ϑ, in a parallel-beam arrangement, and the modulus, $|\mathcal{P}|$, of its Fourier spectrum.

Fourier solution of the inverse problem of transmission imaging, as it reduces it into two basic transforms: (1) the Radon transform in which transmission measurements are presented as ray-sums, with each set of ray-sums providing a projection, and (2) the Fourier transform of the projections. For each projection, a two-dimensional array of Fourier coefficients of the sought solution is generated. The inverse Fourier transforms of these coefficients leads to the sought solution.

With the availability of Fast Fourier Transform (FFT) computing algorithms, the solution of the inverse problem of transmission tomography appears simple and straightforward. There are, however a number of complications in applying the central slice theorem of Eq. (12.10). First, $\mathcal{P}$, is a one-dimensional transform, as Eq. (12.9) indicates, while $\mathcal{C}$ is a two-dimensional one, defined by Eq. (12.8). Secondly, these transforms are provided in the polar, (r, ϑ), coordinates system, while the solution is sought in Cartesian, (x, y), coordinates, where the image pixels are formed. Thirdly, this is essentially an analytical solution that requires complete and continuous projections of the measurement ray-sums, in order to be able to perform the Fourier transforms and subsequent inversion. In practice, however, no matter how large the number of radiation measurements recorded is, one always has a finite number of projections. This results in a discrete frequency space that may contain "gaps" i.e. the absence of some frequencies needed in the construction of the two-dimensional set of coefficients. Interpolating between the one-dimensional projection coefficients then becomes necessary, to be able to estimate the absent values. The discretized nature of measurements and image parameters also necessitates replacing the integrals in the above equations with summations and limiting the domain of summation to finite values. Fourth, in practice, the projections in Eq. (12.9), are replaced by measured values with associated uncertainties. Fifth, an originally well-posed problem will become ill-posed when solved by the Fourier transform method, because it is in effect a discretization method in the frequency domain due to the finite number of projections. These numerical aspects of the problem are discussed in the ensuing sections.

12.2.2 Completeness and Continuity

A continuous function can be constructed from a set of values sampled at a closely spaced discrete points, if the function is band-limited, i.e. its Fourier transform is nil outside a finite frequency range. If the bandwidth of the function is $2k_{max}$, it covers the frequency range: $|k| \leq k_{max}$. Then, a band-limited function has a non-zero Fourier transform only when $|k| \leq k_{max}$, where k_{max} is the maximum frequency. The sampling theorem dictates that the sampling frequency be equal to or greater than the bandwidth, in order to exactly reconstruct a function from its discrete samples. Therefore, if M_s measurements are acquired in a transmission projection, at equally spaced intervals each of width, Δr, i.e. a sample frequency (samples per unit length) of $\frac{1}{\Delta r}$, then:

$$\frac{1}{\Delta r} \geq 2k_{max}; \quad k_{max} \leq \frac{1}{2\Delta r} \tag{12.11}$$

is required for complete sampling of the projection profile. The condition of Eq. (12.11) is known as the Nyquist condition. It is the only restriction needed to utilize the central slice theorem of Eq. (12.10) to reconstruct continuous images in transmission tomography, as it permits interpolation between projection values without loss of information due to sampling. However, it should be kept in mind, as indicated in Section 8.3.1, that the condition of Eq. (12.11) cuts off terms in the integral of Eq. (12.5) for $|k| > k_{max}$. Abrupt changes in material (sharp edges) can produce such high frequencies. Not allowing the sampling of these frequencies may result in their reappearance disguised as lower frequencies; resulting in an aliasing artifact. The elimination altogether of these high frequencies leads to the Gibbs phenomenon: an artifact that produces an overshoot, or ringing, at sharp boundaries (a low-density ring can appear artificially inside a high-density boundary (Brooks and Di Chiro, 1976)).

Bandwidth limiting makes an originally well-posed problem ill-posed, when solved by the Fourier transform method. This is because it excludes frequencies outside the band from contributing to the image. In addition, the chosen width of the band determines the domain of solution. Therefore, the image obtained by the Fourier transform method is not unique, which violates one of the conditions of well-posedness. Consequently, an inverse problem solved by the Fourier transform method will still remain susceptible to error. An error in a measurement directly alters the values of the Fourier coefficients, hence the solution.

12.2.3 Discretization

The sampling of measurements in a projection is a discrete process. Let us consider the case of a beam of width w rastering an object along a line to produce a projection profile. The beam width determines the maximum spatial frequency (cycles per unit length):

$$k_{max} = \frac{1}{w} \tag{12.12}$$

According to the Nyquist condition of Eq. (12.11), one needs to sample at least two measurements per beam width, so that:

$$\Delta r = \frac{w}{2} \tag{12.13}$$

is the smallest uniform-sampling width that allows continuous construction of the projection profiles from discrete sampling. Therefore, the number of measurements acquired in a projection, M_r, is given by:

$$M_r = \frac{2R}{\Delta r} = \frac{4R}{w} \tag{12.14}$$

where R is the radius within which the imaged section is enclosed. Then the bandwidth, $2k_{max}$, in the Fourier space can be divided into equal frequency intervals, each with width Δk so that:

$$\Delta k = \frac{2k_{max}}{M_r} = \frac{2}{w} \times \frac{w}{4R} = \frac{1}{2R} \tag{12.15}$$

The many projections required in transmission tomography can be acquired by rotating the parallel beam by an angle, $\Delta\vartheta$, around the central axis of the imaged section. Then one acquires, with uniform sampling,

$$M_\vartheta = \frac{\pi}{\Delta\vartheta} \tag{12.16}$$

projections over $0 \leq \vartheta \leq \pi$, since in parallel-beam transmission tomography projections within $\pi \leq \vartheta \leq 2\pi$ are redundant,[3] as they provide the information given by the projections within $0 \leq \vartheta \leq \pi$. Each projection rotation will produce a change, Δk, in frequency, since the maximum frequency, k_{max} should not be exceeded. Therefore, one should chose the angular interval, Δ_ϑ, such that:

$$\Delta\vartheta = \frac{\Delta k}{k_{max}} = \frac{w}{2R} = \frac{\Delta r}{R} = \frac{\pi}{M_\vartheta} \tag{12.17}$$

One has $\Delta r = R\Delta\vartheta$, which is equal to the arc length corresponding to $\Delta\vartheta$ at radius R, and consequently $\Delta r = \frac{\pi R}{M_\vartheta}$ and $M_\vartheta = \frac{\pi R}{\Delta r}$.

The above described discretization of the Radon space (the spatial domain) and the corresponding Fourier space (frequency domain) are depicted in Figure 12.3.

[3] Non-redundant measurements within $\pi \leq \vartheta \leq 2\pi$ can be obtained by off-setting the source or detector positions so that opposing beams interleave (Goldman, 2007).

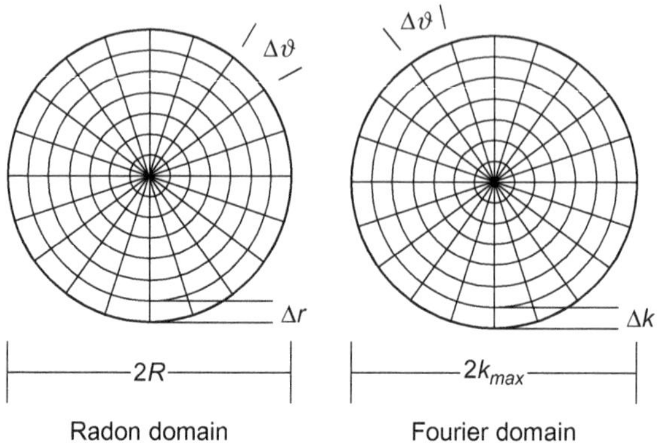

Figure 12.3 Discretization of samples in the Radon and Fourier spaces.

The total number of measurements, M, required to satisfy the sampling condition is then:

$$M = M_r M_\vartheta = \frac{2R}{\Delta r} \times \frac{\pi R}{\Delta r} = \frac{\pi}{2}\left(\frac{2R}{\Delta r}\right)^2 \tag{12.18}$$

The M measurements can be used to image up to N squares pixels in a circle of radius R enclosing the imaged section so that:

$$N = \frac{\pi R^2}{w^2} = \frac{\pi R^2}{4(\Delta r)^2} = \frac{M}{8} \tag{12.19}$$

That is, the number of knowns (measurements) is eight times the number of unknowns (pixel parameters). This is a much higher degree of overdetermination than that typically acquired in practice (from two to three), see Section 9.1.1. This is because satisfying the above conditions leads to sufficient number of pixels from which a continuous image, not affected by the discrete nature of the measurements, can be constructed. This high degree of overdetermination required for image reconstruction by direct inverse Fourier transform is hardly used in image reconstruction. Its study, however, enables one to appreciate the approximate nature of the other analytical methods discussed in Section 12.3.

The fact that the number of available projections is always finite necessitates replacing the continuous integrals in Fourier transforms by summations over discrete frequencies. The Fourier transform, $\mathcal{P}(k, \vartheta)$, of a projection, $p(\vartheta)$, consisting of M_r equispaced intervals of width Δr, can be expressed in terms of M_r discrete points using the discrete Fourier transform:

$$\mathcal{P}(\vartheta, m\Delta k) = \frac{1}{M_r} \sum_{n=-\frac{M_r}{2}}^{\frac{M_r}{2}} p(n\Delta r, \vartheta) \exp\left[-\frac{2\pi inm}{M_r}\right] \tag{12.20}$$

with Δk given by Eq. (12.15) and the appearance of $m\Delta k$ signifies the fact that discrete values are used in Eq. (12.20). It should be kept in mind, however, that $\mathcal{P}(\vartheta, m\Delta k = k) \neq \mathcal{P}(\vartheta, k)$, i.e. a sampled Fourier transform as evaluated from Eq. (12.20) is not equal to the value sampled from a continuous presentation of the projection, $p(r, \vartheta)$, if it were available. In fact, discrete sampled values correspond to shifted (at intervals) values of the continuous (ideal) Fourier transform. Therefore, the continuous transform is obtained approximately by the summation of $\mathcal{P}(m\Delta k, \vartheta)$ (Stark, 1979):

$$\mathcal{P}(k, \vartheta) \approx \sum_{m=-\frac{M_r}{2}}^{\frac{M_2}{2}} \mathcal{P}(\vartheta, m\Delta k) \operatorname{sinc}\left[\frac{k - m\Delta k}{\Delta k}\right] \tag{12.21}$$

where $\operatorname{sinc}(x) = \frac{\sin(\pi x)}{\pi x}$, with $\operatorname{sinc}(x) = 1$ at $x = 0$, so that $\int_{-\infty}^{\infty} \operatorname{sinc}(x) dx = \pi$. Recall that $\operatorname{sinc}(k)$ is the Fourier transform of a rectangular function, $\operatorname{rect}(x)$, of unit width and unit height, i.e. $\operatorname{rect}(x) = 1$ for $|x| \leq 0.5$ and $\operatorname{rect}(x) = 0$ when $x > 0.5$. Therefore, the appearance of the $\operatorname{sinc}(x)$ function in Eq. (12.21) is due to the discrete sampling of projections at finite widths. The approximation in Eq. (12.21) is due to the fact that the exact value requires summation from $-\infty$ to ∞. Equation (12.21) amounts to the filtering of $\mathcal{P}(\vartheta, m\Delta k)$ with $\operatorname{sinc}\left(\frac{k-m\Delta k}{\Delta k}\right)$.

The Fourier transform in Eq. (12.21) is conducted in the polar coordinates, to accommodate projections at various orientations. Equating these projection transforms to the solution transforms, in accordance to the sampling theorem (see Section 12.2.2), produces transforms, $\mathcal{C}(k, \vartheta)$, also in the polar coordinates. Interpolation is required for transforming from the polar mesh to the rectangular grid over which the image is formed. Various interpretation methods are discussed below guided by Stark et al. (1981).

12.2.4 Interpolation

If the conditions of the sampling theorem are met and the projects are sampled uniformly, then exact interpolation can be performed to provide the value of $\mathcal{C}(k, \vartheta)$. Let us assume that the projections are sampled over discrete angles: ϑ_l, $l = 1, 2 \ldots M_\vartheta$ over 2π, with M_ϑ selected to satisfy the condition of Eq. (12.11) of the sampling theorem. Then each projection gives a corresponding Fourier transform of the solution in the polar coordinates, $\mathcal{C}(k, \vartheta_l)$, and the exact value of $\mathcal{C}(k, \vartheta)$ can be evaluated an any angle, ϑ, based on the so-called circular sampling theorem (Stark, 1979), as follows:

$$\mathcal{C}(k, \vartheta) = \sum_{l=0}^{M_\vartheta - 1} \mathcal{C}\left(k, \frac{2\pi l}{N}\right) \frac{\sin\left[\frac{N}{2}\left(\vartheta - \frac{2\pi l}{N}\right)\right]}{\sin\left[\frac{1}{2}\left(\vartheta - \frac{2\pi l}{N}\right)\right]} \tag{12.22}$$

Since the Fourier frequency, k, assumes positive and negative values, including zero, $k = 0, \pm 1, \pm 2, \ldots, \pm k_{max}$, then $M_\vartheta = 2K + 1$ is the smallest value of M_ϑ permitted by the sampling theorem, where in accordance to Eq. (12.11):

$$K = \frac{R}{w} + 1 \tag{12.23}$$

In parallel-beam transmission tomography, projections are obtained over $0 \le \vartheta \le \pi$, since a projection, p, at $\pi + \vartheta$ is equal to that at ϑ. On the other hand, one can view each projection at angle $\vartheta \le \pi$ as having two samples of $\mathcal{C}(k, \vartheta)$ so that the entire range of the polar direction is covered. Therefore, the total number of projections in parallel-beam transmission tomography over $0 \le \vartheta \le 2\pi$ is an even number. The smallest even number for M_ϑ projections is $2K + 2$, i.e. $K + 2$ is the smallest number of projections over $0 \le \vartheta \le \pi$. Therefore, Eq. (12.22) can be rewritten as:

$$\mathcal{C}(k, \vartheta) = \sum_{l=0}^{2K+1} \mathcal{C}\left(k, \frac{\pi l}{K+1}\right) \frac{\sin\left[(K+1)\left(\vartheta - \frac{\pi l}{K+1}\right)\right]}{\sin\left[\frac{1}{2}\left(\vartheta - \frac{\pi l}{K+1}\right)\right]} \tag{12.24}$$

Applying the central slice theorem, Eq. (12.10), $\mathcal{C}(k, \vartheta) = \mathcal{P}(k, \vartheta)$, and using Eqs. (12.21) and (12.24), one obtains:

$$\mathcal{C}(k, \vartheta) = \sum_{l=0}^{2K+1} \mathcal{P}\left(k, \frac{\pi l}{K+1}\right) \frac{\sin\left[(K+1)\left(\vartheta - \frac{\pi l}{K+1}\right)\right]}{\sin\left[\frac{1}{2}\left(\vartheta - \frac{\pi l}{K+1}\right)\right]} \tag{12.25}$$

$$= \sum_{l=0}^{2K+1} \sum_{m=-\frac{M_r}{2}}^{\frac{M_r}{2}} \mathcal{C}\left(m\Delta k, \frac{\pi l}{K+1}\right) \operatorname{sinc}\left[\frac{k - m\Delta r}{\Delta k}\right] \times \frac{\sin\left[(K+1)\left(\vartheta - \frac{\pi l}{K+1}\right)\right]}{\sin\left[\frac{1}{2}\left(\vartheta - \frac{\pi l}{K+1}\right)\right]} \tag{12.26}$$

Equation (12.26) shows that $\mathcal{C}(k, \vartheta)$ can be calculated at (k, ϑ) from discrete sampled values.

12.2.5 Direct Fourier Inversion

Using Eq. (12.25), the continuous function of $\mathcal{C}(k, \vartheta)$ can be constructed from discrete values of the transform of the sampled projections. With the relationships of Eq. (12.7), $\mathcal{C}(k_x, k_y)$ can be computed over a Cartesian grid. Then, the image parameter, $c(x, y)$, can be found at different pixels using the Fourier integral of Eq. (12.5), obtained with the aid of an inverse fast Fourier transform algorithm. This latter approximation, along with the approximation of Eq. (12.21), results in truncation and aliasing errors in the image. Therefore, interpolating Fourier transforms is an ill-posed problem (Lannes et al., 1994), since a small error in interpolation at one frequency will affect other frequencies. Some methods are proposed to reduce these errors, see for example

Stark et al. (1981) and Marèchalt et al. (2000). However, direct Fourier inversion is less commonly used due to the better accuracy produced by the Fourier-frequency filtering method, discussed below.

12.2.6 Fourier-Frequency Filtering

So far, we have focused on analyzing the Fourier transforms of the projections and solution parameters. Let us now return to the fundamental integral of Eq. (12.5) and transform it to the polar coordinates system, (r, ϑ). The purpose of this exercise, guided by Kak and Slaney (2001), is to show that the Fourier transform can be used to filter the projections, and descramble their fundamental components, to arrive at the solution itself. Using the relationships of Eq. (12.7), the Fourier integral of Eq. (12.5) can be expressed in polar coordinates as:

$$c(x, y) = \int_{0}^{\pi} \int_{-\infty}^{\infty} \mathcal{C}(k, \vartheta) \exp[2\pi ikr]|k| dk\, d\vartheta \tag{12.27}$$

Notice that ϑ covers half the polar angular range due to the redundancy of measurements in the other half in parallel-beam transmission tomography, i.e. because $\mathcal{C}(k, \vartheta + \pi) = \mathcal{C}(-k, \vartheta)$; hence the use of $|k|$ instead of k. Using the central slice theorem, $\mathcal{C}(k, \vartheta)$ can be replaced by the corresponding projection $\mathcal{P}(k, \vartheta)$ in Eq. (12.27) to give:

$$c(x, y) = \int_{0}^{\pi} \left[\int_{-\infty}^{\infty} \mathcal{P}(k, \vartheta)|k| \exp[2\pi ikr] dk \right] d\vartheta = \int_{0}^{\pi} q(r, \vartheta)\, d\vartheta \tag{12.28}$$

where

$$q(r, \vartheta) = \int_{-\infty}^{\infty} \mathcal{P}(k, \vartheta)|k| \exp[2\pi ikr]\, dk \tag{12.29}$$

is a "frequency-filtered" form of the projection, $\mathcal{P}(k, \vartheta)$, with a filter frequency equal to $|k|$. This is a ramp-filter in the frequency domain, due to the linear nature of $|k|$. Equation (12.28) indicates that the summation of these filtered projections over the polar angles of the original projections gives the sought solution, $c(x, y)$. This simple, but remarkable observation, is behind the wide use of the filtered backprojection method discussed in Section 12.3.2.

12.3 Backprojection

The essence of this method in its simple form is to project directly measured or filtered values of each projection back into the voxels in the image that contribute to a measurement. With all projections at all recorded polar angles projected back into the image voxels, one would obtain an "imprint" of the image. The objective of filtering the projections prior to backprojecting them is to suppress unwanted artifacts, as shown below after discussing backprojection without filtering.

12.3.1 Direct

Backprojecting direct (unfiltered) measurements is mathematically processed as:

$$\tilde{c}(x,y) = \sum_{m=1}^{M_\vartheta} p(r_m, \vartheta_m)\Delta\vartheta = \sum_{m=1}^{M_\vartheta} p(x\cos\vartheta_m + y\sin\vartheta_m, \vartheta_m)\Delta\vartheta \tag{12.30}$$

where $\Delta\vartheta$ is given by Eq. (12.16), and $\tilde{c}(x,y)$ is a relative value, not the true absolute value, of $c(x,y)$. In other words, Eq. (12.30) gives the value of $c(x_i, y_j)$ for the voxel at (x_i, y_j) in comparison to the value at other points in the image.

Equation (12.30) is a backprojection in the spatial domain, its analogous form in the frequency domain is quite revealing. Replacing $p(r_m, \vartheta_m)$ by its Fourier integral:

$$p(r_m, \vartheta_m) = \int_{-\infty}^{\infty} \mathcal{P}(r,\vartheta)\exp[2\pi kr]dk \tag{12.31}$$

the continuous form of Eq. (12.30) becomes:

$$\tilde{c}(x,y) = \int_0^{\pi}\int_{-\infty}^{\infty} \mathcal{P}(r,\vartheta)\exp[2\pi kr]dk\, d\vartheta =$$

$$\tilde{c}(r,\vartheta) = \int_0^{\pi}\int_{-\infty}^{\infty} \frac{\mathcal{P}(r,\vartheta)}{|k|}\exp[2\pi kr]|k|dk\, d\vartheta \tag{12.32}$$

The introduction of $|k|$ is required to express the integral in the polar coordinates, as in Eq. (12.27). The Fourier transform of Eq. (12.32) is then: $\tilde{\mathcal{C}}(x,y) = \frac{\mathcal{P}(r,\vartheta)}{|k|}$, which together with the center slice theorem, Eq. (12.10), shows that:

$$\tilde{\mathcal{C}}(x,y) = \frac{\mathcal{P}(r,\vartheta)}{|k|} = \frac{\mathcal{C}(r,\vartheta)}{|k|} \tag{12.33}$$

This expression indicates that $\tilde{\mathcal{C}}(x,y)$ is equal to the exact Fourier transform, $\mathcal{C}(x,y)$, divided by the magnitude of the spatial frequency, $|k|$. This has two implications for direct backprojection of Eq. (12.30): (1) $\tilde{\mathcal{C}}(x,y)$ is a scaled (relative) value of the actual Fourier transform of the image parameter, $\mathcal{C}(x,y)$, and (2) high-frequency components are suppressed in the reconstructed image as the solution emphasizes low-frequency components. In addition, simple backprojection associates non-zero values to empty (zero density) voxels. As a result, it produces a foggy background, and in some cases a star effect (an outward spread) may be associated with high density points (Brooks and Di Chiro, 1976).

Equation (12.33) also provides the foundation for Fourier filtered backprojection, as it indicates that multiplying the Fourier transform of the projections, $\mathcal{P}(r, \vartheta)$, by $|k|$, then backprojecting it, will lead to the Fourier transform of the solution, $\mathcal{C}(r, \vartheta)$, as was also indicated by Eq. (12.28).

12.3.2 Fourier Filtered

In this process all projections are filtered with the magnitude of the frequency, in accordance to Eq. (12.28). The resulting values are then backprojected to obtain the required image parameters. This amounts to implementing Eq. (12.27) in a discrete form:

$$\tilde{c}(x, y) = \sum_{m=1}^{M_\vartheta} q(x \cos\vartheta_m + y \sin\vartheta_m, \vartheta_m)\Delta\vartheta \tag{12.34}$$

Notice that this equation is similar to the direct backprojection equation, Eq. (12.30), except for the use of q instead of p, where q is given by Eq. (12.29). This approach, however, not only produces the exact value of the image parameters, $c(x, y)$, but also eliminates the foggy and star artifacts associated with direct backprojection. The latter observation is due to the fact that, in accordance to Eq. (12.33), the Fourier transform of the direct backprojection is divided by the magnitude of the frequency, $|k|$, allowing the contribution of high-frequency components to decrease in proportion to their frequency, while providing the opposite effect for low frequency components.

Aside from the discretization error associated with the summation in Eq. (12.34), the implementation of this method requires the availability of all values of $r_m = x \cos\vartheta_m + y \sin\vartheta_m$. Since $p(r, \vartheta_m)$, and consequently $q(r, \vartheta_m)$, are calculated at a finite number of discrete points, interpolation is required for the r values that are not sampled, i.e. for points not directly in the path of a sampled radiation beam. This can be accomplished by simply using the value of the nearest neighbor, interpolation between adjacent values, or using exact interpolation in accordance with the sampling theorem as discussed in Section 12.2.4. The latter method is hardly used as it requires a large number of measurements, and is computationally demanding. The above mentioned discretization and interpolation approximations result in some loss in image resolution, particularly near edges where streaks in the image may be observed (Brooks and Di Chiro, 1976).

Fourier filtered backprojection is quite attractive because of its simplicity, ease of numerical implementation and the ability to perform it immediately once a projection is recorded, without awaiting to acquire and store all projections. Moreover, in this method, interpolation can be performed more accurately in the spatial domain, unlike in the direct Fourier inversion method, discussed in Section 12.2.5, which requires a more complicated interpolation process (Kak and Slaney, 2001).

12.3.3 Radon Filtered

According to the Fourier transform theory, the Fourier transform of a product of functions, e.g. $\mathcal{F}(k) \times \mathcal{G}(k)$, is equal to the convolution[4] of their individual Fourier transforms: $f(r) * g(r)$. This is known as the convolution theorem. Realizing that Eq. (12.29) consists of the Fourier transform of the product $\mathcal{P}(k,\vartheta)|k|$, and that the Fourier transform of $\mathcal{P}(k,\vartheta)$ is $p(r,\vartheta)$ and that of $|k|$ is $-\frac{1}{2\pi^2 r^2}$, applying the convolution theorem to Eq. (12.29) results in (Brooks and Di Chiro, 1976):

$$q(r,\vartheta) = -\frac{1}{2\pi^2}\int_{-\infty}^{\infty} \frac{p(r',\vartheta)}{(r-r')^2}dr' = \frac{1}{2\pi^2 r^2}\int_{-\infty}^{\infty} \frac{\frac{\partial p(r',\vartheta)}{\partial r'}}{r-r'}dr' \tag{12.35}$$

where the last step involved integration by parts. Equation (12.35) represents a filtering of the derivatives of the projections with the inverse distance from the point of filtering. Once this filtering process is applied, the filtered values given by $q(r,\vartheta)$ are backprojected, in accordance to Eq. (12.30) with $q(r,\vartheta)$ replacing $p(r,\vartheta)$. Since the filtering of Eq. (12.35) involves the projections themselves, not their Fourier transforms, this method is known as Radon filtering.

Although the implementation of Eq. (12.35) appears at first glance to be attractive, because it deals directly with the projections and avoids Fourier transforms, the availability of Fast Fourier transforms alleviates this apparent advantage. Moreover, numerically approximating the derivatives and the singularity at $r' = r$ makes this method susceptible to numerical error propagation. However, Lanzavecchia et al. (1999) utilized the Radon transform, in three-dimensional image reconstruction, to devise a special filter to correct inconsistencies in measurements and for interpolation to fill gaps in the Radon transform.

12.3.4 Shepp-Logan Filtered

Let a Fourier projection, $\mathcal{P}(k,\vartheta)$, be filtered by some weighting function, $\mathcal{W}(k)$, so that:

$$q(r,\vartheta) = \int_{-\infty}^{\infty} \mathcal{P}(k,\vartheta)\mathcal{W}(k)\exp[2\pi ikr]dk \tag{12.36}$$

[4] A convolution operator relates one point to another by their separation distance, so that: $f(r) * g(r) = \int_{-\infty}^{\infty} f(r')g(r-r')dr' = \int_{-\infty}^{\infty} g(r')f(r-r')dr'$.

Applying the convolution theorem to Eq. (12.36) results in:

$$q(r,\vartheta) = \int_{-\infty}^{\infty} p(r',\vartheta)W(r-r')dr' \tag{12.37}$$

where $W(r)$ is the inverse Fourier transform of $\mathcal{W}(k)$. The formulation of Eq. (12.37) requires $W(r)$ to be an even function, i.e. $W(+r) = W(-r)$, and $\mathcal{W}(k)$ be bandwidth limited (Shepp and Logan, 1974). Applying the filtering function of Eq. (12.37) to Eq. (12.34) gives a solution in which a projection, $p(r',\vartheta)$, is weighted by $W(r'-r)$, with $r-r'$ being the distance from a point in the image r to the point r' at which the projection is recorded. Interpolation between neighboring projections can be used if r' does not correspond to a measured projection. Shepp and Logan (1974) proposed the following weight function:

$$W(0) = \frac{M_R}{\pi r^2} = \frac{4}{\pi(\Delta r)}; W(r) = -\frac{M_R^2}{\pi(M_R^2 r^2 - R^2)} = W(n\Delta r) = -\frac{4}{\pi(\Delta r)^2(4n^2-1)} \tag{12.38}$$

with M_r being the number of ray-sums in a projection, and R is the radius enclosing the object. In the discrete form, $W(r)$ is linear between sampled projections. The Fourier transform of Eq. (12.38) is:

$$\mathcal{W}(k) = \left|4k_{max}\sin\left(\frac{\pi k}{2k_{max}}\right)\right| \mathrm{sinc}^2\left(\frac{\pi k}{2k_{max}}\right) \tag{12.39}$$

This is known as the Shepp-Logan filter (Shepp and Logan, 1974). Notice here, when $k << k_{max}$, one has $W(k) \approx |k|$, and this filter becomes identical to the Fourier filter of to Eq. (12.28), but as $k \to k_{max}$, the value of $W(k)$ exceeds $|k|$.

The weight function of Eq. (12.38) can be directly applied in the spatial domain, which makes it quite simple to numerically implement, see the computer program in Shepp and Logan (1974). It is, however, an approximation that does not accommodate higher frequency components well.

12.3.5 Convolution Filtered

Since k has a maximum value of k_{max}, the filtering operation in Eq. (12.29) should be limited to $\pm k_{max}$, so that the Fourier transform of $|k|$ should be expressed as:

$$|\mathcal{K}| = \int_{-k_{max}}^{k_{max}} |k|\exp[2\pi ikr]dk = \frac{k_{max}\sin(2\pi k_{max}r)}{\pi r} - \frac{\sin^2(\pi k_{max}r)}{\pi^2 r^2} \tag{12.40}$$

With this expression, one can repeat the exercise of using the convolution theorem, as done in Section 12.3.3. Then, for a filtered projection, one obtains:

$$\begin{aligned} q(r,\vartheta) &= \int_{-\infty}^{\infty} p(r',\vartheta)\left[\frac{k_{max}\sin[2\pi k_{max}(r-r')]}{\pi(r-r')} - \frac{\sin^2[\pi k_{max}(r-r')]}{\pi^2(r-r')^2}\right]dr' \\ &= \int_{-\infty}^{\infty} p(r',\vartheta)\left[2k_{max}^2\mathrm{sinc}[2\pi k_{max}(r-r')] - \frac{\sin^2[\pi k_{max}(r-r')]}{\pi^2(r-nr')^2}\right]dr' \\ &= k_{max}p(r,\vartheta) - \int_{-\infty}^{\infty} p(r',\vartheta)\frac{\sin^2[\pi k_{max}(r-r')]}{\pi^2(r-r')^2}dr' \end{aligned} \tag{12.41}$$

Use was made here of $\Delta r = \frac{k_{max}}{2}$, in accordance to Eqs. (12.12) and Eq. (12.13), and the fact that $\mathrm{sinc}(2\pi k_{max}(r-r'))$ is equal to unity everywhere except at $r' = r$ where it is equal zero. The first term in Eq. (12.41) is the inverse Fourier transform of the rectangular function, while the second term is that of a triangle function. Therefore, the filter of Eq. (12.41) expresses the difference between a rectangular and a triangle centered around $k = 0$, which avoids the singularity at $r = 0$. This is the Ram-Lak filter, named after Ramachandran and Lakshiminarayanan (1971).

In a discrete form, one has:

$$q(r_i,\vartheta_m) = \frac{p(r_i,\vartheta_m)}{2\Delta r} - \frac{1}{\pi^2\Delta r}\sum_{j=1,\ i-j=\mathrm{odd}}^{M_r}\frac{p(r_j,\vartheta_m)}{(i-j)^2} \tag{12.42}$$

Since $\sin^2[\pi k_{max}(r-r')] = \sin^2(\pi k_{max} n\Delta r) = \sin^2(\frac{n\pi}{2})$ is zero when $n = 0$ or is even, and is equal to one when the integer n is even, only odd values of $i-j$ contribute to the summation in Eq. (12.42).

Limiting the frequency to $|k| \leq k_{max}$ avoids the divergence in the Fourier transform of the unbounded $|k|$; which is equal to $-\frac{1}{2\pi^2 r^2}$ causing it to diverge as $r \to 0$. Given the convolution theorem discussed above, the Fourier integral of Eq. (12.29) will also tend to diverge, if $|k|$ is not bounded. This bandwidth limiting shows its effect in the appearance of the sinc function in the integral of Eq. (12.41), which eliminates frequencies greater than k_{max} in a convolution integral (Brooks and Di Chiro, 1976).

The filter of Eq. (12.42) is known as the convolution filter, because it includes the convolution kernel: $i-j$, which relates one beam i to a neighboring beam j, in parallel-beam transmission tomography, by the separation distance between them. This convolution term appears in the subtraction part of the filter. In effect, it permits the neighbors of a beam in a projection to contribute to its backprojection, with a degree of influence decreasing with the square of the separation distance. The even values of $i-j$ are excluded from the filter, since this filter represents a squared sine-wave originating at beam i that peaks at the next neighbor (Brooks and Di Chiro,

1976). The first term in the filter of Eq. (12.42) is the estimated highest spatial rate of change of the original projection. It has the effect of dampening the measured projections, which further reduces its over-contribution in the backprojection process.

Convoluted-filtered backprojection is quite attractive because it avoids the use of Fourier transforms in frequency filtering, and does not require the derivatives as in Radon filtering. However, it has the disadvantage of band-limiting, discussed in Section 12.2.2, with its associated artifacts of aliasing and the Gibbs phenomenon.

12.3.6 Noise Filtering

Signal noise can produce frequencies greater than k_{max}. To suppress such noise, filtered projections are further multiplied by a smoothing filter: the Shepp-Logan filter (the sinc(x) function), or a low-pass filter. A low-pass filter takes the general form:

$$q_{lp} = \begin{cases} \alpha + (1-\alpha)\cos\left(\frac{\pi k}{k_{max}}\right) & \text{for} \quad k \leq k_{max} \\ 0 & \text{for} \quad k > k_{max} \end{cases} \tag{12.43}$$

With $\alpha = 0$, one has a cosine filter. The Hann (Hanning) filter uses $\alpha = 0.5$, while the Hamming filter has $\alpha = 0.54$ (Hamming, 1977).

12.3.7 Error Propagation

Let σ^2 be the statistical variance associated with each measurement in a projection p. For the sake of simplification, we will assume σ^2 is the same for all measurements. Since backprojection, Eq. (12.28), is linear, the variance of a reconstructed parameter $\tilde{c}(x,y)$ is simply:

$$\sigma_{\hat{c}}^2 = \left(\frac{2\pi}{M_\vartheta}\right)^2 \sum_{m=1}^{M_\vartheta} \sigma_q^2 = \frac{4\pi^2\sigma_q^2}{M_\vartheta} \tag{12.44}$$

assuming an equal variance of σ_q^2 for all projections. Expressing q in terms of a weight, W, as in Eq. (12.37), enables the estimation of σ_q^2 as:

$$\sigma_q^2 = \sum_j^{M_r} \left(\frac{2R}{M_r}\right)^2 \sigma^2 W_j^2 = \frac{4R^2\sigma^2}{M_r^2} \sum_j^{M_r} W_j^2 \tag{12.45}$$

and

$$\sigma_{\hat{c}}^2 = \frac{(4\pi R)^2\sigma^2}{M_\vartheta M_r^2} \sum_j^{M_r} W_j^2 = \frac{\pi\sigma^2}{2Rk_{max}^3} \sum_j^{M_r} W_j^2 \tag{12.46}$$

where in the last expression use is made of Eq. (12.14) and $M_\vartheta = \frac{\pi R}{\Delta r} = 2\pi R k_{max}$; which is consistent with the Nyquist condition $\Delta r = \frac{1}{2k_{max}}$ of the sampling theorem,

Eq. (12.11). With $\int_{-\infty}^{\infty} W^2(u)du = \Delta r \sum_j^{M_r} W_j^2 = \frac{1}{2k_{max}} \sum_j^{M_r} W_j^2$, Eq. (12.46) can be expressed as:

$$\sigma_{\hat{c}}^2 = \frac{\pi\sigma^2}{Rk_{max}^2} \int_{-\infty}^{\infty} W^2(u)du \tag{12.47}$$

and the integral can be evaluated in either the spatial or the frequency domain, since u in the integral is a dummy variable.

Equation (12.47) demonstrates the importance of choosing a weighting function, hence a filter, that reduces the statistical variance. However, it is essential to select a filter based on sound analytical reasoning, so that it produces the exact solution in ideal (noise free) measurements. Equation (12.47) also reflects the fact that a higher sampling frequency reduces the propagation of statistical fluctuations.

12.4 Fan-Beam Transmission Tomography

The methods discussed in Sections 12.1, 12.2, and 12.3 apply to parallel-beam geometry. Most two-dimensional transmission imaging systems employ, however, fan-beam geometries. In a fan beam, radiation emanating from a point at a tip spreads in a plane at different directions, which complicates the mathematical formulation of the analytical problem of backprojection. One approach to dealing with fan-beam projections is to convert them into equivalent parallel-beam ones, in a process known as rebinning. The other approach is to modify the backprojection process itself to accommodate fan beams. These two approaches are discussed below.

12.4.1 Rebinning

A fan-beam projection can be viewed as a set of parallel-beam projections accumulated at various polar angles. Let us consider a fan beam with its central ray making an angle $\frac{\pi}{2} + \vartheta_f$ with the x-axis, as shown in Figure 12.4. Notice here that the detectors are assumed to be arranged over the curve of the fan beam (in the curved direction along r_f); the situation in which detectors are located along a line (along r_l) is discussed in Section 12.5. Let us consider a subbeam (i_f, j_f) (a ray in a fan beam) that provides a projection at a radial distance r_{i_p} and a polar angle ϑ_{j_p} in the polar coordinates, but records a projection at a radial distance r_{i_f} along the arc of detectors measuring the projections from the parent fan beam. Then, assuming that the source-detector assembly is rotated around the source-detector midpoint, the polar coordinates for the subbeam (r_i, ϑ_{j_p}) are related to the fan-beam geometric coordinates $(r_{i_f}, \vartheta_{j_f})$ by:

$$r_{i_p} = |so| \sin\gamma \tag{12.48}$$

$$\vartheta_{j_p} = \vartheta_{j_f} + \gamma \tag{12.49}$$

The task now is to convert the fan-beam geometric coordinates $(r_{i_f}, \vartheta_{j_f})$, where i and j are integers, to the parallel-beam coordinates $(r_{i_p}, \vartheta_{j_p})$, while obtaining integer values

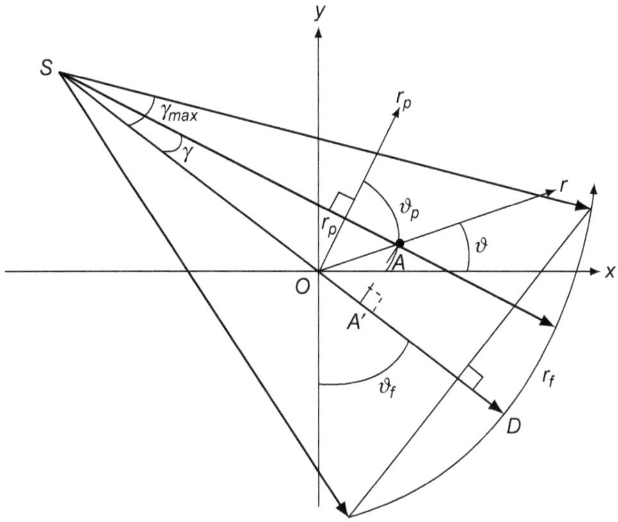

Figure 12.4 A schematic of a cone beam of a half-angle γ_{max} at S, a centerline SOD, and a subbeam at angle $\gamma = \frac{r_f}{|SD|}$, with subscripts f and p referring to fan-beam and parallel-beam geometries, respectively.

for i and j so that the discretized formation for parallel-beam becomes applicable. This requires a two-stage interpolation process (Barrett, 1981). With Δr_f and $\Delta\vartheta_f$ being the discrete sampling intervals in the fan-beam setup and Δr and $\Delta\phi$ the corresponding intervals in the parallel-beam geometry, one then has:

$$r_{i_f} = i_f \Delta r_f; \quad \vartheta_{j_f} = j_f \Delta\vartheta_f \tag{12.50}$$

$$r_i = i\Delta r; \quad \vartheta_j = j\Delta\vartheta \tag{12.51}$$

At a fixed value of i_f, ϑ_{j_p} and ϑ_{j_f} are related by Eq. (12.49), which in the discrete form gives:

$$\tilde{j}_f = j\,\frac{\Delta\vartheta}{\Delta\vartheta_f} - \frac{i\,\Delta r_f}{|SD|\Delta\vartheta_f} \tag{12.52}$$

This produces a noninteger value for $\tilde{j}_f$. However, between the projections corresponding to the integer values just below and above $\tilde{j}_f$ at the same i_f, one obtains an intermediate projection value in the polar direction of ϑ. These projections are not necessarily at uniform radial intervals in the polar coordinates, necessitating a second interpolation between projections at a fixed value of j. A discrete interval in Eq. (12.48) is given by:

$$\tilde{i}_f = \frac{|SD|}{\Delta r_f}\sin^{-1}\left(\frac{i\,\Delta r}{|SO|}\right) \tag{12.53}$$

Interpolation between projections corresponding to the nearest integers of $\tilde{i}_f$ produces projections at uniform radial intervals. Now with uniform sampling in the radial and polar directions, one can apply any of the methods used for image reconstruction for parallel-beam projections.

12.4.2 Direct Backprojection

In parallel-beam direct backprojection (Section 12.3.1), each ray-sum is projected along the line-of-response of the detector, and the backprojections of rays passing through each point (or pixel) in the image are summed. Let us apply the same logic to a point a: (x, y), or (r, ϑ) in Figure 12.4. Writing Eq. (12.28), keeping in mind that fan-beam projections can be acquired over the entire polar range of 2π without redundancy, then guided by Figure 12.4 and using Eqs. (12.36) and (12.37), one has:

$$c(x,y) = \frac{1}{2}\int_{0}^{2\pi}\int_{-r_{p_{max}}}^{r_{p_{max}}} p(r'_p, \vartheta_p) W(x\cos\vartheta_p + y\sin\vartheta_p - r'_p)\, dr'\, d\vartheta_p \tag{12.54}$$

$$c(r,\vartheta) = \frac{1}{2}\int_{0}^{2\pi}\int_{-r_{p_{max}}}^{r_{p_{max}}} p(r'_p, \vartheta_p) W(r\cos(\vartheta_p - \vartheta) - r'_p)\, dr'\, d\vartheta_p \tag{12.55}$$

where W is a weighting function, the subscript p refers to the parallel-beam geometry, $\pm r_{p_{max}}$ are the edge points in the projection beyond which $p(r'_p, \vartheta_p) = 0$, and q is a filter in the spatial domain. Now, using relationships (12.48) and (12.49) to arrive at an expression for the fan-beam geometry, Eq. (12.55) becomes:

$$c(r,\vartheta) = \frac{1}{2}\int_{-\gamma}^{2\pi-\gamma}\int_{-\sin^{-1}\left(\frac{r_{p_{max}}}{|SO|}\right)}^{\sin^{-1}\left(\frac{r_{p_{max}}}{|SO|}\right)} p\left(|SO|\sin\gamma, \vartheta_f + \gamma\right) \times W\left(r\cos(\vartheta_f + \gamma - \vartheta) - |SO|\sin\gamma\right)|SO|\cos\gamma\, d\gamma\, d\vartheta_f \tag{12.56}$$

where the subscript f refers to the fan-beam geometry, and the integration over γ takes into account that one edge of the first fan beam is below its centerline while for the last beam one edge is above the centerline. Equation (12.56) can now be simplified to:

$$c(r,\vartheta) = \frac{1}{2}\int_{0}^{2\pi}\int_{-\gamma_{max}}^{\gamma_{max}} p(\vartheta_f, \gamma) W\left(r\cos(\vartheta_f + \gamma - \vartheta) - |SO|\sin\gamma\right)|SO|\cos\gamma\, d\gamma\, d\vartheta_f \tag{12.57}$$

where γ_{max} is the upper limit of γ, i.e. the angle at the edge of the fan beam (see Figure 12.4).

Let us now focus on the filter function, $W[r\cos(\vartheta_f + \gamma - \vartheta)]$, and attempt to relate it to the Fourier filter, $|k|$ of Eq. (12.29) (Kak and Slaney, 2001). The argument of this filter can be simplified using the geometry of Figure (12.4):

$$|AA'| = |SA|\sin\left(\frac{\pi}{2} - \vartheta_f + \vartheta\right) = |SA|\sin\gamma' = r\cos(\vartheta_f - \vartheta)$$
$$|OA'| = |SA|\cos\left(\frac{\pi}{2} - \vartheta_f + \vartheta\right) - |SO| = |SA|\cos\gamma' - |SO| = r\sin\left(\vartheta_f - \vartheta\right)$$

where $\gamma' = \frac{\pi}{2} - \vartheta_f + \vartheta$ is the angle the direction of r makes with the subbeam (SOD). Then,

$$\begin{aligned} r\cos(\vartheta_f{+}\gamma - \vartheta) - |SO|\sin\gamma \\ &= r\cos(\vartheta_f - \vartheta)\cos\gamma - r\sin(\vartheta_f - \vartheta)\sin\gamma - |SO|\sin\gamma \\ &= |SA|(\sin\gamma'\cos\gamma - \cos\gamma'\sin\gamma) \\ &= |SA|\sin(\gamma' - \gamma) \end{aligned} \tag{12.58}$$

Therefore, $W(r\cos(\vartheta_f + \gamma - \vartheta) = W(|SA|\sin(\gamma' - \gamma))$. Analogous to the inverse Fourier transform of the Fourier filter, $|k|$, one can develop an expression in terms of $W(|SA|\sin\gamma)$:

$$W(r) = \int_{-\infty}^{\infty} |k|\exp[2\pi ikr]\,dk$$

$$\begin{aligned} W(|SA|\sin\gamma) &= \int_{-\infty}^{\infty} |k|\exp[2\pi ik|SA|\sin\gamma]\,dk \\ &= \left(\frac{\gamma}{|SA|\sin\gamma}\right)^2 \int_{-\infty}^{\infty} |k'|\exp[2\pi ik']\,dk' \\ &= \left(\frac{\gamma}{|SA|\sin\gamma}\right)^2 W(\gamma) \end{aligned} \tag{12.59}$$

$$W(\gamma) = \int_{-\infty}^{\infty} |k_\gamma|\exp[2\pi ik_\gamma\gamma]\,dk_\gamma \tag{12.60}$$

$$k_\gamma = \frac{k|SA|\sin\gamma}{\gamma} \tag{12.61}$$

With Eqs. (12.58) and (12.61), the Fourier filtered form of Eq. (12.57) becomes:

$$c(r,\vartheta) = \int_0^{2\pi} \frac{1}{|SA|^2} \int_{-\gamma_{max}}^{\gamma_{max}} p(\vartheta_f,\gamma)W_f(\gamma' - \gamma)|SO|\cos\gamma\,d\gamma d\vartheta_f \tag{12.62}$$

$$= \int_0^{2\pi} \frac{1}{|SA|^2} \int_{-\gamma_{max}}^{\gamma_{max}} p_f(\vartheta_f, \gamma) W_f(\gamma' - \gamma)\, d\gamma d\vartheta_f \tag{12.63}$$

$$= \int_0^{2\pi} \frac{1}{|SA|^2} Q_f(\vartheta_f)\, d\vartheta_f \tag{12.64}$$

with

$$W_f(\gamma) = \frac{1}{2}\left(\frac{\gamma}{\sin\gamma}\right)^2 W(\gamma) \tag{12.65}$$

$$p_f(\vartheta_f, \gamma) = p(\vartheta_f, \gamma)|SO|\cos\gamma \tag{12.66}$$

$$Q_f(\vartheta_f) = p_f(\vartheta_f, \gamma) * W_f(\gamma) \tag{12.67}$$

given that "$*$" is the convolution operator. Notice the presence of the inverse squared-distance term, $\frac{1}{|SA|^2}$, in Eq. (12.64), which was not present in parallel-beam projections. This term gives more weight to points closer to the source than those farther away, accounting for radiation divergence, see Section 2.6, which is ignored in parallel beams where beam divergence is not considered.

Equation (12.64) presents the formulation for backprojection that can be used for fan beams. The filtering function, W, is modified at each subbeam, γ, to W_f by Eq. (12.65). Moreover, the projections are modified, in accordance to Eq. (12.66), at each point by the distance from the source and the point, $|SO|$, and for each subbeam in the fan beam by $\cos\gamma$. The convolution of the modified filter and projection, Eq. (12.67), is then backprojected to obtain the solution, using Eq. (12.64). The convolution process can be accomplished with the aid of the fast Fourier transform algorithm. The formulation of Eq. (12.64) is analogous to that used in parallel-beam projections, enabling the use of the latter's other filters and features.

12.5 Cone-Beam Transmission Tomography

Three-dimensional projections can be directly acquired on a planar detector using a cone-beam rotating around the object. Image reconstruction can then be accomplished by considering each plane within the cone to correspond to a fan-beam, applying backprojection to each fan beam, and adding the contribution from all fan beams to obtain the final three-dimensional image.

A pencil beam in 3D is identified by its direction in cylindrical coordinates:

$$\hat{s}_p = \cos\vartheta_p \sin\varphi_p\, \hat{x} + \sin\vartheta_p \sin\varphi_p\, \hat{y} + \cos\varphi_p\, \hat{z}$$

where $\hat{x}$, $\hat{y}$, and $\hat{z}$ designate unit vectors, and ϑ_p and φ_p are, respectively, the angle the ray makes with the x-axis (polar angle) and z-axis (azimuthal angle). The pencil beam is then the loci of points $\vec{l} + s\hat{s}_p$ where $-\infty \le l \le \infty$ is a distance along the beam and $\vec{l}$

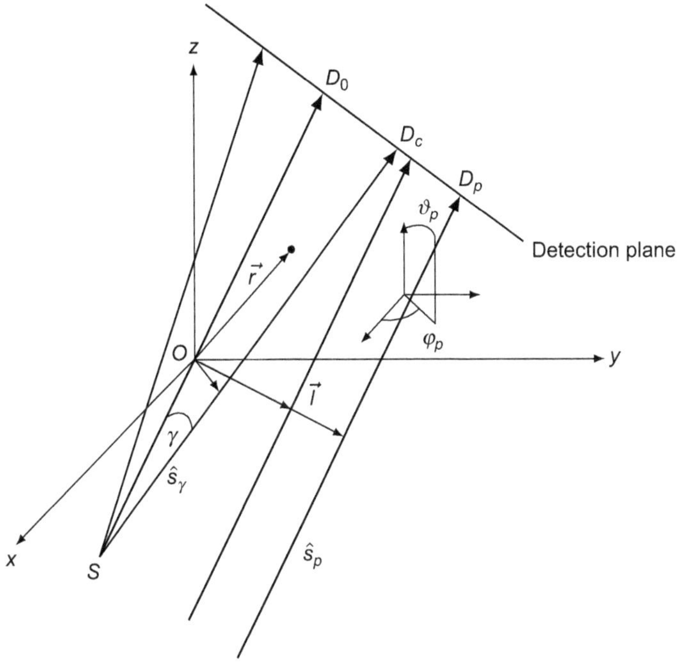

Figure 12.5 A schematic of rays in three dimensions.

is the radial vector from the point of origin (center of beam rotation), normal to $\hat{s}_p$, see Figure 12.5. A set of parallel beams can be formed in three dimensions by maintaining the direction of $\hat{s}_p$ and changing the source location on a plane normal to $\hat{s}_p$, while a cone-beam is formed by fixing the source position while changing the direction of the ray. The corresponding two-dimensional cases are then formed by fixing the azimuthal angle, φ_p, to a certain value corresponding to the plane of interest, e.g. $\varphi_p = \frac{\pi}{2}$ for the x-y plane. A transmission projection for any pencil beam in three-dimensions is then given by:

$$p(\vec{l}, \hat{s}) = \int_{-\infty}^{\infty} c(\vec{l} + s\hat{s})\, ds \tag{12.68}$$

The direction of integration in Eq. (12.68) for a parallel beam is along $\hat{s} = \hat{s}_p$, while that for a subbeam in a cone-beam (making an angle, γ, with the centerline of the cone (along $\hat{s}_0$) and an angle, φ, with the z axis) is along $\hat{s} = \hat{s}_\gamma$. To relate $\hat{s}_\gamma$ to that of the parallel-beam arrangement, we will consider the set of beams parallel to $\hat{s}_0$. Then, as can be seen in Figure 12.5:

$$\hat{s} = \hat{s}_\gamma = \frac{|SO|\hat{s}_0 + \vec{l}}{\sqrt{|SO|^2 + |l|^2}} \tag{12.69}$$

where $\vec{l}$ is the radial vector corresponding to the considered ray at (γ, φ).

The solution in three dimensions for cone-beams can now be expressed by generalizing Eq. (12.55), so that (Denton et al., 1979):

$$c(x,y,z) = c(\vec{r}) = \int_0^{4\pi} \frac{1}{8\pi^3} \frac{|SO|^3}{(|SO| + \vec{r}\cdot\hat{s}_0)^3}\, d\Omega_p \times \int \frac{|SO|}{\sqrt{|SO|^2 + |l|^2}} p(\vec{l}, \hat{s}) W\left(\frac{|SO|[(\vec{r}\cdot\hat{s}_0)\hat{s}_0]}{|SO| + \vec{r}\cdot\hat{s}_0} - \vec{s}\right) d\vec{l} \tag{12.70}$$

where Ω_p is a solid angle within the cone beam, $d\Omega = d\cos\vartheta_p d\varphi_p$ and "$\cdot$" designate a dot product. Note that $|SO| + \vec{r}\cdot\hat{s}_0 - \vec{s}$ is a distance along $\vec{l}$, hence the integration is over $\vec{l}$, i.e. over all subbeams passing by a point in the image at $\vec{r}$. The other weight functions are due to geometric considerations, similar to those considered in Section 12.4, and account for radiation divergence. The filter function in Eq. (12.70) is an appropriate convolving function, see for example (Denton et al., 1979).

Three-dimensional image reconstruction can then be conducted by performing the integrations in Eq. (12.70) in three steps. First, the projections are modified to $\frac{|SO|}{\sqrt{|SO|^2+|l|^2}} p(\vec{l}, \hat{s})$. Secondly, the modified projections are convoluted with W, to obtain the second integral in Eq. (12.70). Finally, the projections are backprojected with the weight $\frac{1}{8\pi^3} \frac{|SO|^3}{(|SO|+\vec{r}\cdot\hat{s}_0)^3}$. This process can be easily discretized as shown in (Denton et al., 1979). Feldkamp et al. (1984) presented an approximate but numerically efficient convolution-backprojection formula for 3D cone-beam image reconstruction. The formula is based on the accumulation of contributions of the image parameters from two-dimensional fan-beam projections. Due to its simplicity and its numerical efficiency, this scheme is widely used in cone-beam systems. Such systems typically involve helical (or spiral) scanning process, in which the object is moved through a rotating source-detector gantry while measurements are gathered. The scheme of Feldkamp et al. (1984) is typically formulated for spiral/helical scanning loci with equi-spatial cone-beam projections, but Wang et al. (2001) provided a scheme for equi-angular cone-beam projections.

12.6 Emission Imaging

As indicated in Chapter 5, a measurement in emission imaging is essentially the summation of radiation activities over all points in the field-of-view of a detector. Therefore, an emission measurement can be manipulated in the manner projections are handled in transmission imaging, except for the fact that emission measurements are modulated by radiation attenuation within the object. If the attenuation coefficient is known at all points in an imaged section, then it would be possible to estimate the degree of attenuation from a point in the object to the position of a detector, and incorporate attenuation factors into the backprojection process to correct for the modulating effect of attenuation. In many cases, the detailed distribution of the attenuation coefficient within the interrogated object is not known, but either its average value is given from *a priori* knowledge, or its distribution along various orientations around the object is estimated from single-projection transmission measurements, or from full transmission tomography. Even with a single uniform value of the attenuation coefficient within the object, the degree of attenuation would change with the change in the distance from the source to the object's edge facing the detector. Therefore, correction for attenuation should be considered at each direction, ϑ, of a projection. This necessitates the introduction of a second Fourier transform with respect to the polar direction, in addition to the usual spatial transform along the radial distance.

For simplicity, let us first consider a parallel-beam arrangement (or an arrangement in which radiation divergence is neglected). Starting with attenuation-free measurements, the two-dimensional Fourier transform of the sought solution, c, is (Metz and Pan, 1995):

$$\begin{aligned} \mathcal{C}(k,\psi) &= \int_{-\infty}^{\infty}\int_{-\infty}^{\infty} c(x,y)\exp[-2\pi(ik_x x + k_y y)]\,dxdy \\ &= \int_{0}^{2\pi}\int_{0}^{\infty} c(r,\vartheta)\exp[-2\pi kir(\cos(\psi-\vartheta)]\,rd\vartheta dr \end{aligned} \tag{12.71}$$

with $x = r\cos\vartheta$, $y = \sin\vartheta$ and $\tan\psi = \frac{k_y}{k_x}$.

The inverse Fourier transform of Eq. (12.71) is:

$$c(r,\vartheta) = \int_{0}^{2\pi}\int_{0}^{\infty} \mathcal{C}(x,y)\exp[2\pi kir(\cos(\vartheta-\psi)]\,kdkd\psi \tag{12.72}$$

The transform $\mathcal{C}(k,\psi)$ is periodic in ψ, except at $k = 0$, with a period of 2π. As such it can be presented by a Fourier series expansion of the form:

$$\mathcal{C}(k,\psi) = \sum_{-\infty}^{\infty} \mathcal{C}(k,k_\psi)\exp[ik_\psi\psi] \tag{12.73}$$

where,

$$\mathcal{C}(k,k_\psi) = \frac{1}{2\pi}\int_{0}^{2\pi} \mathcal{C}(k,\psi)\exp[-ik_\psi\psi]\,d\psi \tag{12.74}$$

$$\begin{aligned} &= \frac{1}{2\pi}\int_{0}^{2\pi}\int_{0}^{2\pi}\int_{0}^{\infty} c(r,\vartheta)\exp[-2\pi kir\cos(\psi-\vartheta)] \\ &\quad \times \exp[-ik_\psi\psi]\,rd\vartheta dr\,d\psi \end{aligned} \tag{12.75}$$

$$= \frac{1}{2\pi}\int_{0}^{2\pi}\int_{0}^{2\pi}\int_{0}^{\infty} c(r,\vartheta)\exp\left[-2\pi kir\cos\left(\psi' + \frac{\pi}{2}\right)\right]$$

$$\times \exp\left[-ik_\psi\left(\psi' + \vartheta + \frac{\pi}{2}\right)\right] rd\vartheta\, dr\, d\psi'$$

$$= (-i)^{k_\psi} \int_0^{2\pi}\int_0^{\infty} c(r,\vartheta)\exp[-ik_\psi\vartheta]\mathrm{J}_k(2\pi kr)\, rd\vartheta\, dr \tag{12.76}$$

where use is made of Eq. (12.71), $\psi' = \psi - \vartheta - \frac{\pi}{2}$, $\exp[-\frac{\pi}{2}] = -i$, and $\mathrm{J}(2\pi kir)$ is the Bessel function of the first kind or order k:[5]

$$\mathrm{J}_k(2\pi kr) = \frac{1}{2\pi}\int_0^{2\pi} \exp[i(2\pi kr\sin\psi') - k\psi']\, d\psi' \tag{12.77}$$

Ignoring attenuation, an emission measurement at a detector point defined in polar coordinates, (r,ψ), is modeled by:

$$e_0(s,\psi) = \int c(s,t)\, dt \tag{12.78}$$

for a detector at position s that records emissions along a line t, and e_0 is a measurement modeled without considering attenuation. Two-dimensional polar Fourier transform of e_0 is considered here to facilitate the next step of incorporating the effect of attenuation:

$$\mathcal{E}_0(k,k_\psi) = \frac{1}{2\pi}\int_0^{2\pi}\int_{-\infty}^{\infty} e_0(s,\psi)\exp[-ik_\psi\psi]\exp[-2\pi iks]\, dsd\psi \tag{12.79}$$

$$= \frac{1}{2\pi}\int_0^{2\pi}\int_{-\infty}^{\infty}\int_{-\infty}^{\infty} c(x,y)\exp[-2\pi(ik_x x + k_y y)]\exp[-ik_\psi\psi]\, dxdyd\psi$$

$$= \mathcal{C}(k,k_\psi) \tag{12.80}$$

where the last equation is obtained from Eq. (12.75), which verifies the center slice theorem of Eq. (12.10) in the polar Fourier transform.

Assuming an attenuation coefficient, $\Sigma(r,\vartheta)$ at point (x,y) within the object and zero outside it, then aided by Figure 12.6, Eq. (12.78) becomes:

$$e_\Sigma(s,\psi) = \int_{-\infty}^{\infty} c(s,t)\exp\left[-\left(\int_{t(s_0,\psi)}^{0} + \int_0^t\right)\Sigma(s,\psi,t')dt'\right] dt \tag{12.81}$$

$$= \int_{-\infty}^{\infty} c(s,t)\exp\left[-\bar{\Sigma}(s,\psi)\left(\int_{t(s_0,\psi)}^{0} + \int_0^t\right) dt'\right] dt$$

[5] The transform: $\int_0^\infty c(r)\mathrm{J}_k(sr)\, rd\vartheta\, dr$ is known as the kth order Hankel transform of $c(r)$, and is useful for radially symmetric functions.

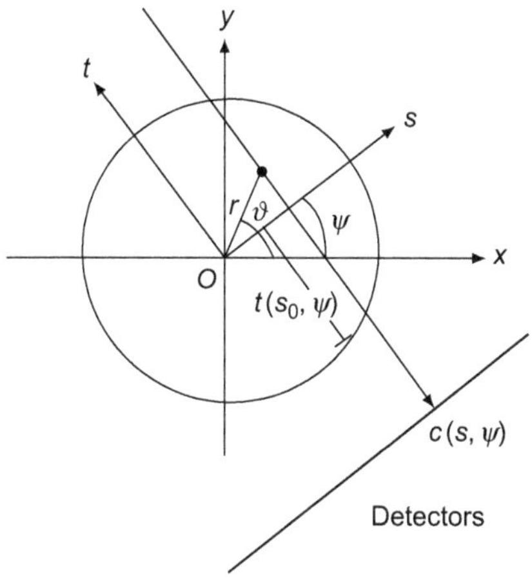

Figure 12.6 A schematic showing a line of sources normal to a detection line making an angle ψ with the x-axis, with the reduced attenuation distance, $t(s_0, \psi)$, measured from the edge of the object to s-axis.

$$= \int_{-\infty}^{\infty} c(s,t)\exp\left[-\bar{\Sigma}(s,\psi)\{t(s_0,\psi)+t\}\right]dt$$

$$= \exp[-\bar{\Sigma}(s,\psi)]t(s_0,\psi)] \int_{-\infty}^{\infty} c(s,t)\exp\left[-\bar{\Sigma}(s,\psi)t\right]dt$$

$$= \exp[-\bar{\Sigma}(s,\psi)t]e_{\bar{\Sigma}(s,\psi)} \tag{12.82}$$

where $t(s_0, \psi)$ is the distance from a point (s_0, ψ) on the s-axis, i.e. toward the detector's position, and $\bar{\Sigma}(s, \psi)$ is the average attenuation coefficient along a ray at (r, ψ), which can be determined *a priori* by a transmission measurement. The transform:

$$e_{\bar{\Sigma}}(s,\psi) = \int_{-\infty}^{\infty} c(s,t)\exp\left[-\bar{\Sigma}(s,\psi)t)\right]dt \tag{12.83}$$

with $\bar{\Sigma}$ being an average attenuation coefficient, is a modified exponentially transformed projection, often called the exponential Radon transform. Without the averaging introduced to produce Eq. (12.83), applying any form of backprojection would be difficult.

The modified projections can be directly backprojected but with an exponential weighting (Tretiak and Metz, 1980):

$$c(r,\vartheta) = \int_{0}^{2\pi} \exp[\bar{\Sigma}t]\frac{|k|}{2}\exp[2\pi iks]\left\{\int_{-\infty}^{\infty} e_{\bar{\Sigma}}(s',\psi)\exp[2\pi iks']\,ds'\right\} dkd\psi \tag{12.84}$$

This, however, like simple direct projection suffers from susceptibility to error propagation.

The two-dimensional Fourier transform of Eq. (12.83) is (Metz and Pan, 1995):

$$\begin{aligned}
\mathcal{E}_{\bar{\Sigma}}(s,\psi) &= \frac{1}{2\pi}\int_{0}^{2\pi}\int_{-\infty}^{\infty} e_{\bar{\Sigma}}(s,\psi)\exp(-2\pi iks)\exp[-ik_{\psi}\psi]\,ds d\psi \\
&= \frac{1}{2\pi}\int_{0}^{2\pi}\int_{-\infty}^{\infty}\int_{-\infty}^{\infty} c(s,\psi)\exp\left[-(\bar{\Sigma}(t,\psi)t + 2\pi iks + ik_{\psi}\psi)\right] ds\,dtd\psi \\
&= \frac{1}{2\pi}\int_{0}^{2\pi}\int_{-\infty}^{\infty}\int_{-\infty}^{\infty} c(r,\vartheta) \\
&\quad \times \exp\left[\bar{\Sigma}(t,\psi)r\sin(\psi-\vartheta) - 2\pi ikr\cos(\psi-\vartheta) - ik_{\psi}\psi\right] ds\,drd\psi \\
&= (-i)^{k_{\psi}}\int_{0}^{2\pi}\int_{0}^{\infty} c(r,\vartheta)\exp[-ik_{\psi}\vartheta]\left\{\frac{1}{2\pi}\times\int_{0}^{2\pi}\exp\left[\bar{\Sigma}(t,\psi)r\cos\psi'\right.\right. \\
&\quad \left.\left. + i(2\pi k\sin\psi' - k_{\psi}\psi')\,d\psi'\right]\right\} rd\vartheta dr
\end{aligned} \tag{12.85}$$

where $\psi' = \psi - \vartheta - \frac{\pi}{2}$. Note that at $\bar{\Sigma} = 0$, Eq. (12.85) becomes identical to Eq. (12.76). Similar to the latter expression, Eq. (12.85) can be written in terms of a Bessel function (Metz and Pan, 1995):

$$\begin{aligned}
\mathcal{E}_{\bar{\Sigma}}(k,k_{\psi}) = (-i)^{k_{\psi}}\int_{0}^{2\pi}\int_{0}^{\infty} c(r,\vartheta)\exp[-ik_{\psi}\vartheta]\left[\frac{k+k_{\Sigma}}{\sqrt{k^2-k_{\Sigma}^2}}\right]^{k_{\psi}} \\
\mathrm{J}_k(2\pi\left(\sqrt{k^2-k_{\Sigma}^2})r\right)
\end{aligned} \tag{12.86}$$

where $k_{\Sigma} = \frac{\bar{\Sigma}}{2\pi}$, and the Bessel function is defined in Eq. (12.77). Then, the Fourier coefficient for unattenuated radiation, $\mathcal{C}(k,k_{\psi})$ of Eq. (12.76), is related to the attenuated one, $\mathcal{E}_{\bar{\Sigma}}(s,\psi)$ by:

$$\mathcal{E}_{\bar{\Sigma}}(k,k_{\psi}) = \left[\frac{k+k_{\Sigma}}{\sqrt{k^2-k_{\Sigma}^2}}\right]^{k_{\psi}} \mathcal{C}\left(\sqrt{k^2-k_{\Sigma}^2}, k_{\psi}\right) \tag{12.87}$$

Or equivalently,

$$\mathcal{E}_{\bar{\Sigma}}\left(\pm\sqrt{k^2+k_{\Sigma}^2}, k_{\psi}\right) = \left[\frac{\pm\sqrt{k^2+k_{\Sigma}^2}+k_{\Sigma}}{k}\right]^{k_{\psi}} \mathcal{C}(k, k_{\psi}) \tag{12.88}$$

Therefore, attenuation shifts a frequency k of the unattenuated Fourier coefficient, $\mathcal{C}(k, k_{\psi})$, to a frequency $\pm\sqrt{k^2+k_{\Sigma}^2}$. For $k_{\psi} > 0$, the amplitude is also increased by $\left[\frac{\sqrt{k^2+k_{\Sigma}^2}+k_{\Sigma}}{k}\right]^{k_{\psi}} > 1$ for the shifted frequency of $\sqrt{k^2+k_{\Sigma}^2}$. On the other hand, the amplitude at the frequency $-\sqrt{k^2+k_{\Sigma}^2}$ is altered by $\left[\frac{-\sqrt{k^2+k_{\Sigma}^2}+k_{\Sigma}}{k}\right]^{k_{\psi}}$ (< 1 for $|k| > k_{\Sigma}$), with a change in sign for odd values of k_{ψ}. The higher the frequency, k, the less pronounced the effect of attenuation, but the opposite is true for increasing values of k_{ψ}. Obviously, the higher the degree of attenuation, the more significant is the influence of attenuation.

From the values of $\mathcal{E}_{\bar{\Sigma}}(k, k_{\psi})$, the unattenuated Fourier transform, $\mathcal{C}(k, k_{\psi})$, can be calculated, using Eq. (12.87). However, this equation produces two values for $\mathcal{C}(k, k_{\psi})$, corresponding to $\pm k$. A linear combination of the two values can be used to estimate $\mathcal{C}(k, k_{\psi})$:

$$\mathcal{C}(k, k_{\psi}) = w\left[\frac{\sqrt{k^2-k_{\Sigma}^2}}{k+k_{\Sigma}}\right]^{k_{\psi}} \mathcal{E}_{\bar{\Sigma}}(k, k_{\psi}) + (1-w)\left[\frac{\sqrt{k^2-k_{\Sigma}^2}}{-k+k_{\Sigma}}\right]^{k_{\psi}} \mathcal{E}_{\bar{\Sigma}}(-k, k_{\psi}) \tag{12.89}$$

where $0 \leq w \leq 1$ can be chosen to minimize noise propagation (Metz and Pan, 1995). Once the equivalent unattenuated values are estimated, they can be equated to the corresponding attenuated projections, in accordance to the central slice theorem, Eq. (12.10). Then any of the methods discussed in this chapter for transmission tomography can be applied.

12.7 Scatter Imaging

In the absence of attenuation, scatter imaging becomes an emission-like problem, with a detector measurement representing the scatter intensity of all voxels in the field-of-view of the detector. Then the two-dimensional Radon transform, discussed in Section 12.6, can be used for image reconstruction in scatter imaging, see also (Truong et al., 2007). Similarly, if constant attenuation is assumed, then the method of the modified projections of Section 12.6 can be applied. However, since the attenuation coefficient (macroscopic cross section) of the incident radiation is lower than that of the scattered radiation, and the latter varies with the energy of scattered radiation, the assumption of constant attenuation is not a very good one. Therefore, analytic methods are not generally adequate for use in scatter imaging.

12.8 Computer Codes

An Internet search will reveal the availability of a number of computer codes for image reconstruction using Fourier-based methods. At the time of writing this book, such a search resulted in the following links:
`http://www.owlnet.rice.edu/~elec431/projects96/DSP/backproject3.html`,
`http://oregonstate.edu/~faridana/preprints/fbp.txt`,
`http://cars9.uchicago.edu/software/idl/tomography.html`,
`http://www.chem.fsu.edu/steinbock/downloads.htm`, and
`http://engineering.purdue.edu/~bouman/software/tomography/`.

12.9 Wavelet Transforms

In Section 8.3.1, it was shown that the linear inverse problem of transmission tomography can be formulated by the superposition of sinusoidal waves (Fourier series). A sinusoidal wave extends indefinitely. A wavelet (a small wave) is confined to a small duration, i.e. it has a starting point and an ending point.[6] This enables the performance of the so-called scale analysis, in contrast to the frequency analysis of Fourier transforms. Scale analysis examines the features of a function (or a signal) at different ranges or resolutions: fine, medium and coarse. While a coarse range provides overall features, a fine range reveals more feature details. This analysis is performed via a "mother wavelet," which is shifted (translated) and dilated (contracted or scaled) and applied to the original signal to provide an approximation. The mother wavelet is then translated, dilated, and applied to the signal again to provide another approximation, and so on until a set of approximations is obtained at different scales. The resulting set of approximations represents the wavelet transforms. An inverse wavelet transform reconstructs a signal from its wavelet transforms. Wavelet transforms tend to be less sensitive to noise because in effect they average the original signal over different ranges or scales. Similar to the Fourier-expansion method, the wavelet method is used for reconstructing linear problems that can be presented by Radon transforms.

With a mother wavelet, $\Psi(x)$, daughter wavelets are formed by a translation with an amount b, and a contraction by a, so that[7]:

$$\Psi(x)_{(a,b)} = \frac{1}{\sqrt{a}}\Psi\left(\frac{x-b}{a}\right) \tag{12.90}$$

[6] Graps (1995) and Strang (1994) provide excellent overviews of the concept of wavelets.

[7] Equations (12.90) to (12.92) are based on: Weisstein, Eric W. Wavelet. From MathWorld–A Wolfram Web Resource. `http://mathworld.wolfram.com/Wavelet.html`

Then a wavelet transform of a function, $c(x)$, is:

$$W_{\Psi(a,b)}c = \int_{\infty}^{\infty} c(x)\Psi(x)_{(a,b)}dx = \frac{1}{\sqrt{a}}\int_{\infty}^{\infty} c(x)\Psi\left(\frac{x-b}{a}\right)dx \tag{12.91}$$

The function, $c(x)$, is then reconstructed in terms of wavelets and the associated scale function by the Calderón's formula:

$$c(x) = K_{\Psi}\int_{-\infty}^{\infty}\int_{-\infty}^{\infty} \langle c(x), \Psi(x)_{(a,b)}\rangle\ \Psi(x)_{(a,b)}a^{-2}\,da\,db \tag{12.92}$$

where the inner product,[8] $\langle c(x), \Psi(x)_{(a,b)}\rangle$, is a wavelet coefficients, and K_{Ψ} is a normalization constant.

In a discrete form, a function, $c(x)$, is defined over several subsequent scales. These functions can be used to define wavelets. Let a function be defined over different successively refined scales (intervals). For N scales, the function, $c(x)$, can be presented over N intervals by an $N \times 1$ vector, $\boldsymbol{c}$, so that:

$$\boldsymbol{c} = \boldsymbol{W}_N\boldsymbol{b} \tag{12.93}$$

where $\boldsymbol{W}_N$ is the wavelet-basis matrix and $\boldsymbol{b} = \boldsymbol{W}_{\Psi}^{-1}\boldsymbol{c}$ is the wavelet in the expansion coefficients determined by the value of $c(x)$ over various scales, and the values of $\boldsymbol{W}_N$ is determined by Eqs. (12.90) and (12.91). For computational convenience, a scaling factor of two is used to proceed from one resolution level to the next.

The above analysis is founded on a predefined mother wavelet. There are a several families of wavelets, see for example Graps (1995). The simplest mother wavelet form is that of Haar (Strang, 1994), which is a square wavelet that is positive over half a unit scale and negative over the other half:

$$\Psi(x) = \begin{cases} 1 & 0 \le x < \dfrac{1}{2} \\ -1 & \dfrac{1}{2} < x \le 1 \\ 0 & \text{otherwise} \end{cases} \tag{12.94}$$

The translation and dilation of this mother function generates the daughter functions:

$$\Psi(x)_{(s,l)} = \Psi(2^s x - l) \tag{12.95}$$

where s is a nonnegative integer and $0 \le l \le 2^s - 1$. In addition to the mother and daughters wavelets, a scaling basis, Φ, is formulated so that an overall integral is defined over the unit scale. The first scale, $N = 1$ considers the entire range (normalized here to one unit), so that $\Phi = \boldsymbol{W}_1 = 1$ in Eq. (12.93) is a 1×1 identity matrix.

[8] $\langle f, g\rangle = \int fg\,dx$.

The second resolution level, has the scaling basis: $\Phi = [1, 1]^{\mathrm{T}}$ and the mother wavelet, $\Psi_{(0,0)} = \Psi(x-0) = [1, 1, -1, -1]^{\mathrm{T}}$; with the two vectors forming $\boldsymbol{W}_2$, a 2×2 matrix. For the third scale resolution, $N = 4$, in Eq. (12.93), the scaling basis over the four intervals will be a vector: $\Phi = [1, 1, 1, 1]^{\mathrm{T}}$. The mother wavelet, $\Psi_{(0,0)} = \Psi(x-0) = [1, 1, -1, -1]^{\mathrm{T}}$, is such that its square wave covers the entire scale. The first dilation produces $\Psi_{(1,0)} = \Psi(2x-0) = [1, -1, 0, 0]^{\mathrm{T}}$, and the subsequent translation produce $\Psi_{(1,1)} = \Psi(2x-1) = [0, 0, 1, -1]^{\mathrm{T}}$. The four vectors together form a 4×4 matrix, $\boldsymbol{W}_4$, and so on. The elements of $\boldsymbol{c}$ in Eq. (12.93) are the average values of $c(x)$ over each of the considered intervals within the unit range $0 \leq x \leq 1$. Rather than inverting the wave matrix in Eq. (12.93) to determine the elements of $\boldsymbol{b}$, the averages and half of the differences of consecutive pairs elements of $\boldsymbol{c}$ are evaluated in a pyramid algorithm that moves the averaged values upward for further averaging and differencing (Strang, 1994). The top averages and differences, along with the bottom differences determine the coefficients.

As far as signal frequency is concerned, the averaging and differencing operations involved in wavelet transforms can be viewed, respectively, as low-pass and high-pass filters of $c(x)$. The fast wavelet transform pyramid algorithm elucidates the multiresolution nature of wavelet transforms in which a signal is resolved at multiple scales by differences that reveal the signal details at a given scale and averages that smooth the signal. Once the wavelet expansion coefficients are determined, the original signal, $c(x)$, can be retrieved in its discrete form by multiplying the wavelet matrix by the coefficients, in accordance to Eq. (12.93).

Retrieving a signal from its wavelet expansion coefficients is a form of filtering, that both smoothes the signal and reveals details, since some of the coefficients correspond to the averages and others to the differences. If the details are not important, coefficients below a certain threshold value can be eliminated before reconstructing the signal. This "denoising" process has the advantage that it is accomplished without smoothing out the sharp edges in the signal (Graps, 1995).

In a multiresolution approximation, separable wavelet basis can be employed in multidimensional wavelet transforms (Mallat, 1989). In two dimensions, (x, y), the scaling function, Φ, and the mother wavelet, Ψ, are combined so that:

$$g(x,y) = \begin{Bmatrix} \Phi(x,y) &= \Phi(x)\Phi(y) \\ \Psi^1(x,y) &= \Phi(x)\Psi(y) \\ \Psi^2(x,y) &= \Psi(x)\Phi(y) \\ \Psi^3(x,y) &= \Psi(x)\Psi(y) \end{Bmatrix} \tag{12.96}$$

to form the wavelet basis (a scaling function and three wavelets).

Wavelet transforms are used in image reconstruction of linear problems from one-dimensional projections in a manner similar to that of Fourier filtered backprojection, discussed in 12.3.2. However, the Fourier filter of Eq. (12.29) is replaced by (Rashid-Farrokhi et al., 1997):

$$q(r,\vartheta)_{2^s} = \int_{-\infty}^{\infty} \mathcal{P}(k,\vartheta)|k|\tilde{\mathcal{G}}_{2^s}(k\cos\vartheta, k\sin\vartheta)\,dk \tag{12.97}$$

where $\tilde{g}(\vec{r}) = g(-\vec{r})$, $\mathcal{G}$ is the Fourier transform of g and $g_2(\vec{r}) = 2^s g(2^s\vec{r})$, with $\vec{r}$ being a distance vectors with (x, y) components. The filtering $|k|\tilde{\mathcal{G}}$ is known as the wavelet ramp filter, in contrast with the ramp filter, k. Rashid-Farrokhi et al. (1997) investigated a number of wavelets for use in computed tomography. For other work on image reconstruction with wallets see, for example, Delaney and Bresler (1995) and Bonnet et al. (2002). The wavelet method is also useful in local (region-of-interest) tomography as discussed in Section 14.2.6.

References

H. H. Barrett, *Radiological Imaging*. New York: Academic Press, 1981.

S. Bonnet, F. Peyrin, F. Turjman, and R. Pros, "Multiresolution reconstruction in fan-beam tomography," *IEEE Transactions on Image Processing*, vol. 11, pp. 169–176, 2002.

B. A. Brooks and G. Di Chiro, "Principles of computer assisted tomography (CAT) in radiographic and radioisotopic imaging," *Physics in Medicine and Biology*, vol. 21, pp. 689–732, 1976.

A. H. Delaney and Y. Bresler, "Multiresolution tomographic reconstruction using wavelets," *IEEE Transactions on Image Processing*, vol. 4, pp. 799–813, 1995.

R. Denton, B. Friedlander, and A. Rockmore, "Direct three-dimensional image reconstruction from divergent rays," *IEEE Transactions on Nuclear Science*, vol. NS-26, pp. 4695–4703, 1979.

L. A. Feldkamp, L. C. Davis, and J. W. Kress, "Practical cone-beam algorithm," *Journal of the Optical Society of America A*, vol. 6, pp. 612–619, 1984.

L. W. Goldman, "Principles of CT and CT technology," *Journal of Nuclear Medicine Technology*, vol. 35, pp. 115–128, 2007.

A. Graps, "An introduction to wavelets," *IEEE Computational Science and Engineering*, vol. 2, pp. 50–61, 1995.

R. W. Hamming, *Digital Filters*. Englewood Cliffs: Prentice-Hall, 1977.

A. C. Kak and M. Slaney, *Principles of Computerized Tomographic Imaging*. Philadelphia: Society of Industrial and Applied Mathematics, 2001.

A. Lannes, E. Anterrieu, and K. Bouyoucef, "Fourier interpolation and reconstruction via Shannon-type techniques; Part I: Regularization principle," *Journal of Modern Optics*, vol. 41, pp. 1537–1574, 1994.

S. Lanzavecchia, P. L. Bellon, and M. Radermacher, "Fast and accurate three-dimensional reconstruction from projections with random orientations via Radon transforms," *Journal of Structural Biology*, vol. 128, pp. 152–164, 1999.

S. G. Mallat, "A theory for multiresolution signal decomposition: The wavelet representation," *IEEE Transactions on Pattern Analysis and Machine Intelligence*, vol. 11, pp. 674–693, 1989.

P. Maréchalt, D. Togane, and A. Cellert, "A new reconstruction methodology for computerized tomography: FRECT (Fourier regularized computed tomography)," *IEEE Transactions on Nuclear Science*, vol. 47, pp. 1595–1601, 2000.

C. E. Metz and X. Pan, "A unified analysis of exact methods of inverting the 2-D exponential radon transform, with implications for noise control in SPECT," *IEEE Transactions On Medical Imaging*, vol. 14, pp. 643–658, 1995.

G. N. Ramachandran and A. V. Lakshiminarayanan, "Three-dimensional reconstruction from radiographs and electron micrographs: Application of convolutions instead of Fourier transforms," *Indian J. Pure Appl. Phys*, vol. 9, pp. 997–1003, 1971.

F. Rashid-Farrokhi, K. J. R. Liu, C. A. Berenstein, and D. Walnut, "Wavelet-based multiresolution local tomography," *IEEE Transactions on Image Processing*, vol. 6, pp. 1412–1430, 1997.

L. A. Shepp and B. F. Logan, "The Fourier reconstruction of a head section," *IEEE Trans. Nucl. Sci.*, vol. 21, pp. 21–43, 1974.

H. Stark, "Sampling theorems in polar coordinates," *Journal of the Optical Society of America A*, vol. 69, pp. 1519–1525, 1979.

H. Stark, J. W. Woods, I. Paul, and R. Hingorani, "Direct fourier reconstruction in computer tomography," *IEEE Trans. Acoust. Speech Signal Processing*, vol. ASSP-29, pp. 237–244, 1981.

G. Strang, "Wavelets," *American Scientist*, vol. 82, pp. 250–255, 1994.

O. Tretiak and C. E. Metz, "The exponential radon transform," *SIAM J. Appl. Math*, vol. 39, pp. 341–354, 1980.

T. T. Truong, M. K. Nguyen, and H. Zaidi, "The mathematical foundations of 3D Compton scatter emission imaging," *International Journal of Biomedical Imaging*, vol. 2007, pp. 1–11, 2007, article ID 92780, doi:10.1155/2007/92780.

B. Wang, H. Liu, S. Zhao, and G. Wang, "Feldkamp-type image reconstruction from equiangular data," *Journal of X-Ray Science and Technology*, vol. 9, pp. 113–120, 2001.

13 Probabilistic Methods

As indicated in Section 8.4, a statistical estimate of the solution of an inverse problem requires *a priori* knowledge of at least the probability distribution, $P(\boldsymbol{c})$, of the solution, $\boldsymbol{c}$. The probability distribution of the measurements, $P(\boldsymbol{e})$, can be taken as a scaling factor in estimating the conditional probability, $P(\boldsymbol{c}|\boldsymbol{e})$, of the inverse mapping from the Bayes' hypothesis, Eq. (8.11), and need not be explicitly known. A solution, $\hat{\boldsymbol{c}}$, can then be estimated in a number of ways, as described in the ensuing sections, starting with the simplest approach of equiprobable distributions.

13.1 Bayesian - Minimum Information

As in the Tikhonov regularization of (10.16), see also Section 10.4.4, one can arrive at a solution without involving *a priori* statistical information. This is accomplished by proclaiming complete ignorance about the statistical nature of the solution, and assuming an equiprobable (uniform) distribution for the elements of the solution vector, $\boldsymbol{c}$. In order to arrive at such distribution, we will normalize the forward mapping $\boldsymbol{e} = \boldsymbol{A}\boldsymbol{c}$, so that (Hussein, 1983):

$$\sum_{i=1}^{N} \bar{A}_{ij} = 1 \tag{13.1}$$

$$\boldsymbol{e} = \boldsymbol{A}\boldsymbol{c} = \bar{\boldsymbol{A}}\bar{\boldsymbol{c}} \tag{13.2}$$

$$\bar{A}_{ij} = \frac{A_{ij}}{\sum_{i=1}^{M} A_{ij}} \tag{13.3}$$

$$\bar{c}_j = \left(\sum_{i=1}^{M} A_{ij}\right) c_j \tag{13.4}$$

where N is the length of vector $\boldsymbol{c}$ and M is that of vector $\boldsymbol{e}$, and $\bar{\boldsymbol{A}}$ and $\bar{\boldsymbol{c}}$ are normalized forms of $\boldsymbol{A}$ and $\boldsymbol{c}$, respectively. The normalization accomplished by Eq. (13.1) enables one to view an element of the matrix $\bar{\boldsymbol{A}}$ as the conditional probability of obtaining a measurement, e_i, given a normalized parameter, $\bar{c}_j$, or correspondingly c_j, i.e.

$$P(e_i|c_j) = P(e_i|\bar{c}_j) = \bar{A}_{ij} \tag{13.5}$$

Computed Radiation Imaging. DOI: 10.1016/B978-0-12-387777-2.00013-6

This enables the establishment of the conditional probability of the forward mapping, $P(\boldsymbol{e}|\boldsymbol{c})$. We also make use of the law of total probability:

$$P(e_i) = \sum_{n=1}^{N} P(e_i|\bar{c}_n)P(\bar{c}_n) \tag{13.6}$$

Then, applying the Bayes' postulate of Eq. (8.11), the *a posteriori* (conditional probability of the inverse mapping), $P(\boldsymbol{c}|\boldsymbol{e})$, is given by:

$$P(c_j|e_i) = P(\bar{c}_j|e_i) = \frac{P(e_i|\bar{c}_j)P(\bar{c}_j)}{P(e_i)} = \frac{P(e_i|\bar{c}_j)P(\bar{c}_j)}{\sum_{n=1}^{N} P(e_i|\bar{c}_n)P(\bar{c}_j)} = \frac{\bar{A}_{ij}P(\bar{c}_j)}{\sum_{n=1}^{N} \bar{A}_{in}P(\bar{c}_n)} \tag{13.7}$$

where use was made of Eq. (13.5). The probability $P(\bar{c}_j)$, with the assumption of equiprobable distribution, is such that:

$$P(\bar{c}_j) = \frac{\bar{c}_j}{\sum_{n=1}^{N} \bar{c}_n} \tag{13.8}$$

Similarly, one can also assume an equiprobable distribution for $P(e_i)$:

$$P(e_i) = \frac{e_i}{\sum_{m=1}^{M} e_m} \tag{13.9}$$

Substituting Eqs. (13.8) and (13.9) in Eq. (13.7):

$$P(\bar{c}_j|e_i) = \frac{\bar{A}_{ij}\frac{\bar{c}_j}{\sum_{n=1}^{N} \bar{c}_n}}{\sum_{n=1}^{N} \bar{A}_{in}\frac{\bar{c}_n}{\sum_{n'=1}^{N} \bar{c}'_n}} = \frac{\bar{A}_{ij}\bar{c}_j}{\sum_{n=1}^{N} \bar{A}_{in}\bar{c}_n} \tag{13.10}$$

Now reapplying the law of total probability in the form:

$$P(\bar{c}_j) = \sum_{i=1}^{M} P(\bar{c}_j|e_i)P(e_i) \tag{13.11}$$

and using Eqs. (13.8), (13.9), and (13.10):

$$\frac{\bar{c}_j}{\sum_{n=1}^{N} \bar{c}_n} = \sum_{i=1}^{M} \frac{\bar{A}_{ij}\bar{c}_j}{\sum_{n=1}^{N} \bar{A}_{in}\bar{c}_n} \frac{e_i}{\sum_{m=1}^{M} e_m}$$

$$\bar{c}_j = \sum_{i=1}^{M} \frac{\bar{A}_{ij}\bar{c}_j e_i}{\sum_{n=1}^{N} \bar{A}_{in}\bar{c}_n} \times \frac{\sum_{n=1}^{N} \bar{c}_n}{\sum_{m=1}^{M} e_m}$$

$$= \sum_{i=1}^{M} \frac{\bar{A}_{ij}\bar{c}_j e_i}{\sum_{n=1}^{N} \bar{A}_{in}\bar{c}_n} \times \frac{\sum_{n=1}^{N}\left(\sum_{n'=1}^{N} A_{nn'}\right) c_n}{\sum_{m=1}^{M} e_m}$$

$$= \sum_{i=1}^{M} \frac{\bar{A}_{ij}\bar{c}_j e_i}{\sum_{n=1}^{N} \bar{A}_{in}\bar{c}_n} \tag{13.12}$$

where use was made of Eq. (13.4), and the fact that

$$\sum_{m=1}^{M} e_m = \sum_{n=1}^{N}\left(\sum_{n'=1}^{N} A_{nn'}\right) c_n$$

The formulation of Eq. (13.12), which is based on proclaiming equiprobable distributions for both the measurements and the solution, can be used to arrive at the simple iterative scheme:

$$c_j^{(k+1)} = c_j^{(k)} \sum_{i=1}^{M} \frac{A_{ij}}{\sum_{n=1}^{M} A_{nj}} \times \frac{e_i}{\hat{e}_i^{(k)}} \tag{13.13}$$

where k is the iteration number and $\hat{e}_i$ is a calculated, according to the forward model, value for the ith measurement, and use was made of Eq. (13.3). This is a successive approximation process that does not rely on matrix inversion. It provides a mild adjustment process at each iteration, and as such does not amplify noise considerably. If the elements of the matrix $\boldsymbol{A}$ are nonnegative, as it is usually the case in radiation imaging, then one can ensure nonnegative values of the elements of $\boldsymbol{c}$ by initiating the iterative process with a nonnegative guess. This method was used for refinement of a solution obtained using another successive approximation process (Hussein et al., 1986). In image restoration, this method is known as the Richardson-Lucy algorithm (Lucy, 1974, Richardson, 1972), see Chapter 16. The next section shows that the iterative process of Eq. (13.13) converges to the most likely solution when the likelihood, $P(\boldsymbol{e}|\boldsymbol{c})$, is governed by Poisson statistics.

13.2 Poisson Distribution

Radiation counting is governed by Poisson statistics, see Section 15.2. Therefore, one would expect both the measurement vector, $\boldsymbol{e}$, and the modeled measurement, $\hat{\boldsymbol{e}} = \boldsymbol{A}\boldsymbol{c}$, to belong to the same distribution. The likelihood, $P(\boldsymbol{e}|\boldsymbol{c})$, can be replaced with equivalent likelihood, $P(\boldsymbol{e}|\hat{\boldsymbol{e}})$. By combining the Poisson probability for each measurement, the conditional probability, $P(\boldsymbol{e}|\hat{\boldsymbol{e}})$, can be expressed as:

$$P(\boldsymbol{e}|\hat{\boldsymbol{e}}) = \prod_{i=1}^{M} \exp\left[-\hat{e}_i\right] \frac{\left(\hat{e}_i\right)^{e_i}}{e_i!} \tag{13.14}$$

where N is the length of the vector, $\boldsymbol{c}$, and M is the number of acquired measurements forming the measurement vector, $\boldsymbol{e}$, and A_{ij} is the $j$the element of $\boldsymbol{A}$.

The most likely solution is the one that maximizes $P(\boldsymbol{e}|\hat{\boldsymbol{e}})$, or equivalently maximizes $\ln P(\boldsymbol{e}|\hat{\boldsymbol{e}})$:

$$\max \ln P(\boldsymbol{e}|\hat{\boldsymbol{e}}) = \max \sum_{i=1}^{M} \left(-\hat{e}_i + e_i \ln \hat{e}_i\right) \tag{13.15}$$

with $\ln e_i!$ removed from maximization, since the solution is sought for a fixed set of measurements. Maximizing Eq. (13.15) requires that:

$$\frac{\partial}{\partial \hat{e}_i} \ln P(\boldsymbol{e}|\hat{\boldsymbol{e}}) = \sum_{i=1}^{M} -1 + \frac{e_i}{\hat{e}_i} = 0 \tag{13.16}$$

which indicates that $\ln P(\boldsymbol{e}|\hat{\boldsymbol{e}})$ is maximum when $\hat{\boldsymbol{e}} = \boldsymbol{e}$, i.e. when the solution is reached. A solution, $\hat{\boldsymbol{c}}$, is obtained when:

$$\frac{\partial}{\partial \hat{c}_j} \ln P(\boldsymbol{e}|\hat{\boldsymbol{e}}) = \sum_{i=1}^{M} -A_{ij} + e_i \frac{A_{ij}}{\sum_{n=1}^{N} A_{in} c_n} = \sum_{i=1}^{M} -A_{ij} + A_{ij} \frac{e_i}{\hat{e}_i} = 0 \tag{13.17}$$

This provides the basis for a likelihood-maximization iterative process of the form (Shepp and Vardi, 1982):

$$\hat{c}_j^{(k+1)} = \hat{c}_j^{(k)} \sum_{i=1}^{M} \frac{A_{ij}}{\sum_{n=1}^{M} A_{nj}} \frac{e_i}{\hat{e}_i^{(k)}} \tag{13.18}$$

where k refers to the iteration number. As the solution converges, $\hat{c}_j^{(k+1)} \to \hat{c}_j^{(k)}$, the likelihood would have reached its maximum. In fact, as k increases, the likelihood can be shown to strictly increase at each step of the iteration, unless it has already reached its maximum value (Shepp and Vardi, 1982). Notice the similarity between Eq. (13.18) and Eq. (13.11), which is based on the Bayesian hypothesis for estimating the *a posteriori* probability, along with an assumed uniform distribution for $\boldsymbol{c}$. With the latter assumption, the term $\ln P(c)$ is irrelevant in determining the logarithm of the *a posteriori*, which then becomes equal to the likelihood. Moreover, maximizing the likelihood of Eq. (13.15) amounts to matching modeled, $\hat{e}_i$, and measured, e_i, values one by one, which is also accomplished by Eq. (13.11). In essence, the Bayesian approach amounts also to maximizing the logarithm of the Poisson-based likelihood.

The expression of Eq. (13.18) can be rewritten as:

$$\begin{aligned}
\hat{c}_j^{(k+1)} &= \hat{c}_j^{(k)} + \frac{\hat{c}_j^{(k)}}{\sum_{n=1}^{M} A_{nj}} \left[\sum_{i=1}^{M} e_i \frac{A_{ij}}{\sum_{n=1}^{M} A_{nj}} - \sum_{n=1}^{M} A_{nj} \right] \\
&= \hat{c}_j^{(k)} + \frac{\hat{c}_j^{(k)}}{\sum_{n=1}^{M} A_{nj}} \frac{\partial}{\partial \hat{c}_j} \ln P\left(\boldsymbol{e}|\hat{c}_j^{(k)}\right)
\end{aligned} \tag{13.19}$$

where use is made of Eq. (13.17) and of the full expansion of $\hat{\boldsymbol{e}}$ in terms of $\hat{c}_j^{(k)}$. In a matrix form, Eq. (13.19) is expressed as:

$$\boldsymbol{c}^{(k+1)} = \boldsymbol{c}^{(k)} + \boldsymbol{D}^{(k)}\, \nabla \ln P\left(\boldsymbol{e}|\hat{c}_j^{(k)}\right) \tag{13.20}$$

where $\boldsymbol{D}^{(k)}$ is diagonal matrix with the jth element equal to $\frac{\hat{c}_j^{(k)}}{\sum_{n=1}^{M} A_{nj}}$ and $\nabla \ln P\left(\boldsymbol{e}|\hat{c}_j^{(k)}\right)$ is the gradient of the logarithm of the likelihood. The formulation of Eq. (13.20) enables the use of a gradient-based iterative algorithm, such as one of those described in Chapter 10. Lange et al. (1987) presented a number of ways to accelerate convergence of the iteration process.

Poisson statistics applies to measurement counts. In both emission and scatter imaging, detector counts are the measurements used in image reconstruction, and the formulations for the likelihood given in this section are directly applicable. In transmission imaging, projections obtained via Radon transforms (see Chapter 4) are employed in the image reconstruction process. Therefore, the formulation for $\ln P(\boldsymbol{e}|\hat{\boldsymbol{e}})$ is slightly different. Since measurements are independent of each other, the Poisson *a priori* is given by:

$$P(\boldsymbol{e}|\hat{\boldsymbol{e}}) = \prod_i \exp(-\hat{e}_i)\frac{\hat{e}_i^{\,e_i}}{e_i!} = \prod_i \exp[-e_{0_i}\exp[p_i]]\frac{(e_{0_i}\exp[-p_i])^{e_i}}{e_i!} \tag{13.21}$$

$$\begin{aligned} \ln P(\boldsymbol{e}|\hat{\boldsymbol{e}}) &= \sum_i -e_{0_i}\exp[-p_i] + e_i \ln(e_{0_i}\exp[-p_i]) - \ln e_i! \\ &= \sum_i -e_{0_i}\exp[-p_i] - e_i p_i + e_i \ln e_{0_i} - \ln e_i! \\ &= \sum_i -e_{0_i}\exp[-p_i] - e_i p_i + \text{constant} \end{aligned} \tag{13.22}$$

The constant is due to the fact that the source strength and measurements are fixed, for a sought set of parameters, $\hat{\boldsymbol{c}}$. Therefore, for transmission, the logarithm of the likelihood, $\ln P(\boldsymbol{e}|\hat{\boldsymbol{e}})$, is given by:

$$-\ln P(\boldsymbol{e}|\hat{\boldsymbol{e}}) = \sum_i \left(e_{0_i}\exp[-p_i] + e_i p_i\right) + \text{constant} \tag{13.23}$$

where $\boldsymbol{e}$ is the vector of the measurements, $\hat{\boldsymbol{e}}$ is the vector of modeled measurements, and $\hat{e}_i = e_{0_i}\exp[-p_i]$ is the ray-sum (projection) corresponding to a measurement e_i, with e_{0_i} being the source intensity.

13.3 Normal Distribution

Appendix 8.A presented a method for maximizing the probability, $P(\hat{\boldsymbol{e}})$, or precisely minimizing $-\ln P(\hat{\boldsymbol{e}})$, assuming that $P(\hat{\boldsymbol{e}})$ is a normal (Gaussian) distribution with a mean, e, and a variance, σ. The use of the normal distribution in radiation counting is justified by the fact that the Poisson distribution, which governs radiation counting, approaches a normal distribution when the number of counts exceeds about 20 counts (Hussein, 2003). The negative of the logarithm of the normal distribution (the likelihood) also lends itself readily to minimization, unlike the Poisson distribution which as shown in Section 13.2 is a more complicated expression. The distribution $P(\hat{e})$ with a mean e is equivalent to the conditional probability $P(e|\hat{e})$, and consequently corresponds to $P(e|\hat{c})$, since $\hat{e} = A\hat{c}$. Therefore, maximizing $P(\hat{e})$ is in effect a maximization of the likelihood, $P(e|c)$.

13.4 Maximum a posteriori (MAP)

The expected solution, c, for a given set of measurements, e, is the one that maximizes the *a posteriori* probability, $P(c|e)$. This, according to the Bayes theorem, Eq. (8.11), is equivalent to:

$$\max \frac{P(e|c)P(c)}{P(e)} \tag{13.24}$$

If nothing is known about the solution, c, except that it arises from some constant (independent of c) distribution, $P(c)$, and if the distribution of e is a fixed distribution, $P(e)$, then the likely solution, $\hat{c}$, is the solution that maximizes also the likelihood $P(e|c)$. However, if an *a priori* distribution, $P(c)$ is known or assumed, c can be treated as a random variable, and one maximizes the *a posteriori*, i.e.,

$$\max \{P(e|c)P(c)\} \tag{13.25}$$

The probability, $P(e)$, in (13.24) is removed from the optimization process, because it is a normalization term for the *a posteriori*. The maximization of the *a posteriori* provides an expected value for the solution, while maximizing the likelihood provides the most likely solution.

Maximizing the *a posteriori* is equivalent to maximizing the logarithm of (13.25):

$$\max \{\ln P(e|c) + \ln P(c)\} \tag{13.26}$$

Now one is maximizing a modified form of the logarithm of the likelihood, by adding the term, $\ln P(c)$. This added term acts as regularizer of the likelihood. One can consider the maximum *a posteriori* method as a likelihood maximization method with an added regularization term. Owing to relationship (13.26), the regularization term is often referred to as the *hyperprior*. Therefore, choosing an *a priori* resembles the choice of the regularization method, discussed in Section 10.4 for matrix-based methods, since regularization is based on some *a priori* knowledge about the nature of the solution.

There is some basic *a priori* knowledge about the solution, $\boldsymbol{c}$, in radiation imaging. For instance, $\boldsymbol{c} \geq 0$ and in linear (or linearized) problems, $P(\boldsymbol{c})$ is additive, i.e. $P(\boldsymbol{c}_1) + P(\boldsymbol{c}_2) = P(\boldsymbol{c}_1 + \boldsymbol{c}_2)$. The latter is due to the fact that, with corpuscular radiation, the higher the value of $\boldsymbol{c}$, the higher the value of the ray-sum in transmission imaging, the radiation intensity in emission imaging and the scattering flux in scatter imaging (when attenuation is constant or ignored in the latter two imaging modalities, and when scattering is ignored in transmission imaging). There is no single form for a positive and additive *a priori* distribution, but in general such distribution tends to take the form (Sivia, 1990):

$$P(\boldsymbol{c}) = K \exp[\alpha u(\boldsymbol{c})] \tag{13.27}$$

where K is a normalization constant, α is some constant, and $u(\boldsymbol{c})$ is the Shannon-Jaynes information entropy defined by Eq. (10.50). In the absence of any other *a priori* information about $\boldsymbol{c}$, one can define the entropy with respect to a uniform distribution, $\beta_i = 1$ in Eq. (10.50). Then, one can express Eq. (10.50) as:

$$u(\boldsymbol{c}) = -\sum_{j=1}^{N} \left(\frac{c_j}{c_{\text{sum}}}\right) \ln\left(\frac{c_j}{c_{\text{sum}}}\right) \tag{13.28}$$

where $c_{\text{sum}} = \sum_{j=1}^{N} c_j$ and the summation over n^+ in Eq. (10.50) is replaced by a summation over the N values of $\boldsymbol{c}$ since they are assumed to be all positive. Then, $\ln P(\boldsymbol{c})$ with the definition of $u(\boldsymbol{c})$ given by Eq. (13.28) is:

$$\ln P(\boldsymbol{c}) = -\sum_{j} \alpha_j \left(\frac{c_j}{c_{\text{sum}}}\right) \ln\left(\frac{c_j}{c_{\text{sum}}}\right) \tag{13.29}$$

with α_j being the value of α associated with c_j. The maximum value of $\ln P(\boldsymbol{c})$ occurs at $c_j = c_{\text{sum}} \exp\left[-\frac{1}{c_{\text{sum}}}\right]$. Therefore, adding the logarithm of this *a priori* to the likelihood in Eq. (13.26) bounds the upper limit of $\boldsymbol{c}$. The entropy *a priori* incorporated into the maximization of (13.26) adds its partial derivative, with respect to c_j, to the partial derivative of the likelihood. The added partial derivative is $-\frac{\alpha_j}{c_{\text{sum}}}\left(\frac{1}{c_{\text{sum}}} + \ln \frac{c_j}{c_{\text{sum}}}\right)$. Since $\frac{c_j}{c_{\text{sum}}} \leq 1$, this partial derivative is always ≥ 0. The larger the value of α, the more influence the *a priori* has on the maximization process of the *a posteriori*, the more pronounced is its upper-bounding effect. Note, however, as $c_j \to 0$, the effect of the *a priori* becomes too strong and needs to be moderated, which can be done via the gamma distribution described below.

Lange et al. (1987) proposed the use of the gamma distribution for the *a priori*, $P(c)$, which assumes that the image voxels are statistically independent. The gamma distribution for a random variable, c, is defined by a shape parameter, α, and a mean $\beta = \alpha c$ (variance $= \frac{\beta^2}{\alpha}$, mode $\frac{\beta}{\alpha}(\alpha - 1)$, skewness $= \frac{2}{\sqrt{\alpha}}$). When α is an integer

$\alpha = K$, the distribution becomes equivalent to the aggregation of K independent exponential distributions, each with a mean of $\frac{\beta}{K}$. The mean of the gamma distribution is, therefore, equal to αc, where c is a gamma distributed random variable. For an image parameter, c_j, the probability density function, $p(c_i)$, for a gamma distribution is:

$$p(c_j; \alpha_j, \beta_j) = \frac{1}{\Gamma(\alpha_j)} \left(\frac{\alpha_j}{\beta_j}\right)^{\alpha_j} c_j^{\alpha_j - 1} \exp\left[-\frac{\alpha_j}{\beta_j} c_j\right] \tag{13.30}$$

where $\Gamma(\alpha)$ is the gamma function.[1] The gamma distribution changes slowly between image voxels and provides some smoothing. It is also quite flexible, as its effect can be controlled by choosing the shape parameter, $\alpha > 1$, allowing it to provide the desired features around $\boldsymbol{c}$. The application of the gamma distribution as an *a priori* probability is facilitated by the fact that $\Gamma(\alpha_j)\left(\frac{\alpha_j}{\beta_j}\right)^{\alpha_j}$ in Eq. (13.30) is independent of c_j and need not to be considered in the maximization process. Therefore, for $\ln P(\boldsymbol{c})$ in Eq. (13.26), one has:

$$\ln P(\boldsymbol{c}) = \sum_j (\alpha_j - 1) \ln c_j - \frac{\alpha_j}{\beta_j} c_j \tag{13.31}$$

The first term, $(\alpha_j - 1)\ln c_j$ steers the value c_j away from zero, while the second term, $-\frac{\alpha_j}{\beta_j} c_j$ prevents c_j from approaching infinity. In Eq. (13.31), $\ln P(\boldsymbol{c})$ is bounded by the distribution's mode (its most likely value), which occurs when $\ln P(\boldsymbol{c})$ is maximum at $c_j^{mode} = \frac{\beta_j}{\alpha_j}(1 - \alpha_j)$. At large values of α: $c_j^{mode} \to \beta_j$, i.e. the mode approaches the mean. On the other hand, at $\alpha = 1$: $c_j^{mode} = 0$, and the log *a priori* of Eq. (13.31) limits the upper value of c_j. The addition of this log *a priori* to the likelihood, $\ln P(\boldsymbol{e}|\boldsymbol{c})$, in Eq. (13.26) biases the maximum of the likelihood towards c_j^{mode}.

The *a priori* gamma distribution is incorporated into the Poisson-based likelihood estimate of Eq. (13.18) by adding its partial derivative, with respect to c_j, to the condition for maximizing the likelihood, Eq. (13.17), so that:

$$\frac{\partial}{\partial \hat{c}_j}\left[\ln P(\boldsymbol{e}|\hat{\boldsymbol{e}}) + \ln(\boldsymbol{c})\right] = \sum_{i=1}^{M} -A_{ij} + e_i \frac{A_{ij}}{\sum_{j=1}^{N} A_{ij} c_j} + \frac{\alpha_j - 1}{\beta_j} - \frac{\alpha_j}{\beta_j} = 0 \tag{13.32}$$

which leads to the iterative process:

$$\hat{c}_j^{(k+1)} = \frac{\hat{c}_j^{(k)}}{\sum_{i=1}^{M} A_{ij} + \frac{\alpha_j}{\beta_j}} \left(\sum_{i=1}^{M} A_{ij} \frac{e_i}{\hat{e}_i^{(k)}} - \frac{\alpha_j - 1}{\beta_j}\right) \tag{13.33}$$

Lange et al. (1987) showed that the addition of the gamma *a priori* enhances the rate of convergence of the iterative process as it approaches the maximum of the *a posteriori*, in addition to bounding the solutions as indicated earlier.

[1] $\Gamma(\alpha) = \int_0^\infty t^{\alpha-1} \exp[-t]\, dt = \int_0^1 \left[\ln\left(\frac{1}{t}\right)\right]^{\alpha-1} dt.$

The inclusion of the *a priori* term, $P(c)$, with the likelihood, $P(\boldsymbol{e}|\boldsymbol{c})$, can be viewed as a form of regularization of the latter. Therefore, the regularization functions of Section 10.4 can also be employed. This is typically done to preserve edges, using the piece-wise functions, discussed in Section 10.4.11. The Gibbs, Huber, and total variation functions of Section 10.4.11 were used for this purpose, see for example Geman and Geman (1984), Chlewicki et al. (2004) and Panin et al. (1999). Alenius and Ruotsalainen (1997) introduced the median root *a priori*, which is based on assuming that the image to be reconstructed is locally monotonic, i.e. $\boldsymbol{c}$ is spatially nonincreasing or nondecreasing in a local region. This is accomplished by penalizing deviations of a voxel value from the median of voxels in the neighborhood (which is the value at the middle of ordered voxels in a neighborhood). In another work, (Alenius and Ruotsalainen, 2002), the same authors generalized this concept by replacing the median by the linear combination of ordered values or weighted averages, to reduce streaking in the reconstructed image.

The concept of nonlinear anisotropic diffusion, introduced by Perona and Malik (1990), can also be used as a form of regularization of the likelihood. This is accomplished by adding a penalty (or a regularization) function to the likelihood so that direction-dependent and $\boldsymbol{c}$-dependent diffusion of the image parameters, $\boldsymbol{c}$, is introduced at each voxel.[2] The diffusion coefficient is introduced via an "edge-shaping" function, $g(|\Delta\boldsymbol{c}|)$, where $|\Delta\boldsymbol{c}|$ is the magnitude of the gradient of $\boldsymbol{c}$. The function, g, is chosen to be a nonnegative monotonically decreasing function such that $g(0) = 1.0$ and $g(b) \to 0$ as $b \to \infty$. The purpose of this anisotropic diffusion process is to smooth noise while preserving the edges (sharp changes) in the image. Perona and Malik (1990) proposed the following formulations for g:

$$g(|\Delta\boldsymbol{c}|) = \exp\left[-\left(\frac{|\Delta c|}{D}\right)^2\right] \tag{13.34}$$

and

$$g(|\Delta\boldsymbol{c}|) = \frac{1}{1+\left(\frac{|\Delta c|}{D}\right)^2} \tag{13.35}$$

where D is a constant that determines the degree of spread of g. The function in Eq. (13.34) tends to favor sharp edges over mild ones, while the function of Eq. (13.35) tolerates wider regions more than narrow ones. Zhu et al. (2006) showed that the following iterative scheme maximizes the likelihood with a nonlinear anisotropic diffusion penalty:

$$\boldsymbol{c}^{(k+1)} = \frac{\boldsymbol{c}^{(k)}}{\boldsymbol{A}^{\mathrm{T}} + \beta\nabla\cdot\left(g\left(|\Delta\boldsymbol{c}^{(k)}|\right)|\Delta\boldsymbol{c}^{(k)}|\right)} \times \frac{\boldsymbol{A}^{\mathrm{T}}\boldsymbol{e}}{\left(\boldsymbol{A}\boldsymbol{c}^{(k)}\right)} \tag{13.36}$$

[2] Linear isotropic (Fickian) diffusion is equivalent to convolving the image with a normal (Gaussian) distribution (Koenderink, 1984).

where k refers to the iteration number, and β is a regularization parameter. Equation (13.36) is applied to one element of the vector $\boldsymbol{c}$ at a time. Rather than using the function $g(\boldsymbol{c})$, Zhu et al. (2006) proposed the use of fuzzy nonlinear anisotropic diffusion coefficients, with the degree of diffusion controlled by fuzzy rules expressed in a linguistic form.

13.5 The Monte Carlo Method

This method relies on the Markov chain of a series of randomly sampled scenarios, governed by appropriate probability distributions. As indicated in Section 11.7, the method of simulated annealing relies on optimizing a cost function iteratively by proceeding from one iteration to another randomly, in accordance to a prescribed probability distribution. Genetic algorithms, discussed in Section 11.6, also rely on random sampling for the selection of a population of parent chromosomes. Direct Monte Carlo sampling can be used to produce a large set of possible solutions of the problem parameters, $\boldsymbol{c}$, and use the mismatch between the corresponding modeled measurements, $\boldsymbol{Ac}$, and the given measurements, $\boldsymbol{e}$, in conjunction with the Bayesian hypothesis, to determine the solution, or subset of solutions, that most likely correspond to the measurements. An outline of this approach is given here, guided by Mosegaard and Tarantola (1995) and Tarantola (2005) (Chapter 2).

The Bayesian hypothesis of Eq. (8.11) can be expressed in terms of the joint probability, $P(\boldsymbol{c},\boldsymbol{e})$, as:

$$P(\boldsymbol{c},\boldsymbol{e}) = P(\boldsymbol{c}|\boldsymbol{e})P(\boldsymbol{e}) = P(\boldsymbol{e}|\boldsymbol{c})P(\boldsymbol{c}) \tag{13.37}$$

The joint probability formulates the random walks of Monte Carlo simulations, which generate the solution sets. Since the *a posteriori* probability, $P(\boldsymbol{c}|\boldsymbol{e})$, is not known, as it corresponds to the sought inverse mapping, one can rely on the right-hand-side of Eq. (13.37) to construct the random walks. With $P(\boldsymbol{e})$ being constant for a given problem, then constructing the joint probability corresponds to constructing the *a posteriori* probability, $P(\boldsymbol{c}|\boldsymbol{e})$. As discussed in Section 13.3, for a normal distribution of measurement uncertainties, minimizing the likelihood, $-\ln P(\boldsymbol{e}|\boldsymbol{c})$, leads to the least-squares solution. Therefore, the likelihood, $P(\boldsymbol{e}|\boldsymbol{c})$, can be replaced by the square root of the term minimized in the least-squares method, see Appendix 8.A, and the *a posteriori* probability can be expressed as:

$$P(\boldsymbol{c}|\boldsymbol{e}) = P_0 \exp[-S(\boldsymbol{c}|\boldsymbol{e})] \tag{13.38}$$

$$S(\boldsymbol{c}|\boldsymbol{e}) = \sum_{i=1}^{N} \frac{((A_{ij}c_j - e_i))^2}{\sigma_i^2} \tag{13.39}$$

where P_0 is a normalization constant, σ_i^2 is the variance in measurement e_i, and $\exp[S(\boldsymbol{c}|\boldsymbol{e})]$ is known as the "misfit function." Therefore, the Monte Carlo method involves two random procedures, one from the *a priori* probability and the second from the *a posteriori* probability.

A random walk would start by randomly sampling a vector, $\boldsymbol{c}_0$, in accordance to $P(\boldsymbol{c})$, if known, or based on some *a priori* information about $\boldsymbol{c}$. In the discussion below, its assumed that $P(\boldsymbol{c})$ is not known, and as such an equiprobale distribution is assumed with a preset lower and upper bounds: $\boldsymbol{c}_{max} \leq \boldsymbol{c} \leq \boldsymbol{c}_{min}$, with the unknown vector having a length N. Then, one can sample a set of N uniformly distributed random numbers, $1 \leq \xi \leq 0, i = 1, 2, \ldots N$ from a uniform distribution. Let these generated random numbers be the elements of a diagonal matrix, $\boldsymbol{X}_k$, then one can perform a random walk from a state $\boldsymbol{c}_k$ to a state $\boldsymbol{c}_{k+1}$ (with $k = 0, 1, 2, \ldots$) by perturbing the former, as follows:

$$\boldsymbol{c}_{k+1} = \begin{cases} \boldsymbol{c}_k + \boldsymbol{X}_k(\boldsymbol{c}_{max} - \boldsymbol{c}_k)\boldsymbol{c}_k, & \text{if } \boldsymbol{Y} \geq 0.5 \\ \boldsymbol{c}_k - \boldsymbol{X}_k(\boldsymbol{c}_k - \boldsymbol{c}_{min})\boldsymbol{c}_k & \text{if } \boldsymbol{Y} < 0.5 \end{cases} \tag{13.40}$$

where $\boldsymbol{Y}$ is a another vector of random numbers whose elements are uniformly distributed within [0, 1]. The above random walk process changes all elements of $\boldsymbol{c}_k$ at once. A milder random walk may change one elements, or a subset of elements, at a time, to ensure that an increase in the likelihood is attained from one random walk to another. The set of sampled random walks is known as the Markov chain, where a new random walk depends only on the one immediately preceding it.

Sampling from the *a priori* probability of Eq. (13.38) is performed after each new random-walk generated set, $\boldsymbol{c}_{k+1}$, by the rejection method, so that:

$$\text{If} \begin{cases} S(\boldsymbol{c}_{k+1}|\boldsymbol{e}) \leq S(\boldsymbol{c}|\boldsymbol{e}) & \text{accept } \boldsymbol{c}_{k+1} \\ S(\boldsymbol{c}_{k+1}|\boldsymbol{e}) > S(\boldsymbol{c}|\boldsymbol{e}) & \text{accept } \boldsymbol{c}_{k+1} \text{ with probability } \exp[-\Delta S(\boldsymbol{c}_{k+1}|\boldsymbol{e})] \end{cases} \tag{13.41}$$

where $[S(\boldsymbol{c}|\boldsymbol{e})]$ is the misfit function of Eq. (13.39), and $\Delta S(\boldsymbol{c}_{k+1}|\boldsymbol{e}) = S(\boldsymbol{c}_{k+1}|\boldsymbol{e}) - S(\boldsymbol{c}_k|\boldsymbol{e})$. The scheme of Eq. (13.41) is the Metropolis rule (Metropolis et al., 1953), which produces an equilibrium distribution as the number of random walks approach infinity. The application of the second step is accomplished by sampling a random number, ξ, from the uniform distribution in the interval [0,1], then if $\xi \leq \Delta S(\boldsymbol{c}_{k+1}|\boldsymbol{e})$, the sampled set is accepted; otherwise the sampled set is rejected and a new step is sampled before proceeding to the new random walk. The random walk, as guided by the Metropolis rule, will reach a nearly stable equilibrium distribution, which is considered as the problem's solution.

This stochastic approach avoids the local minima usually associated with derivative-based optimization methods that attempt to minimize a cost function. The reader will recognize the similarity between this approach and those of genetic evolution and simulated annealing, discussed in Sections 11.6 and 11.7, respectively. In fact, if the misfit of Eq. (13.39) is taken as the energy of the Maxwell-Boltzmann distribution, one has the simulated annealing method. Also, the piecewise regularization with the Gibbs function, discussed in Section 10.4.11, is a form of the Monte Carlo method.

References

S. Alenius and U. Ruotsalainen, "Generalization of median root prior reconstruction," *IEEE Transactions on Medical Imaging*, vol. 21, pp. 1413–1420, 2002.

——, "Bayesian image reconstruction for emission tomography based on median root prior," *European Journal of Nuclear Medicine*, vol. 24, pp. 258–265, 1997.

W. Chlewicki, F. Hermansin, and S. Hansen, "Noise reduction and convergence of Bayesian algorithms with blobs based on the Huber function and median root prior," *Physics Medicine and Biology*, vol. 49, pp. 4717–4730, 2004.

S. Geman and D. Geman, "Stochastic relaxation, Gibbs distributions and the Bayesian restoration of images," *Transactions on Pattern Analysis and Machine Intelligence*, vol. PAMI-6, pp. 721–741, 1984.

E. M. A. Hussein, "Fast neutron scattering method for local void fraction measurement in two phase flow," PhD thesis, McMaster University, Hamilton, ON, Canada, 1983.

——, *Handbook on Radiation Probing, Gauging, Imaging and Analysis: Basics and Techniques*. Dordrecht: Kluwer Academic Pub., 2003, vol. I.

E. M. A. Hussein, D. A. Meneley, and S. Banerjee, "Single exposure neutron tomography of two-phase flow," *International Journal of Multiphase Flow*, vol. 12, pp. 1–36, 1986.

J. Koenderink, "The structure of images," *Biological Cybernetics*, vol. 50, pp. 363–370, 1984.

K. Lange, M. Bahn, and R. Little, "A theoretical study of some maximum likelihood algorithms for emission and transmission tomography," *IEEE Transactions on Medical Imaging*, vol. MI6, pp. 106–114, 1987.

L. B. Lucy, "An iterative technique for the rectification of observed distributions," *Astronomical Journal*, vol. 79, pp. 745–754, 1974.

N. Metropolis, A. Rosenbluth, M. R. A. Teller, and E. Teller, "Equation of state calculations by fast computing machines," *Journal of Chemical Physics*, vol. 21, pp. 1087–1092, 1953.

K. Mosegaard and A. Tarantola, "Monte Carlo sampling of solutions to inverse problems," *Journal of Geophysical Research*, vol. 100, pp. 12 431–12 447, 1995.

V. Panin, G. L. Zeng, and G. T. Gullberg, "Total variation regulated EM algorithm," *IEEE Transaction on Nuclear Science*, vol. 46, pp. 2202–2210, 1999.

P. Perona and J. Malik, "Scale-space and edge detection using anisotropic diffusion," *IEEE Transactions on Pattern Analysis and Machine Intelligence*, vol. 12, pp. 629–639, 1990.

W. H. Richardson, "Bayesian-based iterative method of image restoration," *Journal of the Optical Society of America*, vol. 62, pp. 55–59, 1972.

L. A. Shepp and Y. Vardi, "Maximum likelihood reconstruction for emission tomography," *IEEE Transactions on Medical Imaging*, vol. MI-1, pp. 113–122, 1982.

D. S. Sivia, "Bayesian inductive inference maximum entropy & neutron scattering," *Los Alamos Science*, vol. 19, pp. 180–206, 1990.

A. Tarantola, *Inverse Problem Theory and Methods for Model Parameter Estimation*. Philadelphia: Society for Industrial and Applied Mathematics, 2005.

H. Zhu, H. Shu, J. Zhou, C. Toumoulin, and L. Luo, "Image reconstruction for positron emission tomography using fuzzy nonlinear anisotropic diffusion penalty," *Medical and Biological Engineering and Computing*, vol. 44, pp. 983–997, 2006.

14 Incomplete Problems

14.1 Incompleteness

When the number of measurements is less than the number of unknown image parameters, one has an inverse problem with multiple solutions, i.e. many values of image parameters, $\boldsymbol{c}$, that can correspond to the same set of measurements, $\boldsymbol{e}$. The inverse problem is then said to be incomplete or underdetermined. Even if the problem is well-determined, that is the number of measurements is exactly equal to the number of unknown parameters, the inverse problem can still in effect be incomplete, since statistical fluctuations, systematic measurement errors, modeling errors, and the discretization process itself inevitably produce inconsistencies. To overcome these inconsistencies, one should always seek to overdetermine the problem, either by acquiring more measurements or by decreasing the number of problem unknowns. In imaging, the latter lowers the spatial resolution as it requires increasing the voxel/pixel size, while increasing the number of measurements leads to elevated radiation exposure and prolonged measurement time and/or a more demanding measurement system (in terms of number of detectors and/or scans). There is no rigid rule to determine the number of measurements required to arrive at an acceptable level of overdetermination, but it is usually dictated by practical considerations: in terms of measurement time, available source strength, number and efficiency of detectors, scanning mechanisms, etc. Nevertheless, Section 9.1 presented two methods for determining the number of measurements, one based on the sampling theorem and the other on error minimization.

There are situations though in which overdetermining the inverse problem of radiation imaging, to attain a desired spatial resolution, is not practical or possible. One then has to deal with an incomplete inverse problem, which has potentially many possible solutions. The inverse problem becomes then an "estimation" problem, in which one seeks the most appropriate (most probable, expected, realistic, or optimal) solution amongst all possible solutions. One can also compensate for the "missing" measurements with some *a priori* information or some physical constraints on the solution. Methods for solving incomplete problems are presented in this chapter. We first consider methods discussed in the previous chapters that are suited for solving underdetermined problems.

Computed Radiation Imaging. DOI: 10.1016/B978-0-12-387777-2.00014-8

14.2 General Solution Methods

14.2.1 Least-Squares Solution

The least-squares method, see Chapter 8, inherently accommodates incomplete problems, as it finds the solution with the minimum residual error. That is, it finds the solution that produces modeled measurements, $\boldsymbol{Ac}$, that correspond as much as possible to a given set of measurements, $\boldsymbol{e}$. Using the Euclidean norm to define the difference between $\boldsymbol{e}$ and $\boldsymbol{Ac}$, one has a maximum-likelihood solution, as indicated in Appendix 8.A. This solution amounts in effect to inverting: $[\boldsymbol{A}^{\mathrm{T}}\boldsymbol{A}]\boldsymbol{c} = \boldsymbol{A}^{\mathrm{T}}\boldsymbol{e}$, or the weighted form: $[\boldsymbol{A}^{\mathrm{T}}\boldsymbol{WA}]\boldsymbol{c} = \boldsymbol{A}^{\mathrm{T}}\boldsymbol{We}$, where $\boldsymbol{W}$ is a diagonal matrix whose elements are typically the variance of measurements, to give preference to more statistically reliable measurements, see Eq. (8.3). One can also supplement the least-squares solution with regularization, as discussed in Section. 10.4, to provide a solution with a desired feature. For example, to impose smoothness on the solution, one can choose the solution with maximum information entropy.

14.2.2 Twomey-Phillips Solution

When a reasonable initial estimate of the image to be reconstructed is available, regularization with this solution estimate, see Section 10.4.6, can be used to supplement an incomplete problem. This estimate can also be corrected to better match the measurements using the Twomey-Phillips solution (Twomey, 1963):

$$\hat{\boldsymbol{c}} = \boldsymbol{c}^* + \left[\boldsymbol{A}^{\mathrm{T}}\boldsymbol{A} + \alpha^2\boldsymbol{G}^2\right]^{-1}\boldsymbol{A}^{\mathrm{T}}(\hat{\boldsymbol{e}} - \boldsymbol{Ac}^*) \tag{14.1}$$

where $\boldsymbol{c}^*$ is the initial estimate and the generalized regularization of Eq. (10.17) is implemented. Equation (14.1) can be rearranged as follows:

$$\begin{aligned}
\left[\boldsymbol{A}^{\mathrm{T}}\boldsymbol{A} + \alpha^2\boldsymbol{G}^2\right](\hat{\boldsymbol{c}} - \boldsymbol{c}^*) &= \boldsymbol{A}^{\mathrm{T}}(\hat{\boldsymbol{e}} - \boldsymbol{Ac}^*) \\
\boldsymbol{A}^{\mathrm{T}}\boldsymbol{A}\hat{\boldsymbol{c}} - \boldsymbol{A}^{\mathrm{T}}\boldsymbol{Ac}^* + \alpha^2\boldsymbol{G}^2(\hat{\boldsymbol{c}} - \boldsymbol{c}^*) &= \boldsymbol{A}^{\mathrm{T}}\hat{\boldsymbol{e}} - \boldsymbol{A}^{\mathrm{T}}\boldsymbol{Ac}^* \\
\alpha^2\boldsymbol{G}^2(\hat{\boldsymbol{c}} - \boldsymbol{c}^*) &= \boldsymbol{A}^{\mathrm{T}}(\hat{\boldsymbol{e}} - \boldsymbol{A}\hat{\boldsymbol{c}}) \\
\hat{\boldsymbol{c}} &= \boldsymbol{c}^* + \frac{1}{\alpha^2}\boldsymbol{G}^{-2}(\hat{\boldsymbol{e}} - \boldsymbol{A}\hat{\boldsymbol{c}})
\end{aligned}$$

The last relationship provides an iterative scheme to update the solution estimate:

$$\hat{\boldsymbol{c}}^{(k+1)} = \hat{\boldsymbol{c}}^{(k)} + \frac{\gamma^2}{\alpha^2}\boldsymbol{G}^{-2}\left(\hat{\boldsymbol{e}} - \boldsymbol{A}\hat{\boldsymbol{c}}^{(k)}\right) \tag{14.2}$$

where k is the iteration number with $\hat{\boldsymbol{c}}^{(0)} = \boldsymbol{c}^*$ and γ^2 is a relaxation parameter that controls the rate of change from one iteration to another.

14.2.3 Algebraic Geometric Solution

Thought different least-squares solutions can be obtained with a different weight matrix, $\boldsymbol{W}$, the least-squares solution is essentially governed by the particular algebraic representation of the forward problem, $\boldsymbol{Ac} = \boldsymbol{e}$, which corresponds to a particular set of hyperplanes. Normalizing the forward problem before solving it makes the solution dependent only on the geometry of the hyperplanes, not on their particular algebraic representation. One then has the algebraic reconstruction technique (ART), used in earlier generations of transmission tomography systems. As indicated in Section 10.6.1, Gordon and Mansour (2007) observed that a geometric iterative approach is quite effective in solving strongly underdetermined transmission tomography problems. The same iterative approach is used for solving the tomosynthesis problem discussed in Section 1.2. Normalization was achieved in these solutions by multiplying the system equation, $\boldsymbol{Ac} = \boldsymbol{e}$, by a diagonal matrix, $\boldsymbol{D}_r$, whose elements are the reciprocals of the Euclidean norm of a row vector of $\boldsymbol{A}$. Then, for the resulting system, $\boldsymbol{D}_r\boldsymbol{A} = \boldsymbol{D}_r\boldsymbol{e}$, column normalization was performed by dividing each column of $\boldsymbol{A}$ by the Euclidean norm of the column. The least-squares residual was then minimized in an iterative manner using a conjugate gradient scheme, see Section 10.6.4.

14.2.4 Pseudoinversion

The system matrix, $\boldsymbol{A}$, in an incomplete problem is rank-deficient, similar to a complete problem with noise measurements that makes two rows in a matrix practically equivalent to each other. Therefore, the pseudoinverse (Moore-Penrose inverse) singular-value decomposition method, described in Section 10.2, provides one of the many non-unique solutions of an incomplete problem which matches in part the forward model (i.e. not necessarily matching all available measurements). This method will always produce a solution, but such solution may not be reliable.

14.2.5 Optimization Methods

One of the optimization methods discussed in Chapter 11 can be used to arrive at an optimal solution for incomplete problems, subject to some constraints or *a priori* knowledge. Such constraints will help guide the optimization process to a realistic solution.

14.2.6 Analytic Solutions

In linear problems, using Fourier transforms (Section 12.2), interpolating between known values, or limiting the value of the highest frequency used in image reconstruction, can help overcome the incompleteness of the problem. This comes, however, at the price of reduced spatial resolution.

Wavelets

The multiresolution (scaling) wavelet analysis presented in Section 12.9 can be used to handle incomplete problems, with the view that an incomplete problem for a high-resolution problem is a complete problem at a lower resolution. Therefore, one can reconstruct a low-resolution image over the entire domain, and with the aid of wavelet transform obtain a more detailed resolution within a smaller region-of-interest. This is a local imaging problem, since it focuses only on a localized smaller region.

Wavelet transform is suited for the local problem of imaging when wavelet of vanishing moments are employed (Rashid-Farrokhi et al., 1997).[1] This is due to the fact that the Hilbert transform of a function (see Section 12.1) decays rapidly if it has a large number of vanishing moments. Therefore, selecting wavelets with vanishing moments enables the calculation of wavelet coefficients using local projections. Then with projections from a region-of-interest and a margin of nearby projections, image reconstruction can be performed locally over a small region, see for example Rashid-Farrokhi et al. (1997), Das and Sastry (2002), and Bonnet et al. (2002).

14.2.7 Probabilistic Solution

The Bayesian approach of Chapter 13 with an *a priori* probability distribution, such as the gamma distribution of Section 13.4, augments an underdetermined problem and converts it into an overdetermined problem that can be solved by maximizing the *a Posteriori* probability (Lange et al., 1987). Other methods that can accommodate incomplete problems include the Monte Carlo method discussed in Section 13.5, and other related optimization methods that rely on some form of random sampling, such as genetic evolution and simulated annealing (Ali et al., 1999), see Sections 11.6 and 11.7, respectively. Section 14.4 shows how image features can be incorporated into Markov chain fields. However, the most effective methods for dealing with an incomplete problem is the estimation maximization method discussed below.

14.3 Estimation Maximization

This method deals with problem incompleteness by introducing, based on the statistical behavior of the expected solution, a random set of "unobservable" (virtual[2]) measurements. To explain what is meant by an unobservable measurement, let us consider a single measurement, e_i. This measurement is estimated by the forward model by:

$$\hat{e}_i = \sum_{j=1}^{N_i} A_{ij} c_j \tag{14.3}$$

[1] Orthonormal wavelets with a scaling function with vanishing moments are known as "coiflets."

[2] The term virtual measurements is also used to describe any *a priori* constraints (Rodgers, 1976). Since unobservable measurements are selected based an *a priori* probability distribution, there are in effect virtual measurements.

where N_i is the number of parameters, consequently image pixels/voxels, contributing to e_i. The same measurement, e_i, also corresponds to the summation of a set of random virtual (undetectable) measurements, v_{ij}, so that:

$$\hat{e}_i = \sum_{j=1}^{N_i} v_{ij} \tag{14.4}$$

Comparing Eqs. (14.3) and (14.4), one can state that:

$$\hat{e}_{ij} = A_{ij}c_j$$

is the mean value of the random virtual measurement v_{ij}. Note that zero virtual measurements are trivial and are excluded. Similarly, each other actual measurement can have a corresponding set of virtual measurements. For all available M measurements, one can form a set of virtual measurements. These virtual measurements, along with the M actual measurements, could overdetermine the problem. With iterative updates of the virtual measurements, an acceptable solution can be attained. Of courses, if $M > N$, where N is the number of unknowns, there would be no need to rely on such unobservable (also called latent) measurements.

For virtual measurements, in the from of a vector, $\boldsymbol{v}_i$ corresponding to each actual measurement e_i, one can express the joint probability,[3] i.e. the probability of $\boldsymbol{v}_i$, e_i, and $\boldsymbol{c}$ together, as:

$$P(e_i, \boldsymbol{v}_i, \boldsymbol{c}) = P(e_i|\boldsymbol{v}_i, \boldsymbol{c})p(e_i|\boldsymbol{c})p(\boldsymbol{c}) \tag{14.5}$$

and in turn,

$$P(\boldsymbol{e}|\boldsymbol{v}, \boldsymbol{c}) = \sum_i P(e_i|\boldsymbol{v}_i, \boldsymbol{c}) \tag{14.6}$$

where $\boldsymbol{v}$ is the vector of all virtual measurements. The logarithm of the *a posteriori* that needs to be optimized, Eq. (13.26), can now be expressed as:

$$\ln P(\boldsymbol{e}|\boldsymbol{c}) + \ln P(\boldsymbol{c}) = \ln \sum_i P(e_i|\boldsymbol{v}_i, \boldsymbol{c}) + \ln P(\boldsymbol{c}) \tag{14.7}$$

Notice that Eq. (14.7) requires only knowing the summation of the virtual measurements, not their individual values. The derivatives of Eq. (14.7), with respect to $\boldsymbol{c}$, can be used to drive an iterative process towards the maximum *a posteriori* (MAP). However, obtaining the gradient of Eq. (14.7) is not a trivial task. The method of expectation maximization (EM) can be used instead to estimate and optimize the logarithm of

[3] $P(e,c) = P(e|c)P(c)$.

the joint probability, $\ln P(\boldsymbol{e}, \boldsymbol{v}_i, \boldsymbol{c}|\boldsymbol{c}^{(k)})$, based on a previous estimate $\boldsymbol{c}^{(k)}$ of the sought solution, $\boldsymbol{c}$. Therefore, the estimation (E-step) determines:

$$Q\left(\boldsymbol{c}^{(k+1)}|\boldsymbol{c}^{(k)}\right) = E\left(\ln P(\boldsymbol{e}, \boldsymbol{v}, \boldsymbol{c}|\boldsymbol{c}^{(k)}\right) \tag{14.8}$$

where $E(\cdot)$ designates expected value. Maximization (M-step) $Q(\boldsymbol{c}^{(k+1)}|\boldsymbol{c}^{(k)})$ with respect to $\boldsymbol{c}^{(k)}$ drives the iterative process towards the maximum *a posteriori*. Note that replacing $P(\boldsymbol{e}, \boldsymbol{v}, \boldsymbol{c})$ with $P(\boldsymbol{e}, \boldsymbol{v}|\boldsymbol{c})$ in Eq. (14.8) leads to an estimate of the *a priori*, since $P(\boldsymbol{e}, \boldsymbol{v}, \boldsymbol{c}) = P(\boldsymbol{e}, \boldsymbol{v}|\boldsymbol{c})P(c)$).

14.3.1 Poisson Distribution Likelihood

Emission Imaging

Let us consider the estimation maximization of the logarithm of the likelihood assuming that the virtual measurements are governed by Poisson statistics, as is the case in emission imaging. Then, as in Eq. (13.14), one has:

$$P(\boldsymbol{v}|\boldsymbol{c}) = \prod_i \prod_j \exp\left[-\hat{e}_{ij}\right] \frac{(\hat{e}_{ij})^{v_{ij}}}{v_{ij}!} = \prod_i \prod_j \exp\left[-A_{ij}c_j\right] \frac{(A_{ij}c_j)^{v_{ij}}}{v_{ij}!} \tag{14.9}$$

where $\hat{e}_{ij}$ is a modeled measurement corresponding to the virtual measurement v_{ij}. The logarithm of the likelihood converts the above products into summations:

$$\ln P(\boldsymbol{v}|\boldsymbol{c}) = \sum_i \sum_j -A_{ij}c_j + v_{ij} \ln A_{ij}c_j - \ln v_{ij}! \tag{14.10}$$

The estimation maximization method provides a means for estimating the expected value of the virtual measurements with respect to the actual available measurements, then maximizes this expectation to provide a more likely estimate of the solution (Lange and Carson, 1984). Let us assume that some estimate, $\boldsymbol{c}^{(k)}$, is given, along with a measurement vector, $\boldsymbol{e}$. Then, from Eq. (14.9), one can express the conditional expectation of the virtual measurements with respect to the actual measurements and the estimated solution as:

$$\begin{aligned} Q\left(\boldsymbol{c}^{(k+1)}|\boldsymbol{c}^{(k)}\right) &= E\left(\ln P(\boldsymbol{e}, \boldsymbol{v}|\boldsymbol{c})|\boldsymbol{c}^{(k)}\right) \\ &= \sum_i \sum_j -A_{ij}c_j + E\left(v_{ij}|\boldsymbol{e}, \boldsymbol{c}^{(k)}\right) \ln A_{ij}c_j - E\left(\ln v_{ij}!\right) \end{aligned} \tag{14.11}$$

There is no need to determine $\sum_i \sum_j E\left(\ln v_{ij}!\right)$, since it has no effect on the maximization step, as shown by Eq. (14.13) below. The expected value of $\left(v_{ij}|\boldsymbol{e}, \boldsymbol{c}^{(k)}\right)$ can be obtained using the definition of v_{ij} as being the contribution of voxel j to a measurement e_i. This contribution is determined by the element A_{ij} of matrix $\boldsymbol{A}$, which relates e_i to c_i. Therefore, one can express $E\left(\boldsymbol{e}, v_{ij}|\boldsymbol{c}^{(k)}\right)$ as:

$$E\left(\boldsymbol{e}, v_{ij}|\boldsymbol{c}^{(k)}\right) = \frac{A_{ij}c_j^{(k)}}{\sum_n A_{in}c_n^{(k)}}\, e_i \tag{14.12}$$

Note that the summation of $E\left(\boldsymbol{e}, v_{ij}|\boldsymbol{c}^{(k)}\right)$ of all contributing voxels satisfies Eq. (14.4).

The maximization of $E\left(\ln P(\boldsymbol{e}, \boldsymbol{v}|\boldsymbol{c}|\boldsymbol{c}^{(k)})\right)$ requires:

$$\frac{\partial}{\partial c_j} E\left(\ln P(\boldsymbol{e}, \boldsymbol{v}|\boldsymbol{c}|\boldsymbol{c}^{(k)})\right) = \sum_i -A_{ij} + \frac{A_{ij}c_j^{(k)}}{\sum_n A_{in}c_n^{(k)}}\, \frac{e_i}{c_j} = 0 \tag{14.13}$$

where use is made of Eqs. (14.11) and (14.12), and assuming that A_{ij} is not a function of c_j, i.e. the problem is linear, or quasi-linear in an iterative successive approximation process. Notice also that $E\left(\ln v_{ij}!\right)$ does not involve c_j, because even if it were explicitly formulated it would have been estimated based on $c_n^{(k)}$. With $\frac{\partial^2}{\partial c_j^2} E\left(\ln P(\boldsymbol{e}, \boldsymbol{v}|\boldsymbol{c}|\boldsymbol{c}^{(k)})\right)$ being negative, Eq. (14.11) is concave and Eq. (14.13) produces the maximum value of Eq. (14.11), in the form of an updated estimate:

$$c_j^{(k+1)} = \frac{c_j^{(k)}}{\sum_i A_{ij}}\, \frac{A_{ij}e_i}{\sum_n A_{in}c_n^{(k)}} = \frac{A_{ij}c_j^{(k)}}{\sum_i A_{ij}}\, \frac{e_i}{e_i^{(k)}} \tag{14.14}$$

with $e_i^{(k)}$ being the modeled measurement at iteration k. The expected value of Eq. (14.12) is then updated and maximized, until the iterative process converges at a solution that maximizes Eq. (14.12).

The expressions of Eq. (14.14) is identical to that of Eq. (13.18) which maximizes the likelihood based on only observable measurements. The difference between the two is that when the problem is incomplete, the exact value of the likelihood is not known. With the known data, the EM method finds *a posteriori* estimates of the probabilities of the values of virtual measurements. For each $\boldsymbol{v}$, the EM method calculates an expected value of the likelihood, given $\boldsymbol{e}$ and the estimated $\boldsymbol{c}^{(k)}$. The EM method iterates over the value of the likelihood. As such, the EM method gives a conditional probability distribution of the virtual measurements, as well as an estimate of the solution, $\boldsymbol{c}$. Note that the gradient-based formation of Eq. (13.20) is also applicable here.

Transmission Imaging

In transmission imaging, a virtual measurement, $v_{i,j}$, can be taken as the intensity of radiation entering a voxel j and contributing to detector i (Lange and Carson, 1984). Then the intensity of radiation leaving a voxel j depends on the intensity of radiation entering the voxel, the radiation attenuation coefficient of the material within the voxel, c_j, and the distance, $r_{i,j}$, radiation travels through the voxel towards detector i.

Then, the intensity of radiation leaving the voxel is governed by a binomial distribution[4] of a success probability $\exp[-c_j r_{i,j}]$ and a number of trials $v_{i,j}$. Note that a measured radiation intensity, e_i, is the intensity of radiation leaving the last voxel along the radiation beam viewed by detector i. Therefore, e_i can be viewed as a member of the complete measurement set that includes all virtual measurements. Therefore, v_{i,N_i} is a random variable for which e_i is the measured (mean) value, with N_i being the last voxel seen by detector i. The source (incident radiation) intensity, e_{0_i} of beam i is Poisson-distributed, e_i and $v_{i,j}$ follow a Poisson distribution. Therefore, one has:

$$P(e_{0_i}) = \exp[-e_{0_i}]\frac{e_{0_i}^{v_{1,i}}}{v_{i,1}!}; \quad P(v_{i,j}) = \exp[-\bar{v}_{i,j}]\frac{\bar{v}_{i,j}^{\bar{v}_{i,j}}}{v_{i,j}!}; \quad P(e_i) = \exp[-e_i]\frac{e_i^{v_{i,N_i}}}{v_{i,N_i}!} \tag{14.15}$$

where $\bar{v}_{i,j}$ is the expected value of $\bar{v}_{i,j}$. The expected (mean) value of $v_{i,j}$ is given by:

$$E(v_{i,j}) = e_{0_i} \exp\left[\sum_{j=1}^{N_i} c_j r_{i,j}\right] = \bar{v}_{i,j} \tag{14.16}$$

A virtual measurement resulting from radiation exiting a voxel, $v_{i+1,j}$, is related to that entering the same voxel (both in the same ray i) by the binomial distribution with a success probability $\exp[-c_j r_{i,j}]$ for $v_{i,j}$ number of trials. Therefore, over a beam i, starting from the source and ending at the detector, one can write the conditional probability:

$$\begin{aligned} P(e_i, \boldsymbol{v}_i|\boldsymbol{c}) &= \exp[-e_{0_i}]\frac{e_{0_i}^{v_{1,i}}}{v_{i,1}!} \\ &\times \prod_{j=1}^{N_i-1} \frac{v_{i,j}!}{(v_{i,j} - v_{i+1,j})!e_i!}\left(\frac{\bar{v}_{i+1,j}}{\bar{v}_{i,j}}\right)^{v_{i+1,j}}\left(1 - \frac{\bar{v}_{i+1,j}}{\bar{v}_{i,j}}\right)^{v_{i,j}-v_{i+1,j}} \end{aligned} \tag{14.17}$$

where $\boldsymbol{v}_i$ is the vector of virtual measurements corresponding to beam i. Given that $v_{j+1,i} = v_{j,i}\exp[-c_j r_{i,j}]$ and based on Eq. (14.17), one can formulate the logarithm of the likelihood as:

$$\begin{aligned} \ln P(\boldsymbol{e}, \boldsymbol{v}|\boldsymbol{c}) = &-e_{0_i} + v_{1,i}\ln e_{0_i} - \ln v_{i,1}! \\ &+ \sum_{j=1}^{N_i-1} \{[\ln v_{i,j}! - \ln(v_{i,j} - v_{i+1,j})! - \ln e_i!] \\ &\quad + v_{i+1,j}\ln(\exp[-c_j r_{i,j}]) \\ &\quad + (v_{i,j} - v_{i+1,j})\ln(1 - \exp[-c_j r_{i,j}])\} \end{aligned} \tag{14.18}$$

[4] For a binomial distribution: $P(x) = \frac{n!}{(n-x)!x!}p^x(1-p)^{n-x}$, where x is a random variable with a success probability p and n is the number of trials.

Therefore,

$$
\begin{aligned}
E(\ln P(\boldsymbol{e}, \boldsymbol{v}|\boldsymbol{c})) = \sum_i -e_{0_i} + \bar{v}_{1,i} \ln e_{0_i} - E(\ln v_{i,1}!) \\
+ \sum_{j=1}^{N_i - 1} E\Big(\big\{[\ln v_{i,j}! - \ln(v_{i,j} - v_{i+1,j})! - \ln e_i!]\Big) \\
+ \bar{v}_{i+1,j} \ln(\exp[-c_j r_{i,j}]) \\
+ (\bar{v}_{i,j} - \bar{v}_{i+1,j}) \ln(1 - \exp[-c_j r_{i,j}])\big\}
\end{aligned}
\tag{14.19}
$$

The maximization of this estimation requires:

$$
\frac{\partial E(\ln P(\boldsymbol{e}, \boldsymbol{v}|\boldsymbol{c}))}{\partial \boldsymbol{c}} = 0 \tag{14.20}
$$

$$
\sum_i \bar{v}_{i+1,j} r_{i,j} + (\bar{v}_{i,j} - \bar{v}_{i+1,j}) \frac{r_{i,j} \exp[-c_j r_{i,j}]}{1 - \exp[-c_j r_{i,j}]} = 0 \tag{14.21}
$$

Solving this equation to estimate c_j is not straightforward due to the nonlinearity of Eq. (14.21). However, a number of approximations have been proposed (Lange and Carson, 1984, Lange et al., 1987). Assuming that $c_j r_{i,j} << 1$, which is a reasonable assumption in small voxels, low-density material, and/or high radiation energy, then[5]:

$$
\frac{\exp[-c_j r_{i,j}]}{1 - \exp[-c_j r_{i,j}]} = \frac{1}{\exp[c_j r_{i,j}] - 1} \approx \frac{1}{c_j r_{i,j}} - \frac{1}{2} + \frac{1}{12(c_j r_{i,j})} \tag{14.22}
$$

Taking only the first term in the approximation gives an upper estimate of c_j. On the other hand, the negative second term in the approximation is likely to have the largest magnitude in the expansion, and as such the first two terms produce a lower bound of c_j. The three terms of the approximation taken together provide a more accurate, but more computationally demanding expression, for c_j. Therefore, Eq. (14.21) produces the following approximations for the maximization of Eq. (14.19) (Lange and Carson, 1984), with the virtual measurements estimated at the previous iteration, $k-1$:

First-order approximation:

$$
cj^{(k)} = \frac{\sum_i \bar{v}_{i,j} - \bar{v}_{i+1,j}}{\sum_i \bar{v}_{i+1,j} r_{i,j}} \tag{14.23}
$$

[5] Let $x = c_j r_{i,j} << 1, z = \exp[x] - 1, z + 1 << 1$; $\frac{x}{\exp[x]-1} = \frac{\ln(z+1)}{z} = \frac{z + \frac{z^2}{2} + \frac{z^3}{3} + \cdots}{z} = 1 - \frac{z}{2} + \frac{z^2}{x}$. With $z = \exp[x] - 1, \approx x + \frac{x^2}{2}$, the expression in Eq. (14.22) is arrived at after neglecting third and higher order terms.

Second-order approximation:

$$cj^{(k)} = \frac{\sum_i \bar{v}_{i,j} - \bar{v}_{i+1,j}}{\sum_i \frac{1}{2}(\bar{v}_{i,j} + \bar{v}_{i+1,j}) r_{i,j}} \tag{14.24}$$

Third-order approximation:

$$0 = (cj^{(k)})^2 \sum_i (\bar{v}_{i,j} - \bar{v}_{i+1,j}) \frac{r_{i,j}^2}{12} - cj^{(k)} \sum_i (\bar{v}_{i,j} + \bar{v}_{i+1,j}) \frac{r_{i,j}}{2} + \sum_i (\bar{v}_{i,j} - \bar{v}_{i+1,j}) \tag{14.25}$$

Devising a gradient-driven iterative scheme, such as that of Eq. (13.20), is not as straightforward for the case of transmission imaging. However, in analogy with the latter scheme, Lange et al. (1987) suggested the use of:

$$\boldsymbol{c}^{(k+1)} = \boldsymbol{c}^{(k)} \boldsymbol{D}^{(k)} \nabla \ln P\left(\boldsymbol{e} | \hat{c}_j^{(k)}\right) \tag{14.26}$$

with the jth term in the diagonal matrix $\boldsymbol{D}^{(k)}$ in Eq. (14.26) as:

$$D_{jj} = \frac{c_j^{(k)}}{\sum_i e_i r_{ij}} \tag{14.27}$$

So that:

$$\begin{aligned} c_j^{(k+1)} &= c_j^{(k)} + \frac{c_j^{(k)}}{\sum_i e_i r_{ij}} \frac{\partial}{\partial c_j} \ln P\left(\boldsymbol{e} | \hat{c}_j^{(k)}\right) \\ &= c_j^{(k)} + \frac{c_j^{(k)}}{\sum_i e_i r_{ij}} \left\{ \sum_i \bar{v}_{i+1,j} r_{i,j} + (\bar{v}_{i,j} - \bar{v}_{i+1,j}) \frac{r_{i,j} \exp[-c_j^{(k)} r_{i,j}]}{1 - \exp[-c_j^{(k)} r_{i,j}]} \right\} \end{aligned} \tag{14.28}$$

14.3.2 Scatter Imaging

No formulation is given in the literature for estimation maximization with virtual measurements in scatter imaging. We will developed here a formulation that combines those used in emission and transmission imaging. In scattering, a virtual measurement, $v_{i,j}$, can be taken as the intensity of radiation scattered from voxel j that contributes to detector i. A virtual measurement depends, therefore, on the intensity of radiation entering the voxel, the type and density of material within the voxel, the size of the voxel, and the probability of scattering. The first factor, the intensity of radiation entering a voxel, is Poisson distributed, as in the case of transmission imaging. The density and type of material within the voxel can be simply considered to have a uniform probability distribution (assuming lack of other knowledge) or to have a gamma distribution, see Eq. (13.30). The size of a voxel is obviously not a random variable and is fixed by the desired image resolution. The probability of scattering depends on the type of radiation and its scattering modality, as well as the density and type of material within the voxel. The probability of scattering per unit density for a particular material is a physical property (cross section) that can be determined *a priori*. It can be considered in most cases to be independent of the sought solution, $\boldsymbol{c}$. Let σ_{ij} be the probability of scattering from voxel j to detector i (i.e. the microscopic angular cross section for the reaction). Given these arguments, one can postulate that virtual measurements in scatter imaging are dominantly Poisson distributed, and utilize the same formulations used in emission imaging. However, attention must be paid to the fact that scattering is a nonlinear process, and another iterative process (other than estimation maximization) is need to formulate the matrix, $\boldsymbol{A}$, relating $\boldsymbol{c}$ to $\boldsymbol{e}$. Nevertheless, within each iteration for nonlinearity, the problem can be considered to be quasilinear, enabling the application of estimation maximization.

14.3.3 Ordered Subset Estimation Maximization

In order to accelerate the typically slow expectation minimization iterative process, Hudson and Larkin (1994) proposed an ordered iterative process in which a measurement set is grouped into a number of subsets, and solution updates performed only over one subset at time. This is known as the ordered subsets estimation maximization (OSEM) method. Subsets should be formed in such a way that each voxel has about the same probability of contributing to each subset. The row-action maximum-likelihood algorithm is such an iterative scheme, in which measurements are processed sequentially (Browne and Pierro, 1996). The ordered subset iterative approach typically accelerates iterations toward convergence at early stages, though iteration over the entire set of measurements is needed at the late stages to reach convergence (Leahy and Qi, 2000).

14.4 Markov Random Fields

Adequate solution of an underdetermined problem depends on having reliable *a priori* information. Rather than relying on an estimate of the solution or certain constraints, which can greatly influence the solution, one can take advantage of the fact that most physical objects, and subsequently their images, are composed of clusters of nearly homogeneous regions separated by boundaries or edges. Geman and Geman (1984) described such objects by coupled Markov random fields (MRF), i.e. fields in which a new random event depends only on the one immediately preceding it. Each voxel in an image is designated a clique of neighboring voxels via some unobservable line or a label, $\boldsymbol{l}$, so that each image is modeled by the vector pair $(\boldsymbol{c},\boldsymbol{l})$. This enables expressing the *a priori* information by a probability distribution coupled by some *a priori* energy, Ψ and a partition function, Z:

$$P(\boldsymbol{c},\boldsymbol{l}) = \frac{1}{Z}\exp\left[-\frac{\Psi(\boldsymbol{c},\boldsymbol{l})}{T}\right] \tag{14.29}$$

This is the Gibbs distribution at some temperature, T, used in thermodynamics to describe transitions from one equilibrium state to another, discussed also in Section 10.4.11. The image reconstruction problem can then be formulated as a maximization of the *a posteriori* conditional probability $P(\boldsymbol{c},\boldsymbol{l}|\boldsymbol{e})$, which is equivalent to the minimization of the *a posteriori* "energy": $\|\boldsymbol{Ac}-\boldsymbol{e}\|^2+\Psi(\boldsymbol{c},\boldsymbol{l})$. The energy term, $\Psi(\boldsymbol{c},\boldsymbol{l})$, acts as a regularization functional. Any of the piecewise regularization methods of Section 10.4.11 can be applied here.

Geman and Geman (1984) suggested solving this problem stochastically, rather than using iterative methods that globally and monotonically increase the *a posteriori* probability. The so-called stochastic relaxation (simulated annealing) approach generates a sequences of images that evolve locally, while converging to the optimal solution. Each voxel in the image is designated a clique of neighboring voxels. Starting with a high temperature, T, a local random change, sampled from the conditional probability $P(\boldsymbol{c},\boldsymbol{l}|\boldsymbol{e})$, is made based on the current value of voxels and their cliques. The energy is then lowered, to assure convergence to a solution at the lowest temperature, and the random sampling process is repeated.

Nikolova et al. (1998) viewed an image as consisting of a vector pair $(\boldsymbol{c},\boldsymbol{l})$, with $\boldsymbol{l}$ being an unobservable line or label. With $\boldsymbol{l}$ being a noninteracting Boolean-like process, it can be determined by minimizing $\Psi(\boldsymbol{c},\boldsymbol{l})$, to obtain a value $\hat{\boldsymbol{l}}(\boldsymbol{c})$ for a given $\boldsymbol{c}$. Then the *a posteriori* energy, $\|\boldsymbol{Ac}-\boldsymbol{e}\|^2+\Psi(\boldsymbol{c},\hat{\boldsymbol{l}})$, is minimized. This dual-stage optimization process better accounts for the presence of both uniform clusters and edges, which is the initial premise of this MRF model.

References

F. E. A. F. Ali, Z. Nakao, and Y.-W. Chen, "An evolutionary approach to CT image reconstruction," *Bulletin of the Faculty of Engineering, University of the Ryukyus*, vol. 57, pp. 81–84, 1999.

S. Bonnet, F. Peyrin, F. Turjman, and R. Pros, "Multiresolution reconstruction in fan-beam tomography," *IEEE Transactions on Image Processing*, vol. 11, pp. 169–176, 2002.

J. A. Browne and A. R. D. Pierro, "A row-action alternative to the EM algorithm for maximizing likelihoods in emission tomography," *IEEE Transactions on Medical Imaging*, vol. 15, pp. 687–699, 1996.

P. C. Das and C. S. Sastry, "Region-of-interest tomography using a composite Fourier-wavelet algorithm," *Numerical Functional Analysis and Optimization*, vol. 23, pp. 757–777, 2002.

S. Geman and D. Geman, "Stochastic relaxation, Gibbs distributions and the Bayesian restoration of images," *Transactions on Pattern Analysis and Machine Intelligence*, vol. PAMI-6, pp. 721–741, 1984.

D. Gordon and R. Mansour, "A geometric approach to quadratic optimization: An improved method for solving strongly underdetermined systems in CT," *Inverse Problems in Science and Engineering*, vol. 15, pp. 811–826, 2007.

H. M. Hudson and R. S. Larkin, "Accelerated image reconstruction using ordered subsets of projection data," *IEEE Transactions on Medical Imaging*, vol. 13, pp. 601–609, 1994.

K. Lange and R. Carson, "EM reconstruction algorithms for emission and transmission tomography," *Journal of Computer Assisted Tomography*, vol. 8, pp. 306–316, 1984.

K. Lange, M. Bahn, and R. Little, "A theoretical study of some maximum likelihood algorithms for emission and transmission tomography," *IEEE Transactions on Medical Imaging*, vol. MI6, pp. 106–114, 1987.

R. M. Leahy and J. Qi, "Statistical approaches in quantitative positron emission tomography," *Statistics and Computing*, vol. 10, pp. 147–165, 2000.

M. Nikolova, J. Idier, and A. Mohammad-Djafari, "Inversion of large-support ill-posed linear operators using a piecewise Gaussian MRF," *IEEE Transactions on Image Processing*, vol. 7, pp. 571–585, 1998.

F. Rashid-Farrokhi, K. J. R. Liu, C. A. Berenstein, and D. Walnut, "Wavelet-based multiresolution local tomography," *IEEE Transactions on Image Processing*, vol. 6, pp. 1412–1430, 1997.

C. D. Rodgers, "Retrieval of atmospheric temperature and composition from remote measurements of thermal radiation," *Reviews of Geophysics and Space Physics*, vol. 14, pp. 609–624, 1976.

S. Twomey, "On the numerical solution of Fredholm integral equations of the first kind by the inversion of the linear system produced by quadrature," *Journal of the Association for Computing Machinery*, vol. 10, pp. 97–101, 1963.

15 Testing

The quality of a reconstructed image depends not only on the formulation and solution of the associated forward and inverse problems, but also on the quality and number and scheme of acquiring required measurements. Therefore, before designing and constructing an actual imaging system for laboratory and field use, the numerical robustness of this inverse problem and the effect of measurements on the problem's solution should be examined. It is desirable also to assess and evaluate various numerical schemes for solving the inverse problem. Moreover, the quality of the reconstructed image needs to be examined.

Numerical testing should be done in steps. The first step is to consider an ideal (error-free) problem that employs modeled measurements (synthetic data) produced by the forward model. From the perspective of solving the inverse problem of image reconstruction, modeled measurements are free of uncertainty. Failure to solve the ideal problem is indicative of failure in implementing the solution strategy, or even of flaws in the methodology itself. After succeeding in solving the ideal inverse problem, one should consider the solution of a "noisy," but otherwise ideal inverse problem, in which random noise is artificially introduced to the modeled measurements. Various levels of noise will demonstrate the extent with which the inverse problem is susceptible to noise in the measurements, and the maximum level of noise beyond which a solution to the problem becomes meaningless. The third step in testing should involve modeled measurements produced by some other method independent of the forward problem, e.g. by Monte Carlo simulations. Such testing will assess the adequacy of the forward model and the effect of its idealizations on the solution of the inverse problems. Following these three steps with modeled measurements, one can proceed with testing using laboratory measurements, assured that the numerical and physical aspects of the problem have been addressed and only practical issues can further affect the solution. These four levels of testing are discussed below, followed in Section 15.5 by definition of parameters for determining and measuring the quality of a reconstructed image.

15.1 Ideal Problem

In the ideal problem, data generated by the forward model is used in place of actual measurements. Therefore, the ideal problem is an exact problem not affected by model approximations, statistical variability, or any other sources of random or systematic

Computed Radiation Imaging. DOI: 10.1016/B978-0-12-387777-2.00015-X errors. Its value is in verifying and debugging numerical algorithms and associated computer programs for solving the inverse problem. Once computer codes are verified, the ideal problem can be used to ensure that physically meaningful results are obtained for different expected image configurations. In the design process, the ideal problem can also be utilized to optimize measurement gathering (or scanning) strategies. These testing aspects are discussed below.

In an ideal problem, the computer program used in image reconstruction should reproduce an exact image of the object for which the modeled measurement data were generated. The most convenient approach perhaps is to start with a uniform object of the same size and spatial resolution as that typically expected in practice. The modeled measurements for such a uniform object should produce predictable trends, without abrupt changes except at boundaries, with magnitudes comparable to those obtained by hand or spreadsheet calculations. If such expected results are not obtainable, the input data as well as the algorithm used for formulating the forward problem and the associated computer coding should be reexamined. If no acceptable image is obtainable at all, the numerical inversion algorithmic itself should be reexamined. Once image reconstruction for a uniform object is successfully accomplished, one should test the inversion process with a variety of nonuniform configurations, similar in nature to those expected in practice, for various spatial resolutions and for a variety of materials. Section 15.6 gives some configurations suited for testing image reconstruction algorithms. Of particular interest are the continuous phantoms formed with the help of analytical shapes, such as polygons and ellipses, since they can be used to study the effect of discretization on the forward and inverse formulations of the problem.

With an ideal problem, one can examine many of the numerical characteristics of the problem, such as existence, uniqueness, continuity, and stability of solution, for a given set of modeled measurements. In an iterative scheme, one can also investigate whether the problem converges to a unique solution, the conditions for such convergence, and its independence of the initial guess. Testing an iterative process should be first conducted with an initial guess identical to the known solution, then the exact solution should be arrived at immediately, without any iteration. Failure to pass this initial test would indicate inadequacy in programming or formulating the iterative process. A number of tests should be conducted to demonstrate that the space within which the physical parameters are reconstructed is complete, by showing that the problem converges within the bounds of the space.

Before proceeding any further, one must be satisfied with the behavior of the ideal inverse problem. If a fully acceptable solution is unattainable under ideal conditions, the problem solution under realistic conditions will be even less acceptable. On the other hand, obtaining a solution with the ideal problem does not ensure an acceptable solution for an actual problem, due to the ill-posedness associated with practical inverse problems. However, some insight into ill-posedness can be obtained from the ideal problem. Sensitivity of the inversion process to numerical errors introduced by truncation of supplied modeled data, and propagation of round-off error during the inversion process, are strong indicators of the problem's ill-posedness. Such small errors should not have a noticeable effect on the solution of the supposedly well-posed ideal problem. By executing a particular test problem under both single and double computing-machine precession, the effect of numerical-error diffusion can be observed. If error diffusion is problematic, the numerical implementation of the inverse problem should be inspected to determine the cause. Once corrected, or if no obvious sources for error diffusion were found, the intrinsic nature of the inverse problem should be considered. If one has a matrix-based system, it should be ensured that the matrix is not rank deficient. The condition number should also be calculated; the larger this number the more inclined the problem to be ill-posed. The potential ill-posedness of a functional inversion process can be explored by observing how far the χ^2 value, left-hand-side of Eq. (8.4), is from zero. In a frequency-based solution, the presence of high-frequency components in a uniform or a nearly-uniform image could be indicative of susceptibility to ill-posedness. In the backprojection approach, the value of the error, as determined by Eq. (8.10), is a measure of how vulnerable the problem is to well-posedness.

If strong ill-posedness is encountered in an ideal problem, the layout of the imaging process must be altered. In other words, one must revisit the strategy of acquiring measurements, i.e. their number, location, independence, etc. With the ideal problem, one can also examine the effect of system parameters, such as source position, energy, direction and collimation, detector location, energy window, width of field-of-view, voxel size, level of details in the forward model modeling, etc.

One must keep in mind that success in solving the ideal inverse problem does not assure by itself correct solution of the actual inverse problem. Such an ideal solution, for all practical purposes, is trivial, leading to what Colton and Kress (1992) called the *inverse crime*. Such situation can be avoided by synthesizing the measurements using a forward model that is different in precession from the forward model inverted in the inversion process, e.g. by generating each model at a different level of discretization.

15.2 Noisy-Ideal Problem

Before testing an inverse problem with actual data, its ill-posedness can be studied by contaminating the ideal data with random noise. This enables examining the numerical aspects of the problem without worrying about modeling errors and other deviations from ideal conditions that are encountered when acquiring measurements in the laboratory. Radiation-counting measurements are governed by the Poisson distribution, because it is a discrete binary process (count/no count) of events that are independent of each other, and arise from a very large population (atoms or their nuclei) with a very small probability (only a very small fraction of the atoms/nuclei produce counts within a measurement time). The Poisson distribution is such that the probability density function of measuring a count of n (say, particles) is:

$$P(n) = \frac{m^n}{n!} \exp(-m) \tag{15.1}$$

where $\sum_{n=0}^{\infty} nP(n) = 1$, with a mean $m = \sum_{n=0}^{\infty} nP(n)$ and a variance $\sigma^2 = \sum_{n=0}^{\infty} n(n-m)^2 P(n) = m$, equal to its mean, i.e. $m = \sigma^2$. When only one measurement count, n_0, is recorded, it can be considered as being the mean value of a Poisson distribution with an estimated sample variance also equal to n_0. In other words, if a measurement is to be repeated many times, one would expect a Poisson distribution with a mean, n_0, and a variance, n_0. It should be noted too that a Poisson distribution approaches a normal (Gaussian) one when m is large. Then the confidence intervals of the normal distribution can also be associated with the Poisson distribution.

In order to simulate statistical fluctuations in forward-model estimated measurements that correspond to radiation counts, one must introduce random variations governed by the Poisson distribution. Forward models, such as those of Eqs. (4.1), (5.1), (5.6), and (6.6), produce modeled measurements that include calibration or system constants that may not be known unless laboratory measurements are recorded. Such constants can, however, be assigned to unity to facilitate testing the image reconstruction software, but then modeled measurements may have values that are much less than unity, which does not agree with the binary nature of the Poisson distribution. However, one usually has an idea of the order of magnitude of the counts of the actual measurements, and can devise a reasonable scaling factor which when multiplied by the modeled measurements would produce values on the order of magnitude of the expected measurements. The scaled values can then be contaminated with Poisson random noise, and the resulting values can be rescaled back to obtain values comparable in magnitude to the original noise-free modeled measurements. To illustrate this process, let us consider a modeled ideal (error-free) measurement, e_i, with a scale factor of 10^r where r is a positive integer, then the scaled measurement, $e_i \times 10^r$, can be contaminated with a Poisson noise generated by an algorithm such as (Atkinson, 1979):

1. Using a uniform distribution in the (0,1] interval,[1] sample some n random numbers: $\xi_1, \xi_2, \ldots \xi_n$, and for each number generate the corresponding random variables from an exponential distribution, with the transformation:

$$\eta_j = -\ln(\xi_j) \quad j = 1, 2, \ldots n$$

2. Continue generating the above random variables, one at a time, until:

$$\sum_{j=1}^{n} \xi_j \geq e_i \times 10^r$$

Then, $n-1$ is a randomly sampled variable from a Poisson distribution of mean equal to $e_i \times 10^r$. In turn, $(n-1) \times 10^r$ corresponds to a molded measurement, e_i, contaminated with Poisson noise. Given that computers generate "pseudo" random numbers that begin with a predesignated "seed," the same sequence of random variables can be regenerated by repeating the problem with the same seed. This enables diagnosis of algorithms without the interference of truly random effects.

At high counts, the Poisson distribution approaches a normal (Gaussian) distribution, with variance and mean equal to each other. Therefore, for a mean value m, the

[1] Zero is excluded from the sampling process to avoid indefinite logarithms. The interval [0,1] can be replaced, for example, by the interval [0.001,1].

standard deviation is $\sigma = \sqrt{m}$, leading to a fractional (relative) standard deviation of $f = \frac{\sigma}{m} = \frac{1}{\sqrt{m}}$. Knowing the expected magnitude of an actual measurement, one can estimate the expected relative error and sample a random variable, χ, from a $(1,f^2)$ normal distribution, i.e. with a mean equal to unity and a relative variance, f^2. Then, $\chi \times e_i$ corresponds to a modeled measurement, e_i, contaminated with normally distributed noise (also called Gaussian white noise). The Box-Muller method (Box and Muller, 1958) can be used to sample χ as follows:

1. Sample a random number, ξ from a uniform distribution in $(0,2\pi]$.
2. Sample another random number, η, from a uniform distribution $(0,1]$.
3. Generate the random variables:

$$\chi_1 = \sigma\sqrt{-2\ln\eta}\cos\xi + m, \quad \chi_2 = \sigma\sqrt{-2\ln\eta}\sin\xi + m$$

These random variables are normally distributed with mean, m, and variance σ^2. Either one of the sampled variables, χ_1 and χ_2, can be used to contaminate a modeled measurement.

In a normal distribution, one can associate confidence levels with various values of σ: 68% of the measurements lie within $\pm\sigma$ around the mean, 95% are within $\pm 2\sigma$, 99.7% within $\pm 3\sigma$, and so on. A confidence level of 95% is recommended for use in uncertainty analysis.[2] Since, a Poisson distribution approaches a normal distribution at high counts, the same confidence levels can be utilized in both distributions.

Image reconstruction should be tested at various noise levels. This enables the determination of the highest permissible level of noise in measurements that can be tolerated by the inverse problem while producing acceptable solutions. The same analysis can also be used to determine the source strength that would result in a desired level of confidence in the reconstructed images. Noise levels can easily be varied when sampling from a normal distribution $(1,f^2)$ by varying the value of f. In a Poisson distribution, the noise level can be varied by changing the scaling factor, 10^r.

Failure to accommodate an acceptable level of noise in image reconstruction may indicate the need for regularization (see Chapter 10.4). Varying the level of noise is also useful in studying the effect, and determining the degree, of regularization. If even with regularization, it is found that the image reconstruction scheme is quite vulnerable to noise, then one must reexamine the layout of the imaging system and alter it to provide a more robust (better posed) problem. It should be also kept in mind that if the noise level does not have much effect on the error-free synthetic data, and if the forward model used in generating these modeled measurements is the same as that employed in the inversion process, one may end up with an excessively optimistic solution of the inverse problem, leading to the inverse crime described at the end of Section 15.1. Such situation can be avoided by using a level of noise sufficient to perturb the idealized modeled measurements, and/or by employing a forward model for generating the idealized data that is different in precession (e.g. level of discretization) from that used in solving the inverse problem.

[2] Measurement Uncertainty, Part I, ANSI/ASME PTC 19.1-1985.

15.3 Independently-Simulated-Data Problem

An intermediate step before testing with actual laboratory data should involve testing with numerical data generated independently from those modeled using the forward problem. Such data should include effects ignored in the forward model (e.g. scattering in transmission and emission imaging and multiple scattering in scatter imaging), and those not fully simulated in the forward model (such as x-ray beam hardening). It is also preferable to generate such independent measurements using a continuous model to examine the effect of the discretization normally associated with the forward problem. Testing image reconstruction with such independent data, not only provides a means to assess the susceptibility of the inverse problem to modeling errors, but it is also a form of validating the forward problem itself.

The solution of the Boltzmann transport equation, Eq. (2.11), can provide such independent data. The most practical and most widely used method for solving this transport equation is the Monte Carlo method. This method solves a transport problem by the simulation of random walks of radiation source particles in accordance to random variables sampled from probability distributions based on the transport kernels of the Boltzmann transport equation, see for example Hussein (2007). By averaging the contributions of all random walks, an estimate of a desired measurement is obtained, with an associated standard deviation. With the Monte Carlo method, complex geometries can be simulated, and most radiation interactions can be taken into account. The statistical nature of the Monte Carlo results also introduces variabilities not unlike those encountered in laboratory measurements. However, one should aim at obtaining simulated results with as low statistical variance as possible, so that the problem can be validated without the distraction of statistical variations. A number of computer codes are available, but the MCNP code (`http://mcnp-green.lanl.gov/`) is widely used in radiation transport calculations.

Before proceeding to utilize Monte Carlo generated data in image reconstruction, this data should be compared with the ideal ones provided by the forward model, to assess whether the inexactness of the latter can be tolerated. This is a mutual verification process that can point out mistakes in setting up the forward problem, or modeling errors in the simulation. Simulating a simple problem, say for a uniform object, can help identify the sources of error, if any. If the adequacy of the Monte Carlo process is assured, but no good agreement with the forward model is attained, the source of the discrepancy should be identified, and an effort should be made to improve the forward model. If the difference between the two appears not to be significant, but some of the Monte Carlo simulated measurements had a large statistical variance (not significantly reduced by increasing the number of random walks), this large variability could be inherent in the physical process; e.g. due to secondary radiation that contributes a small but consistent amount to the detector response. If simulated and forward model data agree, except for few measurements, careful examination of those measurements should be conducted one-by-one, to determine the cause of such difference. If no identifiable reason for the difference can be easily found in the forward model, the corresponding measurements should be excluded from the measurement set, as they are likely to cause similar problems in practice.

15.4 Laboratory Problem

When solving an inverse problem with laboratory measurements, one is faced with practical challenges that include:

Background: measured radiation that is not related to the interrogated object. This necessitates the introduction of a background correction process.
Inaccuracy: caused by systematic measurement errors in the translation of laboratory spatial parameters (positions of source, object, and detectors) into the forward model, or inaccuracies in determining the latter parameters. Minimizing this error ensures best setup matching between simulation and laboratory arrangements.
Uncertainty: due to statistical, electronic, or temporal fluctuations in measurements, leading to imprecise results. These uncertainties must be quantified.
Calibration: scaling the forward model to incorporate source strength, detector efficiency and counting period, as well as any other parameters not accounted for in the forward model. Therefore, a calibration process must be introduced.
Inexactness: due to approximations and simplifications of the forward model. After accounting for background radiation and calibration, the forward model should produce results that match, within statistical variability, laboratory measurements.

The above aspects are elaborated upon below.

15.4.1 Background

Background is the radiation recorded by the detection system that does not provide relevant information to imaging. This background can be viewed as consisting of passive and active components. The passive background is any radiation detectable in the absence of both the external radiation source (to be used in imaging) and the interrogated object. The active background component is that measured in the source's presence, but in the absence of the object.

The passive background is due to natural radiation in the surroundings, cosmic rays, radiation leakage from a shielded radioisotopic source, electronic noise, thermal drift, mechanical vibrations (microphonics), afterglow (residual signal after termination of a preceding immediate exposure), etc. Thermal energy can also produce a detectable signal, independent of radiation exposure. This passive component tends to be low in intensity and can be filtered out electronically or subtracted from detected signals.

The active background component consists of direct and indirect components. The indirect component is caused by scattering or secondary emissions from surrounding structures. In transmission imaging, the direct background is the signal measured in the absence of the object, called the flat field. This signal is used in determining the projections by providing the normalization factor, I_0, in Eq. (4.2).

Measurement of the direct radiation background component in transmission tomography is, therefore, essential and is in effect a calibration process. In emission imaging with an embedded source, a stand-alone source of known activity, employed in the absence of the object but placed at its intended location, can be used to provide an estimate of the direct background component, per unit source activity. This component can, in turn, be used to calibrate detectors. In induced emission, the radiation (of the same type used in emission imaging) detected in the presence of the activating source, but in the absence of the object, represents the direct background component. The indirect background component is typically much smaller than the direct one, but it is also irrelevant to image reconstruction, and should be minimized as much as possible.

In transmission imaging and embedded emission imaging, the passive component of radiation background should be subtracted from the active component, before using the latter to calibrate the measurements. The passive component should be recorded for each detector, since no two detectors are exactly identical in their response. In induce-emission and scatter imaging, the total (active and passive) background component should be subtracted, detector-by-detector, from the signals measured in the presence of the object. Attention should also be paid to the subtraction of background radiation from the signals recorded for the interrogated object, since the presence of the object itself can attenuate and/or scatter background radiation, altering the background levels recorded in the absence of the object. Therefore, it is always desirable to reduce the background radiation as much as possible. Methods for background reduction are suggested in Chapter 17 of Hussein (2003).

15.4.2 Inaccuracy

Matching the setup used in the forward model with that of the laboratory is vital for validating both the forward and inverse problems. Systematic errors, if not corrected, can cause mismatching between the two setups. Perhaps the most elementary cause of such systematic errors is the inability to visually see radiation and hence determine its exact path and direction. This makes it difficult to match the assumed model geometric setup with a laboratory setup. A small semiconductor laser beam may be positioned behind the source and detector collimators to observe their fields-of-view, but such laser beams tend to diffuse quite broadly and do not provide a faithful representation of radiation divergence. Moreover, it is difficult to know exactly the source's position (particularly the location of a radioisotope within a shielding container) to properly place the laser beam. Small markers (small diameter objects) can be placed between the source and the detector and moved around the expected region of intersection of the fields-of-view of the source and detector, to determine its boundaries. Such markers can also be used to determine the source's profile (change in intensity across a detector aperture). A small calibration source can also be employed as a marker, and moved in front of a detector's collimator to map the detector's field-of-view. Such information can be conveyed to the modeling process to adjust the model until a reasonable matching is obtained between the simulated and modeled arrangements. In performing such measurements, attention should be paid to the three-dimensionality of radiation transport, i.e. the source and detector fields-of-view should be mapped at two orthogonal planes. This is useful even when only a section of the object along one of the two planes is considered, as it can be used to determine the effective thickness of the imaged section.

When modeling radiation interactions (scattering and induced emission) within an object, the location of the interrogated object must be known. In the laboratory, it is convenient to designate an imaging region between the source(s) and detector(s), within which the interrogated object should be placed. Markers (small-diameter rods, pins, etc.) can be used to ensure that the boundaries of this region are exposed to source radiation and contribute to detectors. Markers can also be placed in the proximity of the interrogated object, e.g. near its surface, and left during the acquisition of image measurements. Reconstructed images of these markers will then help relate spatial locations in the reconstructed image to the corresponding ones in the imaged object.

Systematic error can also result when acquiring actual spatial position measurements in the laboratory. Such measurements should be performed with care and with high-accuracy devices. Care should also be taken in conveying such measurements to the forward model, to avoid human error.

15.4.3 Measurement Uncertainties

As a result of the probabilistic nature of radiation emission and interactions, there is always statistical variability associated with radiation measurements. This statistical variability is governed by the Poisson distribution, discussed in Section 15.2. A recorded count has, therefore, an associated random "error," also called quantum noise. Since image reconstruction is a quantitative process, the uncertainty associated with a recorded count must also be quantified. This is facilitated by the fact that the statistical variance of a count, according to the Poisson distribution, is equal to the count itself. Also at large counts, the Poisson distribution approaches a normal distribution, enabling the quantification of the confidence intervals associated with the range of counts around a recorded (considered to be an estimate of the mean) value. Obviously, statistical variability can be reduced by increasing the counting period, the source strength, detector efficiency and/or size and the size of the source and detector collimator apertures. Estimating the statistical variance also enables weighting of measurements in image reconstruction to reduce the importance of weak (large variance) measurements, see Section 8.

Electronic processing of detected radiation events is also subject to statistical fluctuations because electrons themselves are discrete. The contribution of electronic noise, called shot noise, tends, however, to be much less than that of the random error associated with radiation generation and interaction.

Temporal changes in source emission rate, due to the decay of a radioisotope or instabilities in the power source of a radiation generator, also cause fluctuations in the measurements. Such temporal changes become significant if imaging measurements are gathered within a long period of time. Taking active background measurements before and after acquiring imaging data can be used as an indicator of the change in source strength during the measurement period. Temporal changes can also be caused by the object itself if it is not stationary, as in the case of fluid flow or a living subject (due to respiration, cardiac motion, peristalsis, voluntary or involuntary motion). Overlapping measurements can smooth out the effects of motion.

15.4.4 Calibration

The forward models of emission and scatter imaging, Eqs. (4.1), (5.1), (5.6), and (6.6), contain system constants that need to be determined. These constants depend on parameters that are not affected by imaged objects, and as such in principle can be determined by calibration. This can be done by recording laboratory measurements with low statistical uncertainties for a reference object, and dividing the measurement values by the corresponding normalized (i.e. with the system constant set to unity) computed values. A reference object is typically a uniform object of known material density and composition, representative of the objects to be integrated. Using the terminology of Chapter 1, a system constant, K, is given by:

$$K = \left. \frac{\hat{e}|_{\delta_e << \hat{e}} - B}{(mc)|_{K=1}} \right|_{\text{Reference Object}} \tag{15.2}$$

where B refers here to the measured background radiation. For the same source and for identical detectors, the calculated values of the system constant, K, should be the same for all measurements, irrespective of the nature of the reference object. In reality, however, no two detectors are fully identical in their response, and the strength of an external source can have a non-uniform profile which can alter the value of K as the distance between the detector and the source changes. System constants may, therefore, be calculated for each detector position.

System constants should be independent of the nature of the reference object. This independence should be tested by determining the values of the system constants using two significantly different reference objects. If the system constants change significantly from one reference object to another, their use in image reconstruction will have a considerable effect on the results. The inexactness of forward models is likely the reason the system constants will depend on the reference object, since model simplifications typically ignore object-dependent secondary effects (such as scattering in transmission and emission imaging, absorption in emission and scatter imaging, and multiple scattering in scatter imaging). Calibrating for such object-dependent effects can be a daunting task. Fortunately, however, in many cases, secondary effects are not very significant and can be ignored, or the detection system can be designed to reduce their contribution to the recorded counts (e.g. by energy discrimination to discard undesired scattering). In other cases, secondary effects can be incorporated in the image reconstruction process (as in accounting for radiation attenuation in emission imaging, see Sections 5.1 and 5.2), or considered as a penalty that need to be minimized (as in accounting for multiple scattering in scatter imaging, as discussed in Section 6.3). Nevertheless, we introduce here a calibration strategy that can alleviate the object-dependence of the system constants.

Let us express a laboratory measurement in terms of three components: a modeled one, a component not accounted for in the forward model, and the background component, so that:

$$\hat{e} = K(mc)|_{K=1} + K_s c + B, \quad \delta_e << e \tag{15.3}$$

where K_s is a secondary calibration constant, which when multiplied by the modeled parameter, c, gives the secondary effects ignored in the forward model. Here, we are assuming that the secondary effects are linearly-dependent on c, but if needed higher-order dependence can be added with the aid of more calibration constants. The value of K in Eq. (15.3) can be calculated using Eq. (15.2) for a reference object in which the secondary effects are known to be minimal, e.g. a thin object where undesirable scattering and attenuation effects are not that significant. Alternatively, the value of K for a given source-detector configuration can be obtained using a set of reference objects, and finding the best-fit value that provides an object-dependent K value, since usually $K(mc)|_{K=1} > K_s c$. An estimate for K can be found by minimizing $\left[\frac{\hat{e}-B}{\hat{mc}|_{K=1}} - K\right]^2$, with a least-squares scheme. Two values for the secondary calibration factor can then be determined using two reference objects, one, K_{s_l} with a low value c_l and the other, K_{s_h} with a high c_h value. Then for an imaged object, an average of its c values can be utilized to estimate the secondary calibration constant, K_s, to be used in its calibration:

$$K_s = (1-\alpha)K_{s_l} + \alpha K_{s_h}, \quad \alpha = \frac{c - c_l}{c_h - c_l}$$

In essence, the estimated K_s value would act on a secondary forward mapping of measurements in accordance to Eq. (15.3). This dual-reference object calibration process can be extended to multiple-reference objects, if in Eq. (15.3) higher orders of c are used to correct for secondary effects at a higher level of dependence on c. Also, the secondary calibration can be updated iteratively, during the image reconstruction process as better estimates of c are obtained.

Calibration can also be enabled during each imaging exercise by placing a number of small reference (calibration) objects of known characteristics at the periphery of the field of imaging. The reconstructed image parameters can then be scaled to those of the known calibration objects. The use of more than one calibration object, positioned at various locations, is desired to average out any spatial bias in the imaging process.

15.4.5 Inexactness

Even though the exactness of the forward model can be assessed with Monte Carlo simulations, as discussed in Section 15.3, there are aspects of experimental measurements that are difficult to incorporate or exactly emulate in simulations. For instance, simulated source characteristics in terms of size, self-attenuation, energy distribution, and orientation may not exactly match the true characteristics, while detector size, efficiency, energy response, and direction of incidence may not be fully incorporated or faithfully simulated. Moreover, Monte Carlo simulations are typically performed for a unit source, and in the forward model the detection efficiency is incorporated into a calibration constant along with the source strength. Also, the calibration process is not always a direct linear scaling process, as discussed in Section 15.4.4. Even after correcting for background radiation, minimizing systematic error, evaluating statistical uncertainty, and calibration, comparison between modeled measurements and those measured in the laboratory, when performed over a number of test object, may reveal some peculiarities that may not have been taken into account in the background correction and the calibration processes, or when dealing with systematic error. For instance, one is likely to discover, at least for some test objects, that the background correction and the calibration process may not produce consistent matching between the forward model and the experiments. Objects very transparent or too opaque to radiation may affect the background component and/or the calibration process, and minor systematic errors in determining spatial locations may skew the geometric appearance of some objects toward or away from some detectors. Even when background correction and calibration produce consistent results, the experimental results will always have their own statistical fluctuations, and forward-model calibrated and background-corrected measurements should produce results that lie within the bounds of statistical variability. If such agreement is not reached, one should revisit the experimental processes and introduce necessary corrections.

15.5 Image Quality

A perfect image is a continuous image, while a reconstructed image is discretized over pixels or voxels. The discretization process has an immediate impact on image quality: it smoothes out continually changing features and lumps them into a single value averaged over the area of a pixel or the volume of a voxel. As a result, the *spatial resolution*[3] of the image is reduced, blurring sharp features. This averaging process also reduces the *contrast sensitivity*, i.e. the magnitude of abrupt changes (narrow peaks and valleys, and edges) will be underestimated. Even if an imaged object is uniform, discretization will reduce sharpness and contrast at the boundaries of the image, unless the discretization mesh perfectly matches the edges of the object. Even with perfect voxel matching, no two adjacent pixels/voxels will have exactly the same numerical attribute, due to the uncertainties always associated with measurements employed in image reconstruction and their subsequent propagation through the process. Therefore, uniform objects with curved boundaries (such as disks) provide convenient means for the quantitative determination of an image's quality.

A standard method for measuring the spatial resolution and material contrast is provided by ASTM International.[4] This method is based on examining the reconstructed image of a uniform disk of material. It determines the spatial resolution by the sharpness at the edge of the disk, and computes the contrast sensitivity by the statistical analysis of the reconstructed image at the center of the disk, as well as at the peripheries (see Section 15.5.2). The ASTM method uses the modulation transfer function (MTD) to quantify the spatial resolution, and the contrast discrimination function (CDF) to quantify the contrast sensitivity. These two functions are defined in Sections 15.5.1 and 15.5.2, respectively. Although this standard was developed specifically for transmission-based x-ray and gamma-ray CT systems, the concepts it introduces are equally valid for use in any numerically reconstructed image.

Since numerical values for the image parameter, $\boldsymbol{c}$, are reconstructed, it is natural to examine the quantitative accuracy of these values. However, this is not always a straightforward task, as explained in Section 15.5.4. Methods for determining the

[3] Resolution is the smallest observable change in a quantity.

[4] ASTM Standard E1695 - 95(2006)e1, Standard Test Method for Measurement of Computed Tomography (CT) System Performance, ASTM International, West Conshohocken, PA, www.astm.org.

spatial resolution, contrast resolution, and accuracy are discussed below. It should be noted that images are eventually viewed by the human eye, and therefore both contrast and resolution should be judged by human perception, unless images are presented as tabulated values where absolute pixel/values are individually given.

15.5.1 Spatial Resolution

The spatial contrast of an imaging system is not only determined by the size of voxel/pixel of the reconstructed image, but is also affected by the size of the source's field-of-view and the detector's aperture, the distance between two adjacent detectors, and by the number of available measurements relative to the number of image voxels/pixels.

Let us consider two high contrast small spots far apart in space within an otherwise uniform background, see Section 15.5.2 for definition of contrast. The reason the two spots must have a high contrast is to ensure that they are recognizable beyond the effects of noise and measurement errors in the reconstructed image. If these two spots are brought close to each others in space, they will remain distinguishable until they become too closely spaced. Spatial resolution is the minimum distance between the two spots at which they can still be observed as two distinct spots. One would not expect the spatial resolution to be less than the width of a pixel/voxel, because if the edges of the two spots exist within the same pixel/voxel, the image reconstruction process will smear the attributes of the two spots together into a single value within the shared pixel/voxel.

Point Spread Function

The spatial spread of a true value of a physical parameter, $c(\vec{r})$, during the imaging process, can be mathematically expressed as:

$$\tilde{c}(\vec{r}) = \int_{-\infty}^{\infty} K(\vec{r};\vec{r'})c(\vec{r'})\, d\vec{r'} \tag{15.4}$$

where $\tilde{c}(\vec{r})$ is the observed imaging parameter and $K(\vec{r};\vec{r'})$ is a system-specific spread function, or kernel. A point at position, $\vec{r}_0$, in the image can be represented by an impulse (a Dirac delta) function, $\delta(\vec{r}-\vec{r}_0)$. Therefore, the point spread function (PSF) is given by:

$$P(\vec{r}) = \int_{-\infty}^{\infty} K(\vec{r};\vec{r'})\delta(\vec{r'}-\vec{r}_0)\, d\vec{r'} = K(\vec{r};\vec{r}_0) \tag{15.5}$$

Therefore, PSF provides a means of determining the spread function, $K(\vec{r};\vec{r'})$, one point at a time. Recall that the concept of the point spread function was introduced in Section 9.2.2 to relate an impulse change in an image parameter, c, to a measurement profile.

Even under ideal conditions where there are no physical constraints and no measurement noise that cause spatial spread in the reconstructed image, the discretization process (see Section 7.1), necessary for numerical image reconstruction, will introduce a spatial spread, for a voxel at position $\vec{r}$, in the form of a rectangular function:

$$\text{rect}\left(\frac{\vec{r'}-\vec{r}}{\Delta r}\right)=\begin{cases}1 & \text{if } \left|\frac{\vec{r'}-\vec{r}}{\Delta r}\right|<\frac{1}{2}\\ 0 & \text{if } \left|\frac{\vec{r'}-\vec{r}}{\Delta r}\right|>\frac{1}{2}\end{cases} \tag{15.6}$$

where Δr is the voxel width. For uniform voxels, this spread function is independent of the position of the voxel, and as such the kernel, $K(\vec{r};\vec{r'})$, in Eq. (15.4) can be represented by a shift-invariant (also called isoplanatic) function: $K(\vec{r}-\vec{r'})$. According to Eq. (15.5), the PSF of discretization is the rectangular function: $\text{rect}\left(\frac{\vec{r}-\vec{r}_0}{\Delta r}\right)$.

In practice, however, system imperfections and measurement noise produce a PSF spread non-uniformly in space that surpasses a voxel's volume. Such spread may be skewed more in one spatial direction than another. The degree of the spread of the PSF defines the spatial resolution. It is typically measured by the full width at half maximum (FWHM). The lack of PSF isotropy is determined by the variation of the FWHM when measured in different directions. The skewness (asymmetry) of the PSF at any given spatial direction can also be quantified to indicate any bias in the spread of the PSF in two opposite directions. The PSF can vary within the image, and should be measured at the center of the image, its edges, and in between. One should aim at designing an imaging system that produces a PSF that is isotropic, symmetric, and independent of position.

The PSF can be measured with the aid of a wire of material with an attribute considerably different (much higher or much lower) from that of the surrounding reference medium. In emission imaging, a wire-like (line) source is used. The spatial resolution should be measured at different plane orientations in the image, one plane at a time. The wire should be positioned normal to the selected plane, and the geometric cross section of the wire should be less than that of an image pixel/voxel in the plane, for the wire to be representative of a point. On the other hand, the wire should be larger in length than the thickness of a pixel or the height of a voxel normal to the plane, to ensure that the pixel/voxel is completely covered by the wire material in the direction normal to the plane of measurement. A number of wires positioned in different locations should be simultaneously used to examine the dependence of the spatial resolution on position. The location of wires within pixels/voxels should differ such that in some cases a wire overlaps more than one pixel/voxel, and on average wires occupy random positions within pixels/voxels.

Line Spread Function

Using the point spread function to measure the spatial resolution is limited by the fact that in reality one cannot physically represent a point. A wire no matter how thin still has a finite diameter. Also, if the wire has too small a diameter, it may not appear in the reconstructed image with a sufficient contrast to be distinguishable from the background; rendering it useless as the high contrast medium needed in spatial-resolution determination. It is, therefore, often more practical to measure the line spread function (LSF).

The LSF measures the distribution in a reconstructed image of a high-contrast line of infinitesimal thickness placed within an object. It is the integration of the point spread function along a line. If the PSF at point at position $\vec{r}$ is $P(\vec{r})$, then the LSF along a line in a particular direction $\vec{v}$ is:

$$L(\vec{v}) = \int_{-\infty}^{\infty} P(\vec{r})\, d\vec{v} \tag{15.7}$$

This equation indicates that the LSF can be obtained from the PSF by integrating the latter along the line over which the LSF is defined. However, the LSF can also be measured by placing a line made of a high contrast material (or a line source in emission imaging) within a reference uniform medium. The thickness of the line object should be smaller than the width of voxel/pixel in the imaging system. The LSF can be quantified with the FWHM of the observed PSF at varies locations of the line, with all indicators integrated to provide an overall LSF for a given line. One can also progressively bring two wires close to each other and observe their positions in the reconstructed image. The smallest distance beyond which the image of the two wires overlap each other defines the spatial resolution around the location of the two wires. A number of wires branching out from the tip of a triangle (forming a fan) should be used for measuring the LSF in a plane, since this configuration provides a varying separation distance between the lines, see for example Smith (2003, Chapter 25). Then each wire and the adjacent space define a line pair, and the spatial resolution can be expressed in terms of line pairs per unit distance (e.g., lines per mm: lpm). With such a line-pair gauge, the distribution of the LSF in the plane can be examined at different lpm values. The lpm at which the separation of the wire is barely visible in the image determines the spatial resolution of the image.

It should be noted the LSF's depends on the direction of the line. Consequently, though a LSF in a given direction can be calculated by integrating the PSF along the line of the LSF, the opposite is not true, i.e. the PSF cannot be calculated from a single LSF, unless the PSF is isotropic.

Edge Response Function

Although a line spread function is easier to measure than a point spread function, measuring the former can still suffer from one of the disadvantages of the latter; namely, the disappearance of the high contrast advantage if the wire used to measure the PSF is too thin to be visible in the reconstructed image. Both the PSF and LSF are based on the system response to an impulse function, at a point or over a line. However, the system response can also be obtained via a step function, which is the premise of the edge response function (ERF). A step function is essentially a line function that ends at the point in space where the step is created. Therefore, an ERF, $S(\vec{v}, v_0)$, that has a non-zero value until location, v_0, on the line direction, $\vec{v}$, is related to the LSF, $L(\vec{v})$, along this line by:

$$S(\vec{v}, v_0) = \int_{-\infty}^{v_0} L(\vec{v'})\, d\vec{v'} \tag{15.8}$$

Consequently,

$$L(\vec{v}) = \frac{dS(\vec{v}, v_0)}{dv} \tag{15.9}$$

The ERF can be measured with the aid of a block of a high contrast material (or a plate source in emission imaging) that has a planar edge. The plane of the edge should be placed perpendicular to the plane within which the spatial resolution is to be determined. The planar edge of the block should be aligned at a slight angle with either the edges of pixel/voxel rows or columns in the measurement plane to magnify the edge response by having the plane edge partially occupying image voxels. This facilitates the calculation of the LSF from the ERF, using Eq. (15.9). The ERF is quantified by the distance within which the edge response rises from 10% to 90% (Smith, 2003, Chapter 25).

Modulation Transfer Function

The spatial resolution, as indicated above, is typically expressed in terms of line pairs per unit distance, which is a spatial frequency. Therefore, the spatial resolution can also be examined in the frequency domain, with the aid of Fourier expansions, as discussed in Section 8.3.1. This is because an image can be viewed as consisting of sinusoidal waves with a range of spatial frequencies with varying amplitudes. A uniform image has low-frequency components, except at its edges where high frequency components are needed to describe the abrupt change of the image attributes to zero outside the image domain. In a more general image with many features, a wide range of frequencies would be present. The amplitude of each frequency component should be sufficiently high to distinguish it from the low frequency of the background. The absence of some frequencies, or a reduction in their amplitude and a shift in their phase, would blur the image and reduce its spatial resolution. Therefore, similar to introducing an impulse into an image to determine its point spread function, one can introduce a frequency component to determine the *modulation transfer function* of the image.

Let us consider a true feature in an image of a frequency component of unit amplitude, $u(\vec{r}) = \exp[2\pi i(\vec{k} \cdot \vec{r})]$, with $\vec{k}$ being the spatial frequency (wave number) vector. According to Eq. (15.4), this feature would be blurred and appear as:

$$\tilde{u}(\vec{r}) = \int_{-\infty}^{\infty} K(\vec{r}; \vec{r'}) \exp[2\pi(\vec{k} \cdot \vec{r'})]\, d\vec{r'} \tag{15.10}$$

If the shape of blurring of a point in an image is independent of the location of the point (i.e. shift invariant), then the kernel, $K(\vec{r};\vec{r'})$ in Eq. (15.4) can be replaced by the convolution kernel: $K(\vec{r}-\vec{r'})$. Substituting $\vec{r''}=\vec{r}-\vec{r'}$ in Eq. (15.10) and using the convolution kernel, one has:

$$\tilde{u}(\vec{r}) = \exp[2\pi(\vec{k}\cdot\vec{r})]\int_{-\infty}^{\infty} K(\vec{r''})\exp[-2\pi(\vec{k}\cdot\vec{r'})]\,d\vec{r''} = u(\vec{r})\mathcal{K}(\vec{k}) \tag{15.11}$$

with the last step arrived at using the definition of the Fourier transform, Eq. (8.6). The use of $K(\vec{r''})$ in Eq. (15.11) indicates that a shift-invariant spread function retains the frequency of a feature in the image, while changing its amplitude and phase shift by the Fourier transform of the spread function, $\mathcal{K}(\vec{k})$. This factor of change is known as the transfer function, as it determines the amount of change (by blurring) in the amplitude and phase shift of a frequency component. Notice that the transfer function is a constant for a given frequency, k, i.e. it is not a function of position as expected for a shift-invariant PSF. The *magnitude* of the transfer function is known as the modulation transfer function (MTF). Recalling from Eq. (15.5) that an impulse image produces the point spread function, $p(\vec{r})$, therefore, the MTF is simply the magnitude of the Fourier transform of a shift-invariant PSF. Consequently, the MTF is the one-dimensional Fourier transform of the LTF, and the latter is found from the ERF by taking its derivative according to Eq. (15.9).

An image consists of the superposition of many frequency components. Therefore, an image can be represented as:

$$c(\vec{r}) = \int_{-\infty}^{\infty} \exp[2\pi(\vec{k}\cdot\vec{r})]\mathcal{C}(\vec{k})\,d\vec{k} \tag{15.12}$$

where $\mathcal{C}(\vec{k})$ is the Fourier transform of the original image parameter, $c(\vec{r})$, at various locations, $\vec{r}$, in the image:

$$\mathcal{C}(\vec{k}) = \int_{-\infty}^{\infty} \exp[-2\pi(\vec{k}\cdot\vec{r})]c(\vec{r})\,d\vec{r} \tag{15.13}$$

Using the fact that the amplitude and phase shift of each frequency component is modulated by the transfer function, $\mathcal{K}(k)$, then the image is blurred to:

$$\tilde{c}(\vec{r}) = \int_{-\infty}^{\infty} \mathcal{K}(\vec{k})\exp[2\pi(\vec{k}\cdot\vec{r})]\mathcal{C}(\vec{k})\,d\vec{k} \tag{15.14}$$

A Fourier transform of Eq. (15.14) yields:

$$\hat{\mathcal{C}}(\vec{k}) = \mathcal{C}(\vec{k})\mathcal{K}(\vec{k}) \tag{15.15}$$

One would arrive also at Eq. (15.15) by the Fourier transform of Eq. (15.4) when the kernel is shift invariant, which affirms the fact that a frequency component in an image is blurred by the MTF without a change in frequency. It should be noted that the exponential term in Eq. (15.14) is a sinusoidal function, which is difficult to introduce in imaging systems. When rectangular waves, or other waves that introduce sharp-edges, are used, the obtained transfer function is no longer exactly an MTF, and should be more rigorously referred to as the contrast transfer function (CTF) (Morgan, 1983).

The line-pair gauge, described in Section 15.5.1, provides different spatial frequencies, in terms of line pairs per unit distance; with the highest frequency being at the tip of the triangle and the lowest at the opposite end. The Fourier transform, performed digitally with a fast Fourier transform, of the LSF, measured with this gauge provides the MTF at various frequencies. Also, the ERF measured with a block inclined with the pixel/voxel lines, can be used to find the MTF at various spatial frequencies, by taking the derivatives of the ERF to find the LSF and then applying the Fourier transform. The MTF can also be measured directly, albeit approximately, by using a star or a sunburst gauge (Morgan, 1983). The spatial frequency at which the MTF is so reduced in value that it is not practically observable defines the spatial resolution of the image. For emission imaging, capillary line sources can be used, see for instance the QC phantoms of the Data Spectrum Corporation, Hillsborough, North Carolina (`http://www.spect.com/products-all.html`).

As indicated at the beginning of this section, the ASTM procedure for measuring the spatial resolution in computed tomography, requires the use of a uniform disk of a high-contrast material. The PSF of a disk would produce a profile in the form a rectangular pulse along a line passing through the center of the disk, provided that the PSF is symmetric and isotropic. The LSF is then the integral of the profile of PSF along a certain direction. In effect, the LSF is the summation of the image parameters in the pixels/voxels, along the selected direction. The profile of the LSF is then obtained by traversing the disk in a direction normal to the direction of the line of the LSF, and adding the pixel/voxel values in the direction of the line. Therefore, for a rectangular function of a uniform disk, the LSF should have a $\sin\vartheta$ profile, with ϑ being the angle with the normal to the line of the LSF as it sweeps from one end of the disk to the opposite end. The edge of the disk also covers pixels/voxels at various rates, providing the ability to determine the ERF at different spatial frequencies, the derivative of which produces the LSF. As indicated earlier, the Fourier transform of the LSF provides the MTF. The derivative of the ERF is also the slope of the PSF point spread function around the circumference of the disk. Therefore, the MTF takes into account the edge response at all angles. To determine whether the PSF of the imaging system is symmetric and isotopic, the LSF should be measured at various directions, at least at two mutually perpendicular directions.

15.5.2 Contrast

Contrast is the ability to distinguish in an image two neighboring features of nearly identical intensity. The contrast of an image is affected by the size and number of voxels/pixels of the image, the strength of the source, the detector efficiency, and the measurement time (factors that control the strength of recorded measurements). The image reconstruction process can also affect contrast, if it results in smoothing of image parameters, which smears small differences between neighbors.

Contrast is defined by the absolute difference between image parameters, Δc, between two similar but not identical physical attributes. If these two features are indistinguishable from each other, they would appear as a single entity. We will differentiate here between visual (subjective) distinguishability and numerical (quantifiable) distinguishability. The former is used in visual sciences, see for example Wandell (1995), to determine the *detection threshold* i.e. the value below which it becomes difficult for human observers to discern the two features. Since the value of c is numerically readily available, it is more objective to quantify the detection threshold as the smallest numerical value of $|\Delta c|$, below which $|\Delta c|$ is not distinguishable from statistical uncertainties. Contrast detection threshold is also known as *contrast resolution* or *contrast discrimination.*

The two features considered in defining contrast can be provided by an insert and a surrounding uniform medium, if the insert and the surroundings are slightly different in their physical attributes. The contrast detection threshold depends on the size of the examined feature, since a large feature is easier to discern from an image's background than a smaller one. Therefore, contrast can be defined by the smallest volume (or number of voxels/pixels) required to distinguish in an image a certain contrast level. Contrast resolution and spatial resolution are naturally related. It is difficult to simultaneously have an image with both a high spatial resolution and contrast resolution; to determine one, the other must be suppressed. As such, in determining the spatial resolution, in Section 15.5.1, we needed small spots with high contrast, while for measuring contrast large spots are employed.

A contrast ratio is typically used to quantify contrast. A number of measures can be utilized to define this ratio (Mantiuk et al., 2006):

$$\text{Simple Contrast} = \frac{c_f}{c_b}$$

$$\text{Weber Fraction} = \frac{|c_f - c_b|}{c_b}$$

$$\text{Logarithmic Ratio} = \log_{10}\left|\frac{c_f}{c_b}\right|$$

$$\text{Signal-to-Noise Ratio} = \frac{|c_f - c_b|}{\sigma_f}$$

where c_f and c_b designate, respectively, the values of the image parameter, c, for the test (insert) feature and its immediate neighboring background, and σ_f is the standard error (variance) in the value of c_f. The $\log_{10}$ is used, instead of natural logarithm, as it is the convention employed in vision science; it results in, when multiplied by ten, a contrast defined in units of decibels (dB). The logarithmic ratio enables the definition of the contrast over a wide range.

While measuring spatial resolution requires the use of a high contrast test object, measuring contrast demands a low-contrast test object, i.e. low c value, where the uncertainties (noise) in c can compete with the value of c itself. Therefore, contrast resolution is often referred to as low-contrast resolution (Kalender, 2005). The contrast threshold necessary for the detection of an intrusion in an otherwise low-contrast uniform image is determined, as recommended by the International ASTM Standard E1695 - 95(2006)e1, by statistical analysis of the reconstructed image of a uniform disk. The standard errors (deviation from mean) for the c values in regions of various sizes should be calculated. Regions of progressively increasing size (number of pixels/voxels) should be considered at the center of the image, as well as its peripheries. The minimum contrast (detection threshold), $|\Delta c|_{min}$, is determined by assigning a certain confidence level; say $3.29\sigma_c$ for a 95% probability of detection (assuming a normal distribution of the c values). A plot of the detection threshold contrast versus the region size (which can be designated in terms of the number of pixels/voxels in each region) forms the so-called contrast discrimination function (CDF). The contrast resolution, value of $|\Delta c|_{min}$, for a given intrusion (insert) size, is the value corresponding to the intrusion size in this empirically constructed CDF function; with the view that the intrusion must produce a c value that exceeds the threshold detection contrast for its size. The Weber fraction at the detection threshold is then the ratio of the quotient of the threshold contrast to the average c value of the background.

While CDF provides a predicted detection threshold contrast, the actual detection contrast can be determined by testing using an object with inserts of various sizes containing a material with attributes slightly different from those of the background material; such as polystyrene in water in transmission and scatter imaging, and too weak but slightly different source activities in emission imaging. The averages of the distribution of the c values, in the pixels/voxels corresponding to each insert in the reconstructed image, are then calculated. The difference between these average c values and the c value for the background medium determines the actual detection contrast, $|\Delta c|_{actual}$, for different insert sizes. The ratio of the actual detection contrast to the threshold detection contrast, $\frac{|\Delta c|_{actual}}{|\Delta c|_{min}}$, defines the so-called visibility factor (von Falck et al., 2008). At a value of one, the visibility factor is the threshold detectability at the confidence level at which the CDF was devised.

15.5.3 Sensitivity

Sensitivity is the magnitude of the rate of change of an output of an operation with respect to its input. Since in imaging the input is a measurement vector, $\boldsymbol{e}$, and the output is a reconstructed image represented by the vector $\boldsymbol{c}$, image sensitivity is defined by a matrix in which the ij element is $\left|\frac{\partial c_i}{\partial e_j}\right|$, where c_i and e_j are, respectively, elements in $\boldsymbol{c}$ and $\boldsymbol{e}$. This matrix can be derived mathematically from the problem's forward model, and it should be examined to determine, and subsequently remove, any measurement that results in poor sensitivity. Experimentally, a number of approaches, each providing a different perspective to sensitivity, can be applied. For example, the sensitivity of a system to statistical fluctuations in measurements can be determined by taking measurements for a certain object at two levels of uncertainties; say by changing the source strength and/or measurement acquisition time to values slightly different from the nominal design operating values, and reconstructing the corresponding images. The voxel-by-voxel differences between the reconstructed images at the two levels of uncertainties relative to the observed change in measurements determine the sensitivity-to-fluctuations matrix. One desires, of course, low sensitivity to fluctuations. Alternatively, one can image two uniform objects of different physical properties. Again, the voxel-by-voxel differences between the reconstructed images and measurements for each object produce a contrast-sensitivity matrix.

Experimentally-determined sensitivity matrices incorporate physical and measurements effects not included in the forward model. Measuring or determining the sensitivity matrix is useful in the design process, while for an operating system measuring the spatial resolution and contrast are sufficient as measures of the system's sensitivity. This is because sensitivity affects both the spatial resolution, through the point spread function, as well as the change in the c values. Therefore, the term *contrast sensitivity* is also used to define contrast. Note that in visual sciences contrast sensitivity is defined as the negative of the logarithm of the Weber fraction at the threshold detection contrast (Wandell, 1995). With this definition, a low threshold detection contrast would indicate large sensitivity, as one would intuitively expect.

15.5.4 Accuracy and Precision

Since in radiation imaging, the constructed image parameters, $\boldsymbol{c}$, are physical quantities, one should question the accuracy of these parameters; i.e. how close the indicated values are to the corresponding "true" values. Accuracy can be measured by comparing the true value of the imaged physical attributes to the reconstructed values. In emission imaging, this is doable, since the true source activity can be directly measured. In transmission and scatter imaging, particularly when imaging with a multienergetic source, such as an x-ray source, the true image parameters should be independently deduced from transmission and scattering measurements (as appropriate). In other words, one should not rely on theoretical values, to avoid uncertainties in determining material density and composition and the effect of the source energy distribution on the imaged physical attributes. Alternatively, one can rely on the calibration of image parameters, by reconstructing the image of an object of known attributes and comparing the results against the theoretical values of these attributes. Such calibration process is also useful in determining the so-called *contrast scale factors* of an imaging system. The contrast factor is the difference in the actual physical attributes of two media relative to the corresponding observed change in a reconstructed image of the two media (Lin et al., 1993). The scale factors can be useful in determining, in transmission imaging with multienergetic radiation source, the effective energy of the source by finding the energy that produces nearly a contrast scale factor of unity.

The precision of an imaging system is determined by the degree of repeatability and reproducibility of $\boldsymbol{c}$, within the anticipated range of statistical fluctuations. However, operating conditions and equipment positioning, if the design is not robust, can have an effect on the acquired measurements and the manner in which they are related to $\boldsymbol{c}$ in the forward model. Therefore, constancy tests for image spatial resolution, contrast and accuracy should be periodically performed to detect any changes in system components.

15.5.5 Systematic Artifacts

Statistical fluctuations in measurements tend, by their nature, to produce random artifacts in the reconstructed image, though the degree of which may depend on the method of imaging and the image reconstruction algorithm. Systematic errors can also produce image artifacts. The most prominent of systematic errors could be the one caused by the simplification of the forward model employed in the image reconstruction process, as discussed in Part I of this book. Such modeling errors have a direct effect on imaging because the forward model is inverted to reconstruct the image. Even if the forward model is reasonably representative of the physical process, the discretization of the image into voxels or pixels results in averaging of the imaged physical property over the volume of each discretized element. If an element happens to encompass more than one material, discretization will provide an average value of the imaged parameter based on the volumetric fraction occupied by each element. The result is an artifact called the *partial-volume effect* that blurs boundaries between image regions. The opposite of this effect occurs if radiation divergence is not accounted for in the forward model, resulting in pixels/voxels outside the intended source/detector fields-of-view being folded into the designated ones, producing an overfilling of those pixels/voxels.

Systematic errors are also caused by mechanical effects (motion and displacement), geometric factors (misalignment and object falling outside designated imaging domain), or equipment failure, as well as bias in the image reconstruction process itself. Also poor measurement sampling produces a systematic artifact. The fact that each detector itself has a physical finite volume, i.e. it is not a point, introduces widening of the point spread function and a systematic reduction in spatial resolution. Drift and instability in detectors and associated electronics result in change in the measurements, hence are also a cause for image artifacts. In addition, detector saturation by very high radiation intensity or susceptibility to electronic noise at low radiation intensity both result in a measurement that does not faithfully reflect the intensity of radiation reaching the detector. Mechanical vibrations and external electric fields can also alter a detector's repose. Some detectors have a hysteresis effect, or afterglow, that contaminate subsequent measurements taken shortly after by the same detector. Detectors in close proximity can also influence each other through cross-talk.

Owing to their systematic nature, systematic artifacts tend to produce certain recognizable patterns in the reconstructed images, such as rings, streaks, banding, cupping, or simply characteristic blurring. Imaging of a uniform test object with low-uncertainty measurements can reveal such artifacts. Other sources of errors that cause artifacts (systematic and otherwise) are given in this book under "error" and/or "artifact."

15.6 Test Objects

For medical imaging, computed tomography (CT) image parameters for sections of human male and female cadaver are made available by the U.S. National Institute of Health through the *Visible Human Project*: `http://www.nlm.nih`

.gov/research/visible/getting_data.html. This section presents some simple section configuration that are suited for the assessment of image reconstruction algorithms for medical and industrial imaging.

15.6.1 Single Inserts

After testing with a uniform object, the next step is to consider an insert within a uniform object. The simplest approach is to introduce, in an otherwise uniform object, inserts of different material attributes. One can start with a linear strip, move to a cubic or a parallelepiped volume, then perhaps to some arbitrary shape. These inserts enable examining an imaging system's ability to distinguish abrupt changes and boundaries, i.e. to test image sharpness. However, particularly in medical imaging, one needs to also examine changes associated with curved shapes.

The simplest curved configuration is that of a sphere. The size and density of a spherical insert can be varied. The spherical shape is equivalent to a circular insert in section (tomographic) imaging. The sphere is analytically easy to generate and incorporate into the forward model to generate modeled data for testing. The location of a voxel ijk extending from point (x_i, y_j, z_k) to point $(x_{i+1}, y_{j+1}, Z_{k+1})$ with respect to a spherical insert of radius R centered at point (x_0, y_0, z_0) is such that:

Voxel is inside sphere, if:

$$\frac{1}{R^2}\left[(x_{i+l}-x_0)^2+(y_{i+m}-y_0)^2+(z_{i+n}-z_0)^2\right] \leq 1 \tag{15.16}$$

for all combinations of $l = 0, 1;\ m = 0, 1; k = 0, 1$.

Voxel is outside sphere, if:

$$\frac{1}{R^2}\left[(x_{i+l}-x_0)^2+(y_{i+m}-y_0)^2+(z_{n+l}-z_0)^2\right] > 1 \tag{15.17}$$

for all combinations of $l = 0, 1; m = 0, 1; k = 0, 1$

Voxel is intersected by sphere, if:

both Eqs. (15.16) and (15.17) are not satisfied. (15.18)

The latter condition would necessitate calculating the volume of the portion of the voxel which lies inside the sphere, so that the voxel attributes can be weighted accordingly in a forward calculation. An approximate estimate of the volume inside the sphere can be obtained by assuming that each voxel's corner-point corresponds to one eighth $\left(\frac{1}{8}\right)$ of the voxel's volume. Then, the number of voxel corners that do not satisfy Eq. (15.16) multiplied by $\frac{1}{8}$ gives the fraction of the voxel volume outside the sphere. Similarly, the number of voxel corners that do not fulfill Eq. (15.17) times $\frac{1}{8}$ gives the fraction of the voxel volume inside the sphere. The two volume portions should complement each other. More accurate volume calculations can be obtained, for voxels partially within the sphere insert, by determining the intersections points of the voxel within the sphere.

A more sophisticated insert is that of an ellipsoid, whose surface can be described by the general equation:

$$\begin{aligned}&\frac{1}{a^2}[(x-x_0)\cos\vartheta\sin\varphi+(y-y_0)\sin\vartheta\sin\varphi+(z-z_0)\cos\varphi]^2\\&\quad+\frac{1}{b^2}[-(x-x_0)\sin\vartheta+(y-y_0)\cos\vartheta]^2\\&\quad+\frac{1}{c^2}[-(x-x_0)\cos\vartheta\cos\varphi-(y-y_0)\sin\vartheta\cos\varphi+(z-z_0)\sin\varphi]^2=1\end{aligned} \tag{15.19}$$

where a, b and c are, respectively, the radii along the X, Y, and Z axes of the spheroid, ϑ is the polar angle the major axis X makes with the x-axis of the coordinates system, and φ is the azimuthal angle the X-axis makes with the z-axis of coordinates. In the formulation of Eq. (15.19), it is assumed that the ellipsoid was not azimuthally rotated, but formulations that consider such rotation can be analogously generated. Of course, when $a = b = c$, one has a sphere. When $a = b < c$, one has a prolate spheroid (North-American football-shape), while $a = b > c$ gives an oblate (door-knob shaped) spheroid. The procedure for locating a voxel within an ellipsoid is similar to that for the sphere with the left-hand-side of Eq. (15.19) replacing the corresponding one in the inequalities of Eqs. (15.16) and (15.17). The same approximate approach, used for a spherical insert for determining the volume of a portion of voxel intersected by the insert, can also be applied for an ellipsoid insert.

15.6.2 Multiple Inserts

While test objects with a single insert enable examination of the ability of an image reconstruction scheme to distinguish between two materials and determine the location and shape of an enclosed cluster, they do not challenge the problem with many features that can mask each other. Multiple inserts do not only provide such a challenge, but they also test for more realistic features. Selective voxel-by-voxel fillings can fulfill this purpose. A multi-layered object can be set to emulate a composite material. On the other hand, a checkerboard formation of voxels with alternating materials can test whether the image reconstruction process can decipher adjacent voxels, or simply smears them into some averaged value. Selective voxel-by-voxel designation can also be used to configure a test object that resembles any desired systematic or random patterns.

In order to accommodate curved geometries, multiple spherical or ellipsoidal inserts can be employed. Determining whether a voxel is fully or partially inside a given curved region can be determined in the same fashion used above for a single insert, by deciding whether the voxel is in the inside of an ellipsoidal surface, outside of it, or is intersected by it. When an ellipsoidal surface overlaps others, each surface can be designated a so-called gray level, so that the attribute of a voxel is the sum of the gray-level values of the surfaces within which it intersects, in addition to partial (volume-weighted) values for the surfaces that intersects it. A surface gray level could be negative to be able to reduce the value of the attributed physical parameter. This is the approach used by Shepp and Logan (1974) for a section of the skull. This test section is widely used in testing medical images, because in addition to being easy to implement, it provides rich and nearly-realistic textual and density changes. MATLAB provides a function for this phantom, along with a modified version. A 3D extension of this function that generates a phantom consisting of an arbitrary number of ellipsoids in 3D is also available via the MATLAB Central File Exchange: `http://www.mathworks.com/matlabcentral/fileexchange/loadFile.do?objectId=9416`. These continuous phantoms can also be used to examine the effect of discretization on image reconstruction.

The Phantom Laboratory, Salem, NY (`http://www.phantomlab.com/`) produces phantoms for testing medical imaging systems. Data Spectrum Corporation, Hillsborough, North Carolina (`http://www.spect.com/products-all.html`) provides a number of phantoms for use in emission imaging (SPECT and PET). The phantom group (Arbeitsgruppe Phantome) of the Institute of Medical Physics Friedrich-Alexander-University Erlangen-Nürnberg, Henkestr, Erlangen, Germany `http://www.imp.uni-erlangen.de/phantoms/`) describes a number of phantoms for testing transmission-based CT systems that can also be utilized in scatter imaging. Imaging Solutions, Sunnybank Hills, Qld, Australia (`http://www.imagingsolutionsaus.com.au/`) supplies phantoms and tissue simulation solutions.

References

A. C. Atkinson, "The computer generation of Poisson random variables," *Applied Statistics*, vol. 28, pp. 29–35, 1979.

G. E. P. Box and M. E. Muller, "A note on the generation of random normal deviates," *Ann. Math. Stat.*, vol. 29, pp. 610–611, 1958.

D. Colton and R. Kress, *Inverse Acoustic and Electromagnetic Scattering Theory*, ser. Applied Mathematical Sciences. Berlin: Springer-Verlag, 1992, vol. 93.

E. M. A. Hussein, *Handbook on Radiation Probing, Gauging, Imaging and Analysis: Applications and Design*. Dordrecht: Kluwer Academic Pub., 2003, vol. II.

——, *Radiation Mechanics: Principles and Practice*. Amsterdam: Elsevier, 2007.

W. A. Kalender, *Computed Tomography: Fundamentals, System Technology, Image Quality, Applications*, 2nd ed. Erlangen: Publicis Corporate Publishing, 2005.

P.-J. P. Lin, T. J. Beck *et al.*, "Specification and acceptance testing of computed tomography scanners," American Association of Physicists in Medicine, New York, Tech. Rep. 39, May 1993, (`http://www.aapm.org/pubs/reports/rpt_39.pdf`).

R. Mantiuk, K. Myszkowski, and H.-P. Seidel, "A perceptual framework for contrast processing of high dynamic range images," *ACM Transactions on Applied Perception*, vol. 3, pp. 286–308, 2006.

C. L. Morgan, *Basic Principles of Computed Tomography*. Baltimore: University Park Press, 1983.

L. A. Shepp and B. F. Logan, "The Fourier reconstruction of a head section," *IEEE Trans. Nucl. Sci.*, vol. 21, pp. 21–43, 1974.

S. W. Smith, *The Scientist and Engineer's Guide to Digital Signal Processing*. San Diego: California Technical Publishing, 2003, (`http://www.dspguide.com`).

C. von Falck, A. Hartung, F. Berndzen, B. King, M. Galanski, and H.-O. Shin, "Optimization of low-contrast detectability in thin-collimated modern multidetector CT using an interactive sliding-thin-slab averaging algorithm," *Invest Radiology*, vol. 43, pp. 229–235, 2008.

B. Wandell, *Foundations of Vision*. Sunderland, MA: Sinauer Associates, Inc., 1995.

16 Post-Processing: Image Enhancement

Images are reconstructed over discrete pixels/voxels. An imaged object tends, however, to be composed of regions of continuous material with interfacing boundaries. Within each region, discretization can conceal small changes from one point to another. Even if a region is uniform, measurement uncertainties may give the appearance of fluctuating image parameters. Discretization also blurs abrupt changes at boundaries between regions, since pixel/voxel boundaries are unlikely to coincide with natural boundaries, and measurement noise tends also to smear out the boundaries. One is, therefore, tempted to apply some form of image processing on reconstructed images to accentuate their features, e.g. by smoothing out noise or sharpening the appearance of edges.

Image enhancement aims at producing a better quality image from an image of lesser quality (raw image). It is, therefore, a mapping of one image to another. As such, image enhancement can be viewed as an inverse problem for which the input is the raw image. This mapping is, however, a one-to-one correspondence between the pixel/voxels of the enhanced image and those of the raw image; unlike the inverse problem of image reconstruction where a measurement usually corresponds to many image parameters. As a result, the inverse problem of image enhancement deals with pixels/voxels that are not directly correlated to each other. On the other hand, because in image reconstruction measurements present some form of correlation between pixels/voxels, it is advisable to incorporate some image enhancement measures within the image reconstruction process itself through regularization, see Section 10.4. Post-processing of reconstructed images becomes then a means to enhance the visibility (by the eye or any other visual aid) of reconstructed images. Therefore, direct image enhancement is considered here to be concerned with removing spatial blurring and faintness, in order to provide the human eye with appealing crisper and sharper images. Image processing, enhancement, and restoration methods are widely available for use with digital photographs, video frames, and astronomical images. Here, an overview of image enhancement methods is given, addressed in a fashion similar to that used for solving the inverse problem of image reconstruction. More details can be found in textbooks on digital image processing or signal processing e.g. Castleman (1996), Gonzalez and Woods (2008), Proakis and Manolakis (2007), and Smith (2003). The neural networks optimization method (Section 11.8), and genetic algorithms

Computed Radiation Imaging. DOI: 10.1016/B978-0-12-387777-2.00016-1

(Section 11.6), have been also suggested for this purpose, see Tsoy and Spitsyn (2006) and Paulinas and Ušinskas (2007).

The emphasis in this chapter is on image enhancement, not on image processing. The latter deals with extracting information from an image by segmenting it to extract particular features, recognize or detect certain patterns or characteristics, or as an image analysis tool. These, as well as image enhancement methods, can be found in a variety of image processing software, such as: MATLAB's Image Processing Toolbox (`http://www.mathworks.com/products/image/`); Quantim, licensed free of charge software, (`www.iup.uni-heidelberg.de/institut/forschung/groups/ts/soil_physics/tools/`); and the Hypermedia Image Processing Reference-HIPR (`http://homepages.inf.ed.ac.uk/rbf/HIPR2/hipr_top.htm`). One should keep in mind that methods for image processing, enhancement, and restoration are rapidly evolving fields due to the continuing improvement in computer technology.

16.1 Image Convolution

Let the spatial distribution of a reconstructed image be denoted by $g(\vec{r})$, which is a function that is known only at discrete intervals defined by the $\boldsymbol{c}$ values of the reconstructed image at various pixels/voxels. We are using here the f-g-h notation, not only because it is commonly used in the literature of image processing, but also to distinguish the inverse problem of image enhancement from that of image reconstruction.

The purpose of image enhancement is to remove as much as possible the effect of the spatial (geometric) spread and faintness in the reconstructed image inherent in the discretization process and the associated uncertainties (noise), i.e. to restore $g(\vec{r})$ to an enhanced image, $f(\vec{r})$. Faintness in an image is caused by fluctuations in arriving at the image parameter, g, which are typically reduced by increasing exposure (intensity of radiation source and/or measurement time). Therefore, the effect of faintness can be accounted for in image restoration by considering the image to contain an additive noise component that needs to be removed. Each point in the restored (after processing) image is assumed to have been blurred by a point spread (blurring) function, $h(\vec{r};\vec{r'})$, that describes the influence of point $\vec{r}$ on a neighboring point $\vec{r'}$.

The concept of the point spread function was discussed in Section 9.2.2 in terms of relating an impulse change in an image parameter, c, to a measurement, e, and in Section 15.5 for relating a true change in c to the corresponding change in the reconstructed image. Here, we consider it from the perspective of relating a reconstructed image (considered to be degraded) to the sought enhanced image (considered to be the true image). Ideally, the true image is the amalgamation of contiguous points, each represented by a Dirac delta function modified by a certain amplitude, so that:

$$f(\vec{r}) = \int \delta(\vec{r} - \vec{r'}) f(\vec{r'}) d\vec{r'} \tag{16.1}$$

where $\delta(\vec{r} - \vec{r'}) = 1$ when $\vec{r'} = \vec{r}$, and is equal to zero otherwise. Spatial degradation alterers the delta function to a point spread function, h. Typically, the influence of a point spread function (PSF) depends on the distance $\vec{r}-\vec{r'}$. Therefore, PSF can be expressed as $h(\vec{r}-\vec{r'})$, and is then said to be shift-invariant (or isoplanatic).

Let us assume for simplicity that h is a spatially linear function. The degraded image, g, can then be represented as

$$g(\vec{r}) = \int h(\vec{r}-\vec{r'})f(\vec{r'})\, dr' + \eta(\vec{r}) = h(\vec{r}) * f(\vec{r}) + \eta(\vec{r}) \tag{16.2}$$

where "$*$" designates a convolution operator, and $\eta(\vec{r})$ is the noise component associated with $f(\vec{r})$. Note that if $h(\vec{r}-\vec{r'}) = \delta(\vec{r}-\vec{r'})$, then the image is degraded only by noise. Equation (16.2) shows that image spatial blurring is a convolution of the true image. Therefore, the enhanced image, $f(\vec{r})$, is restored by deconvolution. The restored image is considered to be a *visually* appealing image, since the deconvolution process removes the effect of the PSF, producing a smoother (less rugged) image structure, from which features are easier to identify. Since the eyes see only a surface image, we will assume that planar (section) images are to be enhanced, so that selected planes in a three-dimensional image are considered one at a time; though the generalization to three dimensions is doable in many cases.

Equation (16.2) is easier to perform in the frequency domain, where convolution is transformed to a multiplication operation. Like a musical tone, an image can be decomposed into a set of sinusoidal waves (albeit in space, rather than in time) of various frequencies and amplitudes. This decomposition is accomplished via a Fourier transform, see Section 8.3.1. The convolution of Eq. (16.2) in the spatial frequency domain is simply the multiplication of the Fourier transforms of h and f ($\mathcal{H}$ and $\mathcal{F}$, respectively), so that:

$$\mathcal{G}(\vec{k}) = \mathcal{H}(\vec{k})\mathcal{F}(\vec{k}) + \mathcal{N}(\vec{k}) \tag{16.3}$$

where $\vec{k}$ is the spatial frequency vector and $\mathcal{N}(\vec{k})$ is the Fourier transform of the noise component at frequency $\vec{k}$. Therefore, the Fourier transform, $\mathcal{F}$, of the restored image, f, is obtained by:

$$\mathcal{F}(\vec{k}) = \frac{1}{\mathcal{H}(\vec{k})}\,\mathcal{G}(\vec{k}) \tag{16.4}$$

provided that $\mathcal{N}(\vec{k}) << \mathcal{G}(\vec{k})$, which is hopefully the case. The restored image is then obtained by the inverse Fourier transform (indicated by $\mathscr{F}^{-1}$):

$$f(\vec{r}) = \mathscr{F}^{-1}\left\{\mathcal{F}(\vec{k})\right\} = \mathscr{F}^{-1}\left\{\frac{\mathcal{G}(\vec{k}) - \mathcal{N}(\vec{k})}{\mathcal{H}(\vec{k})}\right\} \tag{16.5}$$

This Fourier transform is to performed in two dimensions for an image plane, or three dimensions in a volume image. The following section examines the nature of the blurring function, h, and noise.

16.2 Image Degradation

16.2.1 Point Spread Function

Without knowing the inherent point spread function (PSF) that causes image degradation, one can assume some simple PSF to enhance image quality. For example, an image can be smoothed using the following PSF:

$$h(\vec{r}-\vec{r'})|_{\text{Smooth}} = a_0\delta(\vec{r}-\vec{r'}) + a_1\delta(\vec{r}-\vec{r'}+\Delta_1\vec{r}) + a_2\delta(\vec{r}-\vec{r'}+\Delta_2\vec{r}) + \cdots;$$
$$a_0 > a_1 > a_2 > 0, \ldots; \quad \sum a_i = 1 \tag{16.6}$$

where $\Delta_n\vec{r}$ is the distance from a first voxel to its nth neighbor, and the a_i's are smoothing weights. Obviously, a_0 is the weight given to a voxel at $\vec{r'} = \vec{r}$, a_1 is that given to its immediate (first neighbors), a_2 is the weight to second neighbors, and so on. For a two-dimensional image of square pixels with width, ΔL, with four immediate neighbors and four second neighbors, one has: $a_0 = 0.25, a_1 = 0.125, a_2 = 0.0625, a_i = 0, i > 2$, $|\Delta_1\vec{r}| = \Delta L$, and $|\Delta_2\vec{r}| = \sqrt{2}\Delta L$. In the frequency domain, the smoothing filter of Eq. (16.6) is a low-pass filter that removes high spatial frequency components, typically associated with noise or sharp edges.

A high-pass filter in the frequency domain enhances edges, and is equivalent to a second derivative finite-difference (Laplacian) filter. In the spatial domain, its PSF is expressed as:

$$h(\vec{r}-\vec{r'})|_{\text{Sharpen}} = a_0\delta(\vec{r}-\vec{r'}) + a_1\delta(\vec{r}-\vec{r'}+\Delta_1\vec{r}) + a_2\delta(\vec{r}-\vec{r'}+\Delta_2\vec{r}) + \cdots;$$
$$|a_0| = \sum_{i>0} |a_i|, \ldots; \quad \sum a_i = 0 \tag{16.7}$$

For the two-dimensional example above, one can employ a second derivative relying only on the first neighbors so that: $a_0 = 4$, $a_1 = -1, a_i > 0, i > 1$, or incorporate second neighbors: $a_0 = 8$, $a_1 = -1, a_2 = -1, a_i > 0, i > 2$.

A high-pass filter can also be described as the complement of a low-pass filter, i.e.

$$h(\vec{r}-\vec{r'})|_{\text{Unsharp masking}} = \delta(\vec{r}-\vec{r'}) - h(\vec{r}-\vec{r'})|_{\text{Smooth}} \tag{16.8}$$

This PSF is known as *unsharp masking* because it emulates the traditional film photography technique; in which a mask, created from a blurred (out of focus) image, is superimposed on the original image. Since deconvolution with the impulse function, $\delta(\vec{r}-\vec{r'})$, leaves an image's spatial spread unaltered, while a low-pass filter retains low frequency components (associated with spatial uniformity), the filter of Eq. (16.8) allows the high frequency components of abrupt edge changes to remain in the image. This results in a mask showing the contours of edges present in the image. Obviously the filter of Eq. (16.8) will produce a faint image, but when subtracted from (overlaid over) the original (unprocessed) image will give an outline of the edges. This image can then be sharpened by selectively increasing the contrast of pixels near the contours relative to those away from it.

The point spread functions described above assume no knowledge of the nature of the physical process or the system that produced the image. The goal was simply to accentuate some image features by smoothing out any spatial spread inherent in the image formation, or by sharpening edges. If these same operations are performed using the actual system's PSF, or at least an estimate of which, one would expect a more realistically enhanced image. The procedures described in Section 15.5, for measuring the PSF relating image parameters to measurements, can also be utilized to determine the PSF relating actual image parameters (those of the calibration reference object) to the reconstructed image parameters. The PSF for the reference object can be assumed to be the same for other objects, since the spatial spread is typically governed by the physical nature of the imaging setup, the discretization mesh over which the image is reconstructed, and the image reconstruction scheme. Even without experimental measurements, one can arrive at an approximate estimate of the PSF using the available unprocessed image itself and some *a priori* knowledge of the imaged object. For instance, one can recognize some image features, such as edge corners of uniform regions. Then, by selecting a few pixels/voxels in the predictable region and comparing their image parameters to the expected actual ones, one can estimate the PSF, which can be employed everywhere in the image, if it is spatially invariant. Section 16.7 presents a blind approach for estimating the inverse of the point spread function. Another approach is to assume a Gaussian (normal) distribution of a certain variance as the PSF. The rationale for this choice is discussed below.

Blurring (spread) of the ideal impulse (delta) function can be attributed to a number of factors. In radiation reconstructed imaging, the cause of such spatial blurring can be traced from the source of radiation (whether external or embedded) to the detector that records the measured radiation and its associated electronics, and through the inverse problem that reconstructs an image from the measurements. An external source of radiation is never a point (an isotopic source has a finite volume and an x-ray machine has a finite focal spot size). A detector, whether stand-alone or as a part of a flat panel detector array, has also a finite volume. The fixed voxel/pixel size within which an image is reconstructed introduces directly a rectangular spread function within the image. The discretization of the forward problem, see Section 7.1, also causes blurring. Mechanical positioning and motion (object and equipment) influence the spread function. Given the above stated factors, it is not easy to describe analytically a point spread function. However, given that blurring is caused by many factors that are generally independent of each other, one can assume that the spatial spread is a random process described by a certain statistical distribution. The inclination in image restoration is to assume a point spread function described by a Gaussian function of a certain variance (width):

$$h(\vec{r}-\vec{r}')|_{\text{Gaussian}} = \exp\left[-\pi\left(\frac{|\vec{r}-\vec{r}'|}{\Delta_W}\right)^2\right] \tag{16.9}$$

where Δ_W is a measure of the width (window) of the function (= $1.065 \times$ FWHM, with FWHM being the full width of the distribution at half its maximum). The justification for using such a distribution is the same justification used to assess most experimental or numerical errors by a normal distribution. In accordance to the central limit theorem, any set of independent observations from the same system will tend to resemble a normal (Gaussian) distribution, as the number of observations increases. Therefore, a normal distribution for the point spread function, h, can be assumed. Note that direct application of a Gaussian PSF is equivalent to a low-pass filter, while a high-pass filter is implemented with the second derivative of the distribution or using the unsharp masking of Eq. (16.8). The width of the distribution, as defined by its variance, determines the stretch of the PSF in the same manner the number of coefficients in the filters of Eqs. (16.6) and (16.7) determines the number of neighbors used in filtering. A normalized form of Eq. (16.9) is given by (RSICC, 1978):

$$h(\vec{r}-\vec{r}')|_{\text{Gaussian (normalized)}} = \frac{1}{\sqrt{2\pi\sigma'}} \exp\left[-\frac{1}{2}\left(\frac{|\vec{r}-\vec{r}'|}{\sigma'}\right)^2\right]; \quad \sigma' = \frac{\Delta w|\vec{r}'|}{2.355} \tag{16.10}$$

This definition assures that the PSF is normalized to unity, which preservers the value of an image parameter, c.

16.2.2 Noise

As Eq. (16.2) indicates, image enhancement requires knowledge of the amount of noise associated with the to-be-enhanced image. This noise can also be estimated from the statistical fluctuation in the value of an image parameter. In radiation imaging, the source of radiation, whether internal or external, is governed by Poisson statistics, see Section 15.2. The intensity of measured radiation also fluctuates in accordance to Poisson statistics. Instability in the electronic and electric systems used with the detection system (or electronically driven sources) causes fluctuations in the recorded measurements. These can be eliminated by amplitude discrimination, as they tend to be low-amplitude signals, but may still influence the measurements by fluctuations that may be described by a Gaussian (normal) distribution. The statistical spread inherent in the imaging setup is propagated via the image reconstruction process, which is typically an ill-posed inverse problem. The approximations introduced in formulating the forward problem, as discussed in Part I of this book, are also a source of error that affect the quality of a reconstructed image. In addition, image artifacts due to systematic errors cause reconstructed image parameters to differ from actual ones. However, fluctuation in reconstructed image parameters are not entirely random and independent, as they tend to be correlated by the integral nature of the measurements from which they are reconstructed. One can argue that such correlation will tend to be systematic and will not alter much the randomness caused by the source of fluctuations. Even when the source of such fluctuations is Poisson distributed, for 20 observations or more, one can use the statistics of the normal (Gaussian) distribution to approximate the actual distribution. Note here, we are speaking of a distribution of the value of an observed image parameter at a given position, rather than the distribution of the spatial spread which can also be described by a Gaussian distribution, but in space, as indicated at the end of Section 16.2.1.

16.3 Frequency Filtering

Direct application of Eq. (16.4) can be viewed as filtering the Fourier transform of the reconstructed image, $\mathcal{G}$, by $\frac{1}{\mathcal{H}}$, to obtain a restored image: $\mathcal{F} = \frac{\mathcal{H}}{\mathcal{G}}$. In enhancing a reconstructed image, one is intuitively tempted to remove the effect of discretization into pixels/voxels. Let us pursue this premise. Discretization can be seen as introducing a rectangular function, $\text{rect}\left(\frac{\vec{r}}{2\Delta r}\right)$, where Δr is the width of the voxel, which is equivalent to the multiplication of a one-dimensional rectangular function in each of the directions of the spatial coordinates of the image. The Fourier transform of a $\text{rect}\left(\frac{\vec{r}}{2\Delta r}\right)$ function will produce a function: $(2\Delta r)^n \text{sinc}(k\vec{r}) = (2\Delta r)^n \frac{\sin(\vec{k})}{|\vec{k}|}$, when n is the dimension of the spatial space of the image ($= 2$ for a section image and $= 3$ for a volume image). The inverse filter will then involves the term: $\frac{|\vec{k}|}{\sin(\vec{k})}$. Since the sine function periodically goes to zero, there will an infinite set of frequencies, giving rise to an unbounded transform. This inverse filter provides an idealization that may be approximated by assigning: an upper-bound on inverse filtering, a heuristic model (Barrett, 1981), linear interpolation, cubic convolution, cubic spline modeling, or more common statistical modeling. The latter is discussed below (Park and Schowengerdt, 1983).

16.3.1 Gaussian Filter

Let us consider the case in which h is assumed to be a Gaussian statistical distribution, given by Eq. (16.9), then for a two-dimensional image:

$$\mathcal{H}(\vec{k})|_{\text{Gaussian}} = \Delta_g^2 \exp\left[-\pi \Delta_g^2 |\vec{k}|^2\right] \tag{16.11}$$

where Δ_g is the width of the Gaussian distribution, and use is made of the fact that the Fourier transform, $\mathcal{H}$, of a Gaussian is another Gaussian. The inverse Fourier Gaussian filter is simply the reciprocal of Eq. (16.11). Since a reconstructed image is band limited, due to the discretization process, the inverse filter will have a cutoff frequency, $k_{max} \leq \frac{1}{2\Delta r}$, where Δr is the pixel/voxel width, see Eq. (12.11). Therefore, the inverse Fourier Gaussian filter becomes:

$$\frac{1}{\mathcal{H}(\vec{k}-\vec{k}')|_{\text{Gaussian}}} = \begin{cases} \frac{1}{\Delta_g^2} \exp\left[\pi \Delta_g^2 |\vec{k}|^2\right] & \text{when} \quad |\vec{k}| \leq k_{max} \\ 0 & \text{when} \quad |\vec{k}| > k_{max} \end{cases} \tag{16.12}$$

Subsequently, in an image with an inherent Gaussian point spread function, in which the effect of noise is negligible, the inverse Fourier Gaussian will result in an overall effect described by:

$$\frac{1}{\mathcal{H}(\vec{k}-\vec{k}')|_{\text{Gaussian}}}\mathcal{H}(\vec{k}-\vec{k}')|_{\text{Gaussian}} = \begin{cases} 1 & \text{when} \quad |\vec{k}| \leq k_{max} \\ 0 & \text{when} \quad |\vec{k}| > k_{max} \end{cases} = \text{rect}(\vec{k}-\vec{k}') \tag{16.13}$$

where $\text{rect}(\vec{k})$ is the rectangular function. Therefore, when the inverse Fourier transform is applied to the filtered frequency image, one gets: $\mathscr{F}^{-1}\text{rect}(\vec{k})$. The Fourier transform of a rectangular function is a sinc function, which like the rect function is, in a multidimensional space, the product of its values in each dimension. Therefore, the net effect of an inverse Gaussian band limited filter is to introduce a sinc function correction to the enhanced image.

16.3.2 Parzen Filter

A filter that resembles the Gaussian filter, but with a narrower spread range, is given by the Parzen function (Parzen, 1962):

$$\frac{1}{\mathcal{H}(\vec{k})|_{\text{Parzen}}} = \begin{cases} 1-6|\vec{k}|^2+6|\vec{k}|^3 & \text{when} \quad 0 \leq |\vec{k}| \leq \frac{k_c}{2} \\ 2(1-|\vec{k}|)^3 & \text{when} \quad \frac{1}{2} < |\vec{k}| \leq k_c \end{cases} \tag{16.14}$$

where k_c is a cut-off frequency. The advantage of this filter is that it can be designed, by the choice of k_c, to go to zero at the boundaries.

16.3.3 Pseudoinverse/Matched Filter

One disadvantage of the Fourier Gaussian filter is that some frequencies may not exist, or are weak, in the image. This results in zero or very small values for $\mathcal{H}(\vec{k})$, which makes the application of the filter of Eq. (16.4) vulnerable to singularities and to the amplification of noise. This can be overcome by utilizing the *pseudoinverse* filter:

$$\frac{1}{\mathcal{H}_B(\vec{k})} = \frac{\mathcal{H}^\star(\vec{k})}{\mathcal{H}^2(\vec{k})+B} \approx \begin{matrix} \dfrac{1}{\mathcal{H}(\vec{k})} & \text{when } \mathcal{H}^2(\vec{k}) >> B \\ \dfrac{\mathcal{H}^\star(\vec{k})}{B} & \text{when } \mathcal{H}^2(\vec{k}) << B \end{matrix} \tag{16.15}$$

where $\mathcal{H}^\star$ is the complex conjugate of $\mathcal{H}$, and B is an arbitrary constant that suppresses frequencies with zero or low amplitudes. Note that applying $\mathcal{H}^\star$ is equivalent to correlating the reconstructed image with the to-be-restored image,[1] resulting in the

[1] The correlation (or cross-correlation) integral is similar to the convolution integral of Eq. (16.2), except that convolution involves reversing a function, shifting it, and multiplying it by another function, whereas correlation only involves shifting a function and multiplying it by another function without reversing.

so-called *matched* filtering. The disadvantage of the bounded filter of Eq. (16.15) is that it amplifies high frequencies by $\frac{1}{B}$, and accordingly amplifies noise present at high frequency when $B << 1$. This may be overcome by using a frequency-dependent constant, but assigning such a constant will require some judgment based on *a priori* knowledge of the nature of the imaged object.

16.3.4 Wiener Filter

Another filtering approach to suppress noise is to replace the constant B in Eq. (16.15) by the relative noise with respect to the signal, so that Eq. (16.15):

$$\frac{1}{\mathcal{H}_{\text{Wiener}}(\vec{k})} = \frac{\mathcal{H}^{\star}(\vec{k})}{\mathcal{H}^2(\vec{k}) + \left(\frac{\mathcal{N}(\vec{k})}{\mathcal{G}(\vec{k})}\right)^2} \approx \begin{cases} \frac{1}{\mathcal{H}(\vec{k})} & \text{when } \mathcal{H}^2(\vec{k}) << \left(\frac{\mathcal{G}(\vec{k})}{\mathcal{N}(\vec{k})}\right)^2 \\ \mathcal{H}^{\star}(\vec{k}) & \text{when } \mathcal{H}^2(\vec{k}) >> \left(\frac{\mathcal{G}(\vec{k})}{\mathcal{N}(\vec{k})}\right)^2 \end{cases} \tag{16.16}$$

where $\mathcal{N}(\vec{k})$ is the Fourier transform of the noise. Note here that $\frac{\mathcal{G}(\vec{k})}{\mathcal{N}(\vec{k})}$ is the signal-to-noise ratio at frequency $\vec{k}$. The filter of Eq. (16.16) is known as the Wiener, or the Wiener-Helstorm, filter.[2] It requires knowing the noise frequency distribution. If white noise (equal amount at all frequencies) is assumed, then $\mathcal{N}(\vec{k}) = 1$, and if the noise is Poisson distributed the variance of the noise is equal to the signal. In either case, the Wiener filter will act as an inverse filter, $\frac{1}{\mathcal{H}}$, when the signal is strong, and as a correlation filter, $\mathcal{H}^{\star}$, when the signal is weak. One last note about the Wiener filter: it minimizes the mean-squares difference between the given image and the restored image at a given frequency, as can be deduced from Eq. (16.16), and as such it is a least-squares filter applied one frequency at a time.

16.3.5 Power Spectrum Equalization Filter

Filtering can be accomplished using the power spectrum (squared amplitude of each frequency component). For white noise, the power spectrum is equal to its variance. Removing the power spectrum of the noise from that of the available image restores the power spectrum density[3] of the enhanced image. Performing power spectral equalization (PSE) (Cannon, 1976) of Eq. (16.2), one has:

$$S_g(\vec{k}) = \mathcal{H}^2(\vec{k}) S_f(\vec{k}) + S_\eta(\vec{k}) \tag{16.17}$$

[2] Strictly speaking, the filter of Eq. (16.16) is called the Wiener filter when h is the impulse function, i.e. when $\mathcal{H} = 1$.

[3] In the frequency domain, the power spectrum density is equal to the Fourier transform of the autocorrelation function of a signal.

where $S(\vec{k})$ refers to the power spectrum density of its subscripted variables. The PSE inverse filter is then:

$$\frac{1}{\mathcal{H}_{\text{PSE}(\vec{k})}} = \left[\frac{1}{\mathcal{H}^2(\vec{k})S_f(\vec{k}) + \frac{S_\eta(\vec{k})}{S_f(\vec{k})}} \right]^{\frac{1}{2}} \tag{16.18}$$

Notice the similarity with the Wiener filter, Eq. (16.16), where both are reduced to simple inverse filtering in the absence of noise, and both do not amplify noise in the absence of the frequency components, $S_f(\vec{k})$ or $\mathcal{G}(\vec{k})$, as they are cut off to zero. However, if $\mathcal{H}_{\text{PSE}}(\vec{k}) = 0$, the PSE filter is not cut off: a feature that makes it sometimes more effective than the Wiener filter. The PSE filter is also called the homomorphic (Greek for "same shape") filter, as it preserves the overall features of the image by power spectrum equalization. To implement the PSE filter, one replaces S_f in Eq. (16.18) by $\frac{S_f - S_\eta}{\mathcal{H}^2}$ at each frequency, $\vec{k}$; provided of course that S_η can be estimated from the statistical characteristics of the noise.

16.3.6 Metz Filter

Another filter that suppresses high-frequency noise is the Metz (Metz and Beck, 1974) filter, expressed as:

$$\frac{1}{\mathcal{H}_{\text{Metz}(\vec{k})}} = \frac{1 - \left[1 - \mathcal{H}^2\right]^n}{\mathcal{H}(\vec{k})} \tag{16.19}$$

with the order $n > 1$ chosen to minimize the mean-squared difference between g and f. The noise amplification at high frequency, caused by the deblurring term in the denominator of Eq. (16.19), is compensated for in the Metz filter by the numerator. Increasing the order of the Metz filter increases the contribution of high frequency components, hence the noise.

16.3.7 Frequency-Dependent Filters

To emphasize the components of a particular spatial frequency range over the others, one can apply frequency-dependent filters. For example a ramp filter, which increases in amplitude with frequency, would emphasize the high-frequency components of the image, but would also accentuate high frequency noise. It is, therefore, a high-pass filter. The Hann and Hamming filters (Hamming, 1977), on the other hand, are low-pass filters with a dynamic frequency response. They take the form:

$$\frac{1}{\mathcal{H}_{\text{Hann}}(\vec{k})} = \frac{1}{2}\left(1 + \cos\pi\frac{|\vec{k}|}{k_{max}}\right) \tag{16.20}$$

$$\frac{1}{\mathcal{H}_{\text{Hamming}}(\vec{k})} = \frac{1}{2}\left(1.08 + 0.92\,\cos\pi\frac{|\vec{k}|}{k_{max}}\right) \tag{16.21}$$

where k_{max} is a cut-off frequency; usually the Nyquist frequency of the image. Both these filters are low-pass filters. Notice that these two filters are similar, except that the Hamming filter has a non-zero value at the maximum frequency, so as not to filter out this frequency. These two filters are to be used when more reliable image parameter values are desired, at the expense of reduced spatial resolution. If, on the other hand, a higher spatial resolution is desired at the expense of a higher level of noise, the Butterworth filter (Hamming, 1977) should be applied. This filter takes the form:

$$\frac{1}{\mathcal{H}_{\text{Butterworth}}(\vec{k})} = \frac{1}{1+\left(\frac{|\vec{k}|}{k_{max}}\right)^{n}} \tag{16.22}$$

where n is an integer that defines the order of the filter; the higher the order, the faster the rate at which the filter approaches the cut-off frequency. Therefore, a sharper cut-off is attained at high n values. In other words, n determines how fast is the transition between retained and filtered out frequencies. Note that at $n = 0$, the filter has a magnitude of unity at all frequencies, while as $n \to \infty$ the filter has a zero amplitude at all frequencies, except at $|k| = 0$ where the amplitude is unity. The ability to apply such sharp cut-offs is one of the advantages of the Butterworth filter, as it allows the elimination of noise at higher frequencies while retaining image contrast. Two or more filters can be combined into a single filter. For instance, combining the ramp filter with a Hanning, Parzen, or Butterworth window can improve image visualization. The combined filters are band-pass (also called windowed, tapering, or apodized) filters, as they tend to emphasize a certain range of frequencies, rather than being low-pass or high-pass filters. Other frequency filters, including those discussed in Chapter 12, can also be employed. The wavelets methods of Section 12.9 is also utilized in image enhancement.

16.4 Matrix Based

Like any other discrete problem, image enhancement can be represented in a matrix form as:

$$\boldsymbol{g} = \boldsymbol{H}\boldsymbol{f} + \eta \tag{16.23}$$

where $\boldsymbol{g}$ is a vector whose elements contain the image parameters of the to-be enhanced image, $\boldsymbol{f}$ is a vector of the sought image, η is the noise vector, $\boldsymbol{H}$ is the point-spread (blurring) matrix, and η is the noise in the image. If $\boldsymbol{H}$ is known, then the inverse problem of finding $\boldsymbol{f}$ from $\boldsymbol{g}$ can be solved using any of the matrix-based methods discussed in Chapter 10. Note, however, that the matrix $\boldsymbol{H}$ is a square matrix since the vectors $\boldsymbol{f}$ and $\boldsymbol{g}$ have the same dimensions; unless one is collapsing the image $\boldsymbol{g}$ into a coarser image $\boldsymbol{f}$ to obtain more reliable values of $\boldsymbol{f}$ at the expense of resolution. With a square matrix, Eq. (16.23) becomes a fully-determined problem, but with no additional data to compensate for the effect of noise. Therefore, some *a priori* knowledge of the nature of the image can aid in compensating for the effect of noise.

A matrix-based restoration of the image $\boldsymbol{f}$ takes the form:

$$\boldsymbol{f} = \left[\boldsymbol{H}^{\mathrm{T}}\boldsymbol{H} + \alpha \boldsymbol{B}^{\mathrm{T}}\boldsymbol{B}\right]^{-1} \boldsymbol{H}^{\mathrm{T}}\boldsymbol{g} \tag{16.24}$$

where α is a regularization parameter, $\boldsymbol{B}$ is a regularization matrix. Notice that when $\boldsymbol{B}$ is a null matrix, Eq. (16.24) becomes a least-squares (or maximum likelihood) estimate of the solution, which is analogous to the inverse filter of Eq. (16.4). A solution with $\boldsymbol{B} = \boldsymbol{I}$ leads to Tikhonov regularization, Eq. (10.9), or equivalently a maximum *a posteriori* solution that produces a smooth solution (a minimum information solution). If $\boldsymbol{B}$ is chosen as a Laplacian, the filtering of Eq. (16.24) smoothes the second order derivative, making the first derivative nearly constant. This helps in eliminating noise, but at the expense of not preserving edges well. Equation (16.24) is arrived at by minimizing the Euclidean norm (L_q, $q = 2$) of the corresponding cost function. Decreasing the order, q, e.g. by using the L_1 norm (total variation) is more effective in preserving edges (Aubert and Kornprobst, 2006). The latter reference gives an overview of variational approaches based on partial-differential equations for image restoration.

A diagonal weight matrix, $\boldsymbol{W}$, can also be introduced in Eq. (16.24) as $\boldsymbol{B}^{\mathrm{T}}\boldsymbol{W}\boldsymbol{B}$, to introduce smoothing with different weights at different portions of the image; e.g. one can choose a weight of about one in smooth regions and a nearly zero weight near edges. Any of the regularization methods of Section 10.4, or a combination of, can be introduced into Eq. (16.24), sequentially or simultaneously as appropriate. Piecewise (local) regularization, discussed in Section 10.4.11, is particularly attractive for applying near boundaries and edges. If the matrices encountered are too large to be directly inverted, one can rely on some of the iterative methods discussed in Section 10.6.2, such as the steepest descent or the Conjugate Gradient methods.

The blurring matrix can be assembled from one of the point spread functions described in Section 16.2.1. When the image, $\boldsymbol{g}$, is arranged by stacking the image parameters of columns of a section of an image, the blurring matrix, $\boldsymbol{h}$, will take a repetitive block form, such as that of a Toeplitz matrix which has constant values along negative-sloping diagonals.[4] Such a matrix type can be readily diagonalized then inverted (Ng, 2004, Vogel, 2002). Hansen (2002) discussed methods for the regularization of Toeplitz matrices, relating them to deconvolution using fast Fourier transforms, and provided examples for image restoration.

16.5 Statistical Methods

Viewing the blurring matrix, $\boldsymbol{h}$, as the conditional probability of obtaining $\boldsymbol{g}$ given $\boldsymbol{f}$, any of the probabilistic methods of Chapter 13 can be utilized. In the most basic formulation, a pixel/voxel with an image parameter, g_i, is related to the "true" parameter,

[4] Weisstein, Eric W., "Toeplitz Matrix," From MathWorld - A Wolfram Web Resource. `http://mathworld.wolfram.com/ToeplitzMatrix.html`

f_i, in accordance to Eq. (16.23), when neglecting the effect of noise, by:

$$h_i = \sum_j h_{ij} f_j \tag{16.25}$$

where h_{ij} is an element of the blurring matrix, $\boldsymbol{h}$. If the point spread function is such that $\sum_j h_{ij} = 1$, then one is applying a smoothing function, as in Eq. (16.6). One can also use a proper distribution (say Gaussian) normalized to unity. In either case, h_{ij} can be represented as a conditional probability: $P(g_i|f_j) = h_{ij}$. Applying the Bayes' hypothesis, Eq. (8.11), one restores f_j iteratively using a scheme analogous to Eq. (13.12):

$$f_j^{(k+1)} = f_j^{(k)} \sum_i \frac{h_{ij} g_i}{\sum_n h_{jn} f_n} \tag{16.26}$$

where k is the iteration number. In image restoration, the scheme of Eq. (16.26) is known as the Lucy-Richardson algorithm (Lucy, 1974; Richardson, 1972). This iterative scheme tends, however, to amplify noise, which also makes reaching a converged solution, where $f_j^{(k+1)} = f_j^{(k)}$, difficult. The difference between $f_j^{(k+1)}$ and $f_j^{(k)}$ becomes so small that it cannot be attributed to noise and its amplification. It becomes then difficult to decide on the number of iterations required to reach an acceptable solution.

Hunt and Sementilli (1992) developed a maximum *a posteriori* (MAP) scheme (see Section 13.4) assuming that an image parameter at position $\vec{r}$, $g(\vec{r})$, is Poisson-distributed with a mean value determined by the right-hand side of Eq. (16.25). This results in the iterative scheme:

$$f(\vec{r})^{(k+1)} = f(\vec{r})^{(k)} \exp\left[\left(\frac{g(\vec{r})}{f(\vec{r})^{(k)} * h(\vec{r})}\right) \otimes h(\vec{r})\right] \tag{16.27}$$

where "$*$" and "$\otimes$" designate the convolution and correlation operators, respectively, see Footnote 1 of this chapter. This nonlinear scheme is reported to be capable of recovering spatial image frequency components beyond the maximum sampling frequency of the imaging system; resulting in the so-called super-resolution recovery. Other MAP methods, discussed in Section 13.4, can be utilized for image enhancement. The stochastic methods of Section 13.5 can also be employed in image enhancement.

16.6 Optimization

Image enhancement can also be viewed as an optimization problem with the objective of minimizing a measure of $\boldsymbol{g} - \boldsymbol{hf}$, along with constraints, such as image parameters being nonnegative and do not exceed a certain preassigned maximum value, or some other *a priori* constraints. The optimization methods of Chapter 11 can then be applied. However, the optimization problem of image enhancement is disadvantaged by being a fully determined problem, where the number of unknowns is equal to the number of known parameters; unless a coarser image is to be obtained to improve confidence in the values of $\boldsymbol{f}$ at the expense of reduction in resolution. In the presence of noise, and with uncertainty in knowing the blurring matrix, $\boldsymbol{h}$, a fully determined problem becomes in effect an incomplete (underdetermined) problem. Incorporating *a priori* information as constraints becomes then important in the optimization problem of image enhancement, see Chapter 14.

One way to overcome the inherent incompleteness of the image enhancement problem is to add terms to the cost function regularization to smooth the image, preserve edges, and/or ensure a particular image texture. Section 10.4 discusses many regularization methods. Another two regularization methods, particularly useful in accentuating image features, are given here.

For sharp edges, the following regularization term, B, was proposed by Conan et al. (2000) to overcome the ringing artifact usually associated with edges:

$$B_{\text{Sharp Edge}} = \alpha \sum_r \frac{|\Delta f(\vec{r})|}{\beta} - \ln\left[1 + \frac{|\Delta f(\vec{r})|}{\beta}\right] \tag{16.28}$$

where $|\Delta f(\vec{r})|$ is a gradient determined by the Euclidean distance of the finite-difference, α and β are adjustable regularization parameters, and the summation is over the voxels/pixels of the image. This regularization term is actually a first-order approximation of the conditional probability $P(\boldsymbol{f}|\boldsymbol{g})$, assuming a Poisson distribution.

For images containing simple shapes, the texture of that shape can be used as a regularization local (piecewise) function of the form (Jeffs and Pun, 1996):

$$B_{\text{Shape}} = \sum_{i,j} b_{ij}|f_i - f_j|^s \tag{16.29}$$

where i, j are the indices of a pixel in a two-dimensional cross section of the considered shape, b_{ij} is a neighborhood influence parameter, and s is shape parameter. Jeffs and Pun (1996) reported s-values for large objects observed in astronomical images. The same approach can be adopted for industrial and medical radiological imaging by analogy or by trial and error. Shape matching can also be achieved by identifying the occurrence of a particular pattern in an image, determining the locations where $g(\vec{r})$ matches a certain pattern, $f_p(\vec{r})$, and maximizing the correlation between $g(\vec{r})$ and $f_p(\vec{r})$.

16.7 Blind Deconvolution

One can attempt to enhance an image without knowing its blurring (point spread) function. This is called blind or myopic deconvolution. There are two general approaches to addressing this problem. The first approach is to estimate the point spread function from the blurred image itself. The second approach is to simultaneously solve for the point spread function (PSF) and the image parameters of the refined images. Some example methods for both approaches are given here. The review articles of Kundur and Hatzinakos (1996a,b), Yitzhaky et al. (1999), and Jiang and Wang (2003) can be consulted for other methods.

A blurred image contains information on the true image, the blurring function and noise. One can decrease the effect of the noise and the true image by estimating the blurring function as the average of a number of estimates based on segments within a single image. The only consistent influence on the averaging process is that of the PSF, if it is spatially invariant. Therefore, one can obtain an estimate of the PSF by dividing the available image into small segments, which may overlap, but are wide enough to encompass the range of the PSF. For each segment, the power spectral density is calculated. The average of these densities then replaces the denominator of the PSE filter, Eq. (16.18). With the numerator of the same equation evaluated as the average of the power spectral density of the entire image minus that of its noise, one obtains an inverse filter that can be used to deconvolve the image.

Another direct, noniterative, blind deconvolution relies on the so-called whitening process (Yitzhaky et al., 1998), in which each segment of an image is first filtered in the frequency domain with a certain high-pass filter, $\mathcal{W}^{-1}(\vec{k})$, to reach a better estimate of the "true" image. Let the average of the spectral density of the filtered image be, $S_{\tilde{f}}(\vec{k})$, which can be related to the spectral power density of the blurred image by: $S_g(\vec{k}) = \mathcal{H}^2 S_f = \mathcal{H}^2\mathcal{W}^{-2}S_{\tilde{f}}$, with S_f being the spectral power density of the true image and assuming that averaging removes the effect of noise. This enables the determination of the Fourier transform of the PSD, $\mathcal{H}(\vec{k})$, with $S_f = S_g - S_\eta$, where S_η is an estimated spectrum power density of the noise.

The other approach to blind deconvolution is a double iterative process to estimate both the PSF and the enhanced image (Ayers and Dainty, 1988). Prior image constraints, (such as the nonnegativity of both the PSF and the sought solution, the image support (spatial domain), and band width), can be incorporated into this solution process. The first stage of the iterative process starts by choosing an approximate of the refined image that satisfies the constraints, and finding its Fourier transform, $\hat{\mathcal{F}}^{(l)}(\vec{k})$, $l=0$ for the initial estimate. After ensuring that this estimate satisfies the *a priori* constraints, an estimate of the PSF function, $\mathcal{K}(\vec{k})$, is found using a Wiener-like filter:

$$\mathcal{H}^{(l+1)}(\vec{k}) = \frac{\mathcal{G}(\vec{k})\left(\mathcal{F}^{(l)}(\vec{k})\right)^*}{\left(\mathcal{F}^{(l)}(\vec{k})\right)^2 + \alpha^2\left(\frac{1}{\mathcal{H}^{(l)}(\vec{k})}\right)^2} \tag{16.30}$$

with an initial estimate assumed for $\mathcal{H}^{(0)}(\vec{k})$, and α^2 being a regularization parameter that accounts for the energy of the noise. The new estimate of the PSD is then used to update the estimate for the solution:

$$\mathcal{F}^{(l+1)}(\vec{k}) = \frac{\mathcal{G}(\vec{k})\left(\mathcal{H}^{(l)}(\vec{k})\right)^*}{\left(\mathcal{H}^{(l)}(\vec{k})\right)^2 + \alpha^2\left(\frac{1}{\mathcal{F}^{(l)}(\vec{k})}\right)^2} \tag{16.31}$$

At each step, one should ensure that the constraints are satisfied, for both the Fourier transforms and their inverses, and corrective actions taken, if necessary.

Simulated annealing has been also used in blind deconvolution, in which a solution is attained by minimizing an objective cost function with respect to both the solutions and the PSF, see McCallum (1990). Alternating minimization of the cost function, first with respect to the solution, and then with respect to the PSD, or vice versa, has been also proposed by You and Kaveh (1996) for a matrix-based cost function. The nonnegativity and support constraints recursive inverse filtering (NAS-RIF) algorithm of Kundur and Hatzinakos (1998) applies a penalty function that discourages having image parameters with values outside of the image's support domain, and those of negative values. In reconstructed images, the support domain is usually defined by the domain of the voxels within which the image is reconstructed, and the support constraint is not necessary. The recursive inverse filtering feature of this algorithm is due to its use of an approximate inverse filter updated iteratively.

References

G. Aubert and P. Kornprobst, *Mathematical Problems in Image Processing*. New York: Springer Science, 2006.

G. R. Ayers and J. Dainty, "Iterative blind deconvolution and its applications," *Optics Letters*, vol. 13, pp. 547–549, 1988.

H. H. Barrett, *Radiological Imaging*. New York: Academic Press, 1981.

M. Cannon, "Blind deconvolution of spatially invariant image blurs with phase," *IEEE Transactions on Acoustics, Speech and Signal Processing*, vol. 24, pp. 58–63, 1976.

K. R. Castleman, *Digital Image Processing*. Englewood Cliffs: Prentice Hall, 1996.

J.-M. Conan, T. Fusco, L. M. Mugnier, and F. Marchis, "MISTRAL: Myopic deconvolution method applied to ADONIS and to simulated VLT-NAOS images," *ESO The Messenger*, vol. 99, pp. 38–45, 2000.

R. C. Gonzalez and R. E. Woods, *Digital Image Processing*, 3rd ed. Upper Saddle River: Pearson/Prentice Hall, 2008.

R. W. Hamming, *Digital Filters*. Englewood Cliffs: Prentice-Hall, 1977.

P. C. Hansen, "Deconvolution and regularization with Toeplitz matrices," *Numerical Algorithms*, vol. 29, pp. 323–378, 2002.

B. R. Hunt and P. J. Sementilli, "Description of a Poisson imagery super-resolution algorithm," in *Astronomical Data Analysis Software and System*, I. D. M. Worrall, C. Biemserfer, and J. Barnes, Eds. San Francisco: Astronomical Society of the Pacific, 1992, vol. 25, pp. 196–199.

B. Jeffs and W. Pun, "Simple shape parameter estimation from blurred observations for a generalized gaussian MRF image prior used in MAP image restoration," in *Proceedings. of International Conference on Image Processing*, vol. 2. IEEE, September 1996, pp. 465–468.

M. Jiang and G. Wang, "Development of blind image deconvolution and its applications," *Journal of X-Ray Science and Technology*, vol. 11, pp. 13–19, 2003.

D. Kundur and D. Hatzinakos, "Blind image deconvolution," *IEEE Signal Processing Magazine*, vol. 13, pp. 43–64, 1996.

——, "Blind image deconvolution revisited," *IEEE Signal Processing Magazine*, vol. 13, pp. 61–63, 1996.

——, "A novel blind deconvolution scheme for image restoration using recursive filtering," *IEEE Transactions on Signal Processing*, vol. 46, pp. 375–390, 1998.

L. B. Lucy, "An iterative technique for the rectification of observed distributions," *Astronomical Journal*, vol. 79, pp. 745–754, 1974.

B. C. McCallum, "Blind deconvolution by simulated annealing," *Optics Communications*, vol. 75, pp. 101–105, 1990.

C. E. Metz and R. N. Beck, "Quantitative effects of stationary linear image processing on noise and resolution of structure in radionuclide images," *Journal of Nuclear Medicine*, vol. 15, pp. 164–170, 1974.

M. Ng, *Iterative methods for Toeplitz systems*. New York: Oxford University Press, 2004.

S. K. Park and R. A. Schowengerdt, "Image reconstruction by parametric cubic convolution," *Computer Vision, Graphics, and Image Processing*, vol. 23, pp. 258–272, 1983.

E. Parzen, "On estimation of a probability density function and mode," *Annals of Mathematical Statistics*, vol. 33, pp. 1065–1076, 1962.

M. Paulinas and A. Ušinskas, "A survey of genetic algorithms applications for image enhancement and segmentation," *Information Technology and Control*, vol. 36, pp. 278–284, 2007.

J. G. Proakis and D. G. Manolakis, *Digital Signal Processing: Principles, Algorithms, and Applications*, 4th ed. Upper Saddle River: Prentice-Hall, 2007.

W. H. Richardson, "Bayesian-based iterative method of image restoration," *Journal of the Optical Society of America*, vol. 62, pp. 55–59, 1972.

RSICC, *FORIST. Neutron spectrum unfolding code*, Computer Codes for Unfolding the Pulse-Height Distribution of Neutrons Measured with Organic Scintillators, Radiation Safety Information Computational Center, Oak Ridge National Laboratory, 1978, PSR-0092/01.

S. W. Smith, *The Scientist and Engineer's Guide to Digital Signal Processing*. San Diego: California Technical Publishing, 2003, (`http://www.dspguide.com`).

Y. Tsoy and V. Spitsyn, "Digital images enhancement with use of evolving neural networks," in *Parallel Problem Solving from Nature - PPSN IX*, ser. Lecture Notes in Computer Science, T. P. Runarsson, H.-G. Beyer, E. Burke, J. J. Merelo-Guervós, L. D. Whitley, and XinYao, Eds. Springer: Springer, 2006, pp. 593–602.

C. R. Vogel, *Computational Methods for Inverse Problems*. Philadelphia: Society for Industrial and Applied Mathematics, 2002.

Y. Yitzhaky, I. Mor, A. Lantzman, and N. S. Kopeika, "A direct method for restoration of motion blurred images," *Journal of the Optical Society of America A*, vol. 15, pp. 1512–1519, 1998.

Y. Yitzhaky, R. Milberg, S. Yohaev, and N. S. Kopeika, "Comparison of direct blind deconvolution methods for motion-blurred images," *Applied Optics*, vol. 38, pp. 4325–4332, 1999.

Y.-L. You and M. Kaveh, "A regularization approach to joint blur identification and image restoration," *IEEE Transaction on Image Processing*, vol. 5, pp. 416–428, 1996.

Bibliography

B. D. Ahluwalia, *Tomographic Methods in Nuclear Medicine: Physical principles, instruments, and clinical applications*. Boca Raton: CRC Press, 1989.

R. Barrett, M. Berry, T. F. Chan, J. Demmel, J. Donato, J. Dongarra, V. Eijkhout, R. Pozo, C. Romine, and H. V. der Vorst, *Templates for the Solution of Linear Systems: Building Blocks for Iterative Methods*, 2nd ed. Philadelphia: Society for Industrial and Applied Mathematics, 1994, `http://www.netlib.org/linalg/html_templates/report.html`.

J. Bushberg, J. Seibert, E. Leidholdt, and J. Boone, *The Essential Physics of Medical Imaging*. Philadelphia: Lippincott Williams & Wilkins, 2002.

T. M. Buzug, *Computed Tomography: From Photon Statistics to Modern Cone-Beam CT*. Berlin: Springer, 2008.

D. Colton and R. Kress, *Inverse Acoustic and Electromagnetic Scattering Theory*, ser. Applied Mathematical Sciences. Berlin: Springer-Verlag, 1992, vol. 93.

Committee on the Mathematics and Physics of Emerging Dynamic Biomedical Imaging, National Research Council, *Mathematics and Physics of Emerging Biomedical Imaging*. Washington: The National Academies Press, 1996.

T. Dyakowski, R. Mann, and B. Edwards, *Industrial Process Tomography*. Amsterdam: Elsevier, 2000.

H. W. Engl, A. K. Louis, and W. Rundell, Eds., *Inverse Problems in Medical Imaging and Nondestructive Testing*. Wien: Springer, 1997.

H. W. Engl, M. Hanke, and A. Neubauer, *Regularization of Inverse Problems*. Dordrecht: Kluwer Academic Publishing, 2000.

C. Epstein, *Introduction to the Mathematics of Medical Imaging*. Philadelphia: Pearson Education, 2003.

S. Euclid, *Computed Tomography*, 2nd ed. Elsevier: Oxford, 2001.

R. Farr and P. Allisy-Roberts, *Physics for Medical Imaging*. London: Saunders, 1997.

D. N. Ghosh Roy, *Methods of Inverse Problems in Physics*. Boca Raton: CRC Press, 1991.

S. Gindikin, *Applied Problems of Radon Transform*. Providence: American Mathematical Society, 1994.

G. H. Golub, *Matrix Computations*. Baltimore: Johns Hopkins University Press, 1996.

A. C. Kak and M. Slaney, *Principles of Computerized Tomographic Imaging*. Philadelphia: Society of Industial and Applied Mathematics, 2001.

W. A. Kalender, *Computed Tomography: Fundamentals, System Technology, Image Quality, Applications*, 2nd ed. Erlangen: Publicis Corporate Publishing, 2005.

G. R. Liu and X. Han, *Computational Inverse Techniques in Nondestructive Evaluation*. Boca Raton: CRC Press, 2003.

F. Mees, Ed., *Applications of X-ray Computed Tomography in the Geosciences*. London: The Geological Society, 2003.

V. A. Morozov, *Regularization methods for ill-posed problems*. Boca Raton: CRC Press, 1993.

——, *Methods for Solving Incorrectly Posed Problems*. New York: Springer-Verlag, 1984.

F. Natterer and F. Wubbeling, *Mathematical Methods in Image Reconstruction*. Philadelphia: Society for Industrial and Applied Mathematics, 2001.

S. S. Rao, *Engineering optimization: Theory and practice*. New York: John Wiley & Sons, 1996.

G. Roach, Ed., *Inverse Problems and Imaging*. Harlow: Wiley, 1991.

R. Robb, *Three Dimensional Biomedical Imaging: Principles and practice*. New York: Wiley-Liss, 1998.

P. Scally, *Medical Imaging*. Oxford: Oxford University Press, 1999.

J. Selman, *The Fundamentals of Imaging Physics and Radiobiology*, 9th ed. Springfield: Charles C. Thomas, 2000.

K. K. Shung, M. B. Smith, and B. M. W. Tsui, *Principle of Medical Imaging*. San Diego: Academic Press, 1992.

P. Sprawls, *Physical Principles of Medical Imaging*. Rockville: Aspen Publishers, 1987.

G. Stimac, *Introduction to Diagnostic Imaging*. Philadelphia: W.B. Saunders Company, 1992.

A. Tarantola, *Inverse Problem Theory: Methods for Data Fitting and Model Parameter Estimation*. Amsterdam: Elsevier Science, 1987.

——, *Inverse Problem Theory and Methods for Model Parameter Estimation*. Philadelphia: Society for Industrial and Applied Mathematics, 2005.

A. N. Tikhonov, *Solutions of Ill-Posed Problems*. Washington: Halsted Press, 1977.

D. M. Trujillo and H. R. Busby, *Practical Inverse Analysis in Engineering*. Boca Raton: CRC Press, 1997.

C. R. Vogel, *Computational Methods for Inverse Problems*. Philadelphia: Society for Industrial and Applied Mathematics, 2002.

G. R. Walsh, *Methods of Optimization*. London: John Wiley & Sons, 1975.

S. Webb, *The Physics of Medical Imaging*. Bristol: Hilger, 1988.

www.ingramcontent.com/pod-product-compliance
Ingram Content Group UK Ltd.
Pitfield, Milton Keynes, MK11 3LW, UK
UKHW020126290926
15471UKWH00003B/319

9 780323 165297